Bio-Based and Biodegradable Plastics

Bio-Based and Biodegradable Plastics

From Passive Barrier to Active Packaging Behavior

Editors

Alexey Iordanskii
Nadia Lotti
Michelina Soccio

MDPI • Basel • Beijing • Wuhan • Barcelona • Belgrade • Manchester • Tokyo • Cluj • Tianjin

Editors
Alexey Iordanskii
Semenov Institute of Chemical Physics
Russia

Nadia Lotti
University of Bologna
Italy

Michelina Soccio
University of Bologna
Italy

Editorial Office
MDPI
St. Alban-Anlage 66
4052 Basel, Switzerland

This is a reprint of articles from the Special Issue published online in the open access journal *Polymers* (ISSN 2073-4360) (available at: https://www.mdpi.com/journal/polymers/special_issues/food_package_plastics).

For citation purposes, cite each article independently as indicated on the article page online and as indicated below:

LastName, A.A.; LastName, B.B.; LastName, C.C. Article Title. *Journal Name* **Year**, *Article Number*, Page Range.

ISBN 978-3-03936-968-3 (Hbk)
ISBN 978-3-03936-969-0 (PDF)

Contents

About the Editors

Alexey Iordanskii currently works at the Semenov Institute of Chemical Physics, Russian Academy of Sciences. Scientific involvement includes physical concept of transport phenomena in ultrathin fibers, planar films as well as the other micro- and nano-scaled polymer objects in biomedicine, environmental science and packaging. A. Iordanskii graduated Lomonosov Moscow University and then earned Ph.D. and DSc dissertations sequentially in the areas of Polymer Materials Chemistry and Physical Chemistry of biopolymers. Its current project is 'Biodegradable composites and ultrathin fibers for biomedicine and environmental applications. Morphology, diffusion, and molecular dynamics.' Accordingly, his research activities are expressed by 250+ international scientific publications and 1159 citations (via Scopus) with H-index = 17.

Nadia Lotti (Ph.D.) is Associate Professor at the University of Bologna since 2011. She carries out her research in the field of Chemistry and Technology of Polymers. She is engaged in national and international research projects. Collaborates with various national and European institutes. Member of the Editorial Board of Journal of Waste Management and Environmental Issues, Applied Science, Sustainability and Polymers of MDPI. Education: Teaching courses within the SSD CHIM / 07 at the University of Bologna, CdS in Civil, Biomedical and Electronic Engineering for Energy and Information. Supervisor and Co-supervisor of Bachelor's, Master's Degree and PhD's Theses. Member of the Teaching Body of the Research Doctorate in Industrial Chemistry (up to the XXV cycle), of the Research Doctorate in Civil, Environmental and Materials Engineering (up to the XXXIII cycle), currently of the Doctorate in Health & Technology, all activated at the same Athenaeum. Academic Career: - 1989: Degree in Chemistry with 110/110 cum laude from the University of Bologna, - 1989-1992: PhD in Industrial Chemistry carried out at the University of Bologna. - 1991: 6-month internship at the University of Massachusetts (U.S.A.) under the supervision of Prof. R.W. Lenz. 1993: annual scholarship funded by the CNR. - 1994: winner of a place in the qualification of Collaborative Chemist put up for competition by the Usl no 38 of Forlì. - 1995 Researcher at DICASM, Faculty of Engineering 2 of the University of Bologna. - 2011 Associate Professor at DICAM, of the University of Bologna. Current research interests: The research interests are addressed to the synthesis and molecular, solid-state and in melt characterization of polycondensates, and to the chemical modification of commercial polymers. She currently focuses her research on the topic of the synthesis of polymers from renewable sources for packaging applications and that of biopolymers for biomedical use. Publications: 163 publications in international journals (2306+ citations, Hindex = 27) (scopus) and about 110 Communications at Conferences.

Michelina Soccio, Ph.D. is Researcher at Civil, Chemical, Environmental and Materials Engineering of University of Bologna, since 2017. M. Soccio is graduated in Industrial Chemistry at University of Bologna (2004). M. Soccio earned Ph.D in Industrial Chemistry at University of Bologna in 2008. Ph. D. Thesis "Synthesis, characterization and properties-structure correlations of innovative polyesters for special employs". She worked as post-doc at the Department of Applied Chemistry and Science of Materials of University of Bologna (2008–2010); at Institute for the Structure of Matter of CSIC in Madrid, Spain (2011–2013); at Department of Materials Science and Engineering of Penn State University, State Collage, Pennsylvania, USA (2014); at the Department of Applied Chemistry and Science of Materials of University of Bologna (2014–2017). Michelina Soccio's research topics concern the synthesis, modification and characterization of high performance polymeric materials for industrial applications as well as of novel biodegradable and biocompatible polyesters for tissue engineering and packaging. She is co-author of 94 international scientific publications (Hindex = 22; 1334 citations).

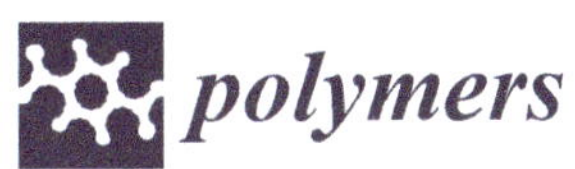 

Editorial

Bio-Based and Biodegradable Plastics: From Passive Barrier to Active Packaging Behavior

Alexey Iordanskii

Semenov Institute of Chemical Physics, Russian Academy of Sciences, 119991 Moscow, Russia; aljordan08@gmail.com

Received: 22 June 2020; Accepted: 6 July 2020; Published: 12 July 2020

Abstract: An overview of the articles has presented for the Special Issue "Bio-Based and Biodegradable Plastics: From Passive Barrier to Active Packaging Behavior". This issue has objective of collecting comprehensive findings regarding structure and functionality of bio-based sustainable polymers performing as multifaceted barrier and packaging in food, cosmetic, and other areas. The content of the collection covers diverse fields of knowledge embracing polymer chemistry, materials science, transport–diffusion phenomena, biodegradation exploring, and others.

Keywords: bio-based polymers; biodegradable packaging; diffusion; permeability; mechanical behavior; biopolymer structure; encapsulation; life cycle analysis

---

The unprecedented energy crisis, triggered by the pandemic, in combination with environmental safety deterioration, has stimulated the accelerated transition from the synthetic bio-stable plastics from hydrocarbon fossils to the bio-based and biodegradable plastics produced from natural renewable resources. In line with the European Strategy for Bioplastics in a Circular Economy, the rational design of biopolymers has to provide them with the desired functionality, sustainability, and nonharmful utility. Mostly affected by this revolutionary trend, biodegradable polymers and their composites must be applied in the areas of packaging manufacturing and especially food packaging implementation. In the framework of this Special Issue the invited authors have presented their significant contributions to augmenting the current trends in barrier and packaging materials with the special characteristics of biodegradable plastics.

As bio-based and biodegradable polymers, polyhydroxyalkanoates, with their basic representative, poly (3-hydroxybutyrate) (PHB), and polylactides (PLA) are the main candidates to replace fossil-based polymer materials especially in the packaging industry. The intention of the comprehensive paper by Siracusa et al. [1] was to manufacture, via blending, fully biodegradable barriers against atmospheric gases and to first combine diffusivity data with blend segmental mobility. A multifaceted approach combining dynamic (permeation techniques and probe-radical ESR methodology) and structural (SEM, DSC, and colorimetry) methods is given to describe the complex behavior of fully bio-based blends with different compositions. Both probe rotation mobility, measured by the ESR method, and atmospheric gas transport characteristics monotonically decreased in the PLA/PHB blend barriers because they were transferred from PLA with lower crystallinity to PHB with a higher one. These findings, in combination with the structure-morphology pattern, could be interpreted as ground barrier packaging characteristics, which are necessary for a coherent interpretation of recently elaborated packaging with active functions such as environmental safety, antimicrobial performance, and the controlled release of food-modifiers.

A comprehensive study considering the physicochemical and mechanical behavior of pristine poly (3-hydroxybutyrate) PHB and copolymers of 3-hydroxybutyrate (HB) with 3-hydroxyvalerate (HV) during long-term enzymatic and hydrolytic biodegradation is presented by Bonartsev et al. [2]. The evolution of principal characteristics such as polymer weight, its average molecular weight, crystallinity, and surface hydrophobicity as function of the HB/HV ratio is carefully described. It is interesting that on the scale of monomer content there is a critical point at ~5.7–5.9% of HV content that could be treated as a transition point for sharp changes in the structural and mechanical properties of the copolymer films.

Along with copolymer synthesis of biodegradable barriers, there is an alternative method of novel packaging design that is based on environmentally safe technology, namely solid-state extrusion without the use of toxic solvents. Rogovina et al. [3] present a prominent biodegradable blend on the base of polylactide (PLA) and PHB. The thermoplastic biopolyesters were obtained by blending under shear deformations and electrospinning methods in the form of films and ultrathin electrospun fibers. The crystallinity and the specific thermal transitions (glass transition, cold crystallization, and melting) for the individual PLA, PHB, and their compositions with polyethylene glycol were analyzed. Thermal and mechanical features of the ternary blends testified the pronounced behavior of PEG as the plasticizer with the noticeable increase in tensile characteristics and with a remarkable decrease in glass-transition temperature.

Chalykh et al. [4] address the topic of water vapor permeability through porous polymeric membranes presenting synthetic and natural transport barriers with a diverse hydrophilic–hydrophobic balance. At hydrolysis and enzymatic attack, water molecules promote macromolecular degradation and therefore affect the functional behavior of barrier and packaging materials. Barrier characterization is also critical for understanding the role of water diffusion and sorption in packaging matrices as the control factors that contribute to bacterial growth and food deterioration. In addition, another important point that water sorption capacity reveals is moisture's impact on gas permeability and selectivity. The prospects of this effect should be thoroughly studied by experts in future publications.

Kildeeva et al. [5] present a circumstantial description of chitosan crosslinked by genipin (Gp) with the aim of depressing polysaccharide solubility and expand the array of functionalities that, along with versatile eco- and bio-compatibility, open up promising prospects in the packaging industry. The Gp as the natural bifunctional reagent improves water resistance and mechanical behavior. Water diffusivity, sorption capacity, and the modeling of isotherms, namely their decomposition into the Langmuir and Flory–Huggins constituents, have shown the complicated pattern of water–crosslinked chitosan interaction. The results of this contribution could be useful for food packaging, edible coatings, or for medical applications to improve tensile parameters and to control moisture absorption.

By an international Spain-UK team, in the manuscript of Quiles-Carrillo et al. [6], the formulation of bio-based high-density polyethylene (BHDPE) with gallic acid (GA) was first melt-compounded and comprehensively investigated. The GA-loaded bio-HDPE films produced by extrusion and cast rolling successively were characterized in terms of their mechanical, morphological, and thermal performance as well as ultraviolet (UV) light stability to evaluate their potential application in food packaging. The encapsulation of the natural antioxidant has essentially improved prospects for the thermal and UV light stability of green BHDPE and, very probably, of other green polyolefins.

Kirsh et al. [7] carefully analyzed the binary (polyethylene (PE) and birch bark extract (BBE)) and ternary (PE, modified starch, and BBE) blends that are designated for novel food packaging. The innovative idea of bend packaging preparation includes the combination of extrusion and ultrasound treating simultaneously. This approach allowed the authors to obtain barrier films with a uniform distribution of components and increased the fluidity of composite blends as well. Embedding the BBE modifies water vapor and oxygen permeabilities. Besides enhancing the antibacterial activity, the natural extract affects the term of biodegradability and simultaneously improves the tensile and elongation characteristics of barrier films.

An ingenious approach to novel composites for food and agriculture packaging is proposed by Mastalygina et al. [8]. In contrast to the mentioned above work of Quiles-Carrillo where composites on the base of bio-HDPE were considered, here, the authors used a kind of "hybrid" material combining synthetic low density PE and natural rubber (NR). A mycological test with fungi and a full-scale soil test clearly showed that 30 wt% of NR affected PE biodegradability via enhancing its hydrophilicity that was affirmed by a water absorption technique. Additional testing by IR spectroscopy and the DSC technique showed the structural evolution of LDPE after exposure to soil and corresponding weight loss by 7.2% for three months of testing. In addition, the composites based on polyethylene with natural rubber additives have satisfactory mechanical and technological properties that determine the suitability of such materials for application as packaging and agricultural films with advanced biodegradability.

To evaluate and measure the environmental impact of the processing, exploitation and disposal of biodegradable materials, in particular PLA bottles, Baldowska-Witos et al. [9] have carefully assessed the life cycle analysis (LCA). At the initial stage, the authors collected technological data (see Table S1) covering rough resources, energy consumption, and emissions to the air, soil, and water. At the next stage, they corrected the environmental impact estimation (Table 2) including greenhouse emissions and radioactive isotopic release (Figure 4). The environmental impact of bottle production from PLA and polyethylene terephthalate was compared. In the conclusion, the authors suggest that for enhancing LCA efficiency it is necessary to carry out an analysis of key data quality and probability in relation to the results; this is presented in the next publication in this Special Issue.

The second paper of Baldowska-Witos et al. [10] is closely related to the previous publication and devoted to the in-depth development of LCA methodology. The authors propose a numerical approach to evaluate the uncertainty and precision of LCA data using stochastic modeling and sensitivity analysis. The immediate applicability of the calculation algorithm to the PLA bottle shaping procedure is based on a number of factors such as (1) the qualified and coherent evaluability of input data in matrix form, (2) the reliability of characteristics and reasonability of conclusions, and (3) the uncertainty degree of input and obtained data estimated via Monte Carlo modeling. By arranging the sequence "input data (materials and energy)—process implementation (PLA preforming, heating, cooling, etc.)—additional impact categories (fossil depletion, global warming, water consumption, etc.)", the authors estimate the key factors that affect the eco-valuable damage category, namely human health, ecosystem quality/resistance, and resource ability.

Funding: This research was funded by the Russian Foundation for Basic Research, grant number 18-29-05017-mk.

Conflicts of Interest: The authors declare no conflict of interest.

References

1. Siracusa, V.; Karpova, S.; Olkhov, A.; Zhulkina, A.; Kosenko, R.; Iordanskii, A. Gas Transport Phenomena and Polymer Dynamics in PHB/PLA Blend Films as Potential Packaging Materials. *Polymers* **2020**, *12*, 647. [CrossRef] [PubMed]
2. Zhuikov, V.A.; Zhuikova, Y.V.; Makhina, T.K.; Myshkina, V.L.; Rusakov, A.; Useinov, A.; Voinova, V.V.; Bonartseva, G.A.; Berlin, A.A.; Bonartsev, A.P.; et al. Comparative Structure-Property Characterization of Poly(3-Hydroxybutyrate-Co-3-Hydroxyvalerate)s Films under Hydrolytic and Enzymatic Degradation: Finding a Transition Point in 3-Hydroxyvalerate Content. *Polymers* **2020**, *12*, 728. [CrossRef] [PubMed]
3. Rogovina, S.; Zhorina, L.; Gatin, A.; Prut, E.; Kuznetsova, O.; Yakhina, A.; Olkhov, A.; Samoylov, N.; Grishin, M.; Iordanskii, A.; et al. Biodegradable Polylactide-Poly(3-Hydroxybutyrate) Compositions Obtained via Blending under Shear Deformations and Electrospinning: Characterization and Environmental Application. *Polymers* **2020**, *12*, 1088. [CrossRef] [PubMed]
4. Anatoly, C.; Pavel, Z.; Tatiana, C.; Alexei, R.; Svetlana, Z. Water Vapor Permeability through Porous Polymeric Membranes with Various Hydrophilicity as Synthetic and Natural Barriers. *Polymers* **2020**, *12*, 282. [CrossRef] [PubMed]
5. Kildeeva, N.; Chalykh, A.; Belokon, M.; Petrova, T.; Matveev, V.; Svidchenko, E.; Surin, N.; Sazhnev, N. Influence of Genipin Crosslinking on the Properties of Chitosan-Based Films. *Polymers* **2020**, *12*, 1086. [CrossRef] [PubMed]
6. Quiles-Carrillo, L.; Montava-Jordà, S.; Boronat, T.; Sammon, C.; Balart, R.; Torres-Giner, S. On the Use of Gallic Acid as a Potential Natural Antioxidant and Ultraviolet Light Stabilizer in Cast-Extruded Bio-Based High-Density Polyethylene Films. *Polymers* **2020**, *12*, 31. [CrossRef] [PubMed]
7. Kirsh, I.; Frolova, Y.; Bannikova, O.; Beznaeva, O.; Tveritnikova, I.; Myalenko, D.; Romanova, V.; Zagrebina, D. Research of the Influence of the Ultrasonic Treatment on the Melts of the Polymeric Compositions for the

Polymers **2020**, *12*, 1537

Creation of Packaging Materials with Antimicrobial Properties and Biodegrability. *Polymers* **2020**, *12*, 275. [CrossRef] [PubMed]

8. Mastalygina, E.; Varyan, I.; Kolesnikova, N.; Gonzalez, M.I.C.; Popov, A. Effect of Natural Rubber in Polyethylene Composites on Morphology, Mechanical Properties and Biodegradability. *Polymers* **2020**, *12*, 437. [CrossRef] [PubMed]

9. Bałdowska-Witos, P.; Kruszelnicka, W.; Kasner, R.; Tomporowski, A.; Flizikowski, J.; Kłos, Z.; Piotrowska, K.; Markowska, K. Application of LCA Method for Assessment of Environmental Impacts of a Polylactide (PLA) Bottle Shaping. *Polymers* **2020**, *12*, 388. [CrossRef] [PubMed]

10. Bałdowska-Witos, P.; Piotrowska, K.; Kruszelnicka, W.; Błaszczak, M.; Tomporowski, A.; Opielak, M.; Kasner, R.; Flizikowski, J. Managing the Uncertainty and Accuracy of Life Cycle Assessment Results for the Process of Beverage Bottle Moulding. *Polymers* **2020**, *12*, 1320. [CrossRef] [PubMed]

Article

Managing the Uncertainty and Accuracy of Life Cycle Assessment Results for the Process of Beverage Bottle Moulding

Patrycja Bałdowska-Witos [1,*], Katarzyna Piotrowska [2], Weronika Kruszelnicka [1], Marek Błaszczak [2], Andrzej Tomporowski [1], Marek Opielak [2], Robert Kasner [1] and Józef Flizikowski [1]

[1] Department of Technical Systems Engineering, Faculty of Mechanical Engineering, University of Science and Technology in Bydgoszcz, 85-796 Bydgoszcz, Poland; weronika.kruszelnicka@utp.edu.pl (W.K.); a.tomporowski@utp.edu.pl (A.T.); robert.kasner@gmail.com (R.K.); fliz@utp.edu.pl (J.F.)
[2] Faculty of Mechanical Engineering, Lublin University of Technology, 20-618 Lublin, Poland; k.piotrowska@pollub.pl (K.P.); m.blaszczak@pollub.pl (M.B.); m.opielak@pollub.pl (M.O.)
* Correspondence: patrycja.baldowska-witos@utp.edu.pl

Received: 30 April 2020; Accepted: 4 June 2020; Published: 10 June 2020

Abstract: Using environmentally friendly materials in the technological process of bottle production fits perfectly into the idea of sustainable development. The use of natural raw materials as well as conscious energy consumption are strategic aspects that should be considered in order to improve the effectiveness of the bottle moulding process. This paper presents a new and structured approach to the analysis of uncertainty and sensitivity in life cycle assessment, one developed in order to support the design process of environmentally friendly food packaging materials. With regard to this "probabilistic" approach to life cycle assessment, results are expressed as ranges of environmental impacts, and alternative solutions are developed while offering the concept of input uncertainty and the effect thereof on the final result. This approach includes: (1) the evaluation of the quality of inputs (represented by the origin matrix); (2) the reliability of results and (3) the uncertainty of results (the Monte Carlo method). The use of the methodology is illustrated based on an experiment conducted with real data from the technological process of bottle production. The results provide insight into the uncertainty of life cycle assessment indicators regarding global warming, acidification and the use of arable fields and farmland.

Keywords: PLA bottle; bio-based and biodegradable polymers; life cycle assessment; environmental impact; Monte Carlo

1. Introduction

The point of reference used in the uncertainty assessment of life cycle assessment (LCA) results is the environmental evaluation of a life cycle, treated as a certain measurement technique based on a given methodology [1]. Its application leads to the "measurement" of two basic elements: environmental aspects that occur in the life cycles of products and subsequent environmental impacts [1–4].

Never can a single value without an uncertainty range represent the true value of an environmental impact because each measurement has uncertainty [5,6]. The process of producing bottles made of polylactide adopted in this paper focuses on the assessment of six technological operations, and concerns primarily the consumption of electricity, water and CO_2 emissions. The implementation of biodegradable polymers into the production process is the result of increasing amounts of residual polymer waste. For years, plastics were produced to obtain durable, environmentally sensitive products. A change in strategy seeking alternative sources of materials has resulted in the development of biodegradable plastics. Therefore, research based on the LCA technique, presenting the potential size of environmental impacts, should also include the impact of uncertainty and the variability of results [7]. This approach is proposed by Lloyd and Riesa [8] at the stage of modeling the uncertainty of characterization test results.

Parameter uncertainty refers to the value of a parameter such as energy or raw material found in processes or products. In contrast, the uncertainty of the model refers to the specific model used, such as the model developed by ReCiPe for life cycle impact assessment (LCIA) by converting emissions and the extraction of resources into a limited number of environmental impact assessments using so-called characterizing factors [9].

The uncertainty analysis of life cycle inventory (LCI) data sets can be divided into quantitative and semi-quantitative analysis. The first is based on statistical methods for the quantitative uncertainty assessment of the LCI database [10–13], and the second on the data quality indicator (DQI) method [9,14–17].

Among the methods of semi-quantitative uncertainty analysis, it is worth considering the DQI semi-quantitative approach based on a qualitative assessment of data quality due to its widespread use in the field of LCA. The DQI approach can be divided into qualitative and semi-quantitative approaches. A qualitative approach evaluates data quality in terms of qualitative descriptors, such as good, fair and poor data quality, which depends on a rather subjective assessment. Usually, the method is subject to a qualitative approach [18,19]. The semi-quantitative approach adopts a numerical rating system, and the assigned quality rating is processed to obtain a single data quality score based on a probability distribution [20,21].

Sonnemann et al. [22] conducted a quantitative analysis of data uncertainty based on the Monte Carlo simulation, indicating its positive cognitive features. Maurice et al. [12] pointed out that quantitative uncertainty analysis is too time consuming. Lloyd and Ries, however, concluded that current quantitative uncertainty analysis does not address significant factors contributing to the uncertainty of LCA results due to the complexity of LCA models [8]. In their research, they also highlighted the problem associated with the long-term and time-consuming process of collecting input data from outside the data library, for example, Ecoinvent available in specific software, e.g., SimaPro or Gabi, which contributes to frequent references to literature. This is one of the reasons why semi-quantitative uncertainty analysis has been used in many studies [9,14,17].

The term uncertainty has a number of interpretations, including those that exclude related terms such as variability and sensitivity [23]. According to the international dictionary of metrological terms, uncertainty is a parameter associated with a measurement result characterizing the scatter of values which can reasonably be attributed to the measured quantity [24,25]. Uncertainty is characterized by a scatter of values (interval size), within which it is possible to place the measured value with a satisfying probability [26,27]. In the case of LCA, uncertainty is contained in the results of "measurement" obtained at different research levels, that is, after the analysis of a set of entries and exits related to environmental aspects (LCI results) and after the environmental impact life cycle assessment (LCIA results) [28].

Considering the complexity of calculations, it is very difficult to present the uncertainty of LCA results in the form of a single equation describing the probability distribution of the values obtained. Therefore, in order to estimate these uncertainties, numeric simulations are conducted. B. Steen [29] indicates that sensitivity analysis enables one to express uncertainty in the form of probability, hence the need to estimate the degree of uncertainty and the distribution of probability. When analysing the LCA method, Kowalski and Kulczycka [30] recommend thorough sensitivity analysis or, if possible, partial uncertainty analysis of selected results and parameters whose uncertainty ranges are known, e.g., by means of the Monte Carlo (MC) simulation. He also indicates that some studies involving the use of LCA are subject to uncertainty, which can result in doubts regarding the value of final indicators, eco-indicators that determine the potential environmental impact of a product or process. The uncertainty of model correctness results from the fact that analysed models are never real. Every LCA is subject to uncertainty which is due to subjective choices made in order to design the model [31].

The existing literature on the theoretical foundations of the mathematical computational structure for performing uncertainty analysis in LCA is still hardly practiced [32]. However, there are various techniques for uncertainty analysis, such as: the theory of possibilities, e.g., [33], fuzzy theory, e.g., [34,35], data quality indicators, e.g., [9,14–17] and expert opinions, e.g., [36,37] or a combination of two or more techniques, e.g., [20,30,31,38]. Despite the wide range of approaches available, the analysis of uncertainty in LCA in the scientific and business process of bottle production remains largely unexplored.

Stochastic modeling, mainly in the form of MC simulations, is the most commonly used approach to analyzing uncertainty among various processes, research areas and industry sectors [39–42]. The usefulness of MC simulation becomes apparent when enough tests have been performed to make a claim regarding data distribution and uncertainty [30]. However, the combination of two elements of stochastic modeling and uncertainty analysis can be problematic. It most often lists three main reasons: (1) there is a need for a larger amount of data to be available than is available [43–45]; (2) there are no clearly described guidelines as to the size of the data set, and this extends the duration of the analysis and the complexity of calculations considerably [46]; (3) the choice of combined research methods seems to be too complex [47]. The most common in the literature are works carried out omitting accuracy analysis [48], or those using error propagation methods, including sampling techniques [49]; much less research is carried out using MC simulation.

The uncertainty of data can be expressed by means of the probability distribution thereof, e.g., the standard deviation or variance. This enables one to specify the range of values they can take. According to R. Heijugns [50], using the MC simulation in iterative LCA is time-consuming, but this disadvantage can be dealt with by calculating the propagation of uncertainty within the LCA methodology. H. Imbeault-Tétrault [51], when comparing the uncertainty propagation calculations with the MC simulation for uncertainty analyses within the LCA methodology, confirmed that the calculation of uncertainty propagation requires less computation time than the MC simulation and suggested that the analytical approach should be used instead. Heijugns and Lenzen [49] describe in detail where uncertainty can be encountered in LCA. MC is time- and cost-efficient, enabling one to determine the confidence level [52].

Uncertainty in studies involving the use of LCA can result in doubts regarding the value of final indicators (eco-indicators) that determine the potential environmental impact of a product or process. There are three types of uncertainty: uncertainty of data, uncertainty related to the correctness (representativeness) of the model applied and uncertainty caused by incompleteness of the model.

In the analysis of data uncertainty conducted by Lewandowska and Fołtynowicz [31], the basic assumption is that the quality of input data should increase with its importance. The authors propose analyzing data quality after having evaluated their environmental impact. As impact assessment is performed for the whole system within precisely specified boundaries, the final result pertains not only to the main data input, but also to the data of all the processes it represents.

Using statistical terms, LCA is used to study and compare systems of products in terms of the same feature, that is, environmental impact [53]. Uncertainty analysis, however, is reduced to determining the feature diversification level for each of the systems. The more diversified the feature is (scatter) the higher the uncertainty it represents [26,52,54].

This article aims to supplement the issues and knowledge about the environmental impact of the bottle shaping process presented in [55] by addressing the quality issues of the required data and the complexity of stochastic modeling for uncertainty and sensitivity analysis, thereby offering a new life cycle assessment model not yet practiced by food sector companies in Poland. The purpose of this paper, therefore, is to propose an evaluation method based on the use of the DQI semi-quantitative approach, stochastic modeling and sensitivity analysis to (1) analyze the uncertainty of beverage bottle production parameters, (2) analyze the uncertainty and accuracy of LCA results for the bottle shaping process and (3) recognize the effects of input parameters on the final results obtained.

2. Materials and Methods

2.1. The Uncertainty and Accuracy Management Procedure

Accuracy is defined as the similarity of the measured or modeled value to the "real one" [56]. Precision, however, represents the quality of repeatability of the results obtained, e.g., each repetition of calculation, of the experiment, and modeling provides a similar result [57,58].

Accuracy and precision are provided by methods which are independent of each other; in other words, a method is imprecise when the results exhibit a large scatter, yet it is accurate, since the mean value of the results corresponds to the real, model or theoretically predicted value [24]. A method is precise but not accurate when the scatter of the results is not large, but the mean value is far from the real, model or theoretically predicted value [59].

In this paper, a procedure for estimating data uncertainty and accuracy based on the known methods for determining the quality and uncertainty of data is proposed. The method combines three approaches: the DQI semi-quantitative approach, the stochastic modeling approach and the global sensitivity analysis. The procedure includes (Figure 1):

- An initial phase, which consists of LCA with contribution analysis,

  Step 1: the semi-quantitative DQI approach,
  Step 2: the stochastic modeling approach with use the MC simulation,
  Step 3: the sensitivity analysis based on the analysis of key issues.

The steps of the procedure are closely related, in such a way that the results of the preceding step are input data for the following step. In this way, it is possible to determine the sensitive points of the LCA along with the determination of the relationships and dependencies of the input data uncertainty with the uncertainty of the results in terms of impact categories and damage areas.

The first initial phase is the LCA allowing the identification of process input data (LCI) and environmental impacts (LCIA). An indispensable element for the proposed procedure is the implementation of contribution analysis to determine the unit processes and categories of impacts with the largest share in the total impact of the PLA bottle shaping production cycle. The LCA of this process was presented in a previous work [55]. The uncertainty analysis involved input data and only relevant impact categories in damage areas. The categories with total contribution equal to 90% were considered relevant. A detailed LCA methodology is provided in Section 2.2.

The first step in assessing data uncertainty is the semi-quantitative DQI approach. It includes the standard quality assessment proposed by Maurice et al. [12] with the use of five DQIs [25,56]: measurement precision (reliability), sample representativeness, appropriate age of data, geographic origin and technological representativeness, and then calculating the aggregated data quality indicator (ADQI) and determining the deviation of input data. A triangular distribution for pro-ecological scenarios of a product life cycle assessment was used in the analysis. The information about distribution obtained in the previous step is used as an input to the second step of the stochastic approach to estimate the uncertainty of significant impact categories using MC simulations. In this way, the impact categories with the greatest uncertainty and the input data of the bottle shaping process that are associated with them, will be identified. As a result of MC simulation, statistical parameters such as average, standard deviation, variance and coefficient of variation are obtained for the distribution of results of significant impact categories. Thereby, it is possible to carry out the key issue analysis using the MC simulation, where the mean and standard deviation will be used as input data for forecasting the sensitivity of the results of the total impact of the bottle shaping process to changes in the impact category value, and thus related changes of inventory data. A detailed methodology of individual stages of the embedded method is presented in Sections 2.1.1–2.1.3.

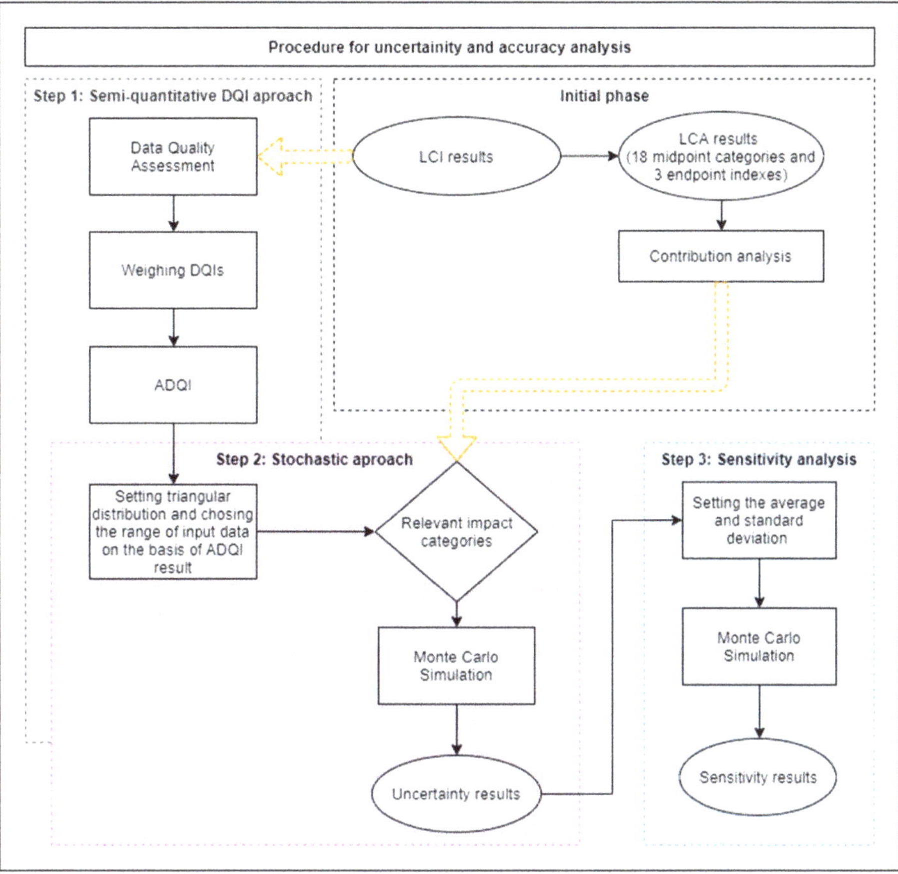

Figure 1. Flow chart of the uncertainty and accuracy management procedure.

2.1.1. Semi-Quantitative DQI Approach

The qualitative data evaluation was carried out using data quality indicators (DQIs). In the analysis the pedigree matrix (Table 1) was used, which is determining five different DQIs: reliability, completeness, time range, geographical range and technological range [18]. First, the input data were assessed in qualitative terms and then in quantitative terms, assigning points on a scale of 1 to 5 for each DQI [25,56]. Depending on the methodology adopted, grade 1 may mean the highest rating (which is consistent with the general method of the matrix of origin [18] and appears in many works, e.g., Baek et al. [20,38], Maurice et al. [12], Ciroth et al. [52]), or, in line with the approach used, for instance, by Lewandowska [60], Canter et al. [15], Kennedy et al. [17], data of the highest quality receive a rating of 5. In this work, we assumed that a score of 1 indicates the lowest quality of data and 5—the highest. DQA was carried out at the level of the unit process, for which each input parameter had the same data source. The selected level of data quality assessment DQA is an attempt to apply the DQI semi-quantitative approach as a practical and simple means to analyze parameter uncertainty.

In the following approach, the so-called aggregated data quality indicator (ADQI), which is the sum of five weighted DQIs, was used [12]. Initially, the weight of each input parameter was obtained by assigning 1 to 5 points to each of the five DQIs. To determine the ADQI value, it is necessary to assume the weight for each DQI. In the simplest terms, it can be assumed that individual DQIs are equivalent. In many works, however, there are methods for determining the weight of individual indicators [12,14,20]. In this work, we adopted weights according to Maurice et al. [12]: 0.25 in relation to the geographical and technological scope and 0.167 in relation to the other three DQIs. Considering the input data of the bottle shaping process (mainly—electricity consumption), it is the geographical and technological scope that will cause the greatest diversity of data, and thus affect the reliability and credibility of LCA results, so it was decided to take the weights as proposed by Maurice et al. [12]. Next, the DQI value was multiplied by the weight of each DQI and the resulting values were summed up.

Table 1. Pedigree matrix [56,60].

DQI	DQI = 5.0	DQI = 4.0	DQI = 3.0	DQI = 2.0	DQI = 1.0
R	Data verified and based on measurements	Verified data based partly on assumptions, or unverified based on measurements	Unverified data based partly on assumptions	Precise estimation	Inaccurate estimation
C	Representative data taken from the appropriate sample and time range	Data collected from a smaller sample, but within a reasonable period of time	Data taken from an appropriate sample, but not in the right time range	Representative data, but from a very small sample	Data not representative from a very small sample or it is unknown
TR	Deviation up to 3 years	Deviation up to 6 years	Deviation up to 10 years	Deviation up to 15 years	Deviation over 15 years or data age unknown
GS	Local scope (or other area envisaged for the purpose of the study)	National scope	Continental range	Global scope	Data of unknown origin
TS	Data on the analyzed process and enterprise	Data on the analyzed process and technology, but from a different source	Data on the analyzed process, but with different technology	Data on similar processes/products with the same technology	Data on similar processes/products with different technology

R—reliability, C—completeness, TR—time range, GS—geographical scope, TS—technological scope.

The next stage of the DQA is to relate the results obtained from origin matrixes to probability distributions [14,17,60]. In the case of point data or very small data sets, reliable determination of the average or standard deviation is not possible [60]. Then, it is necessary to use distributions based on other parameters, e.g., triangular distributions, continuous uniform or beta distributions [17,20]. Triangular and continuous uniform distributions are based on two parameters: minimum and maximum values [60].

Attempts to link DQA results to the distribution of input data were made by Kennedy [17] and Wang [14] using the so-called transformation matrix. In this work, we combined the ADQI results with the levels of deviations of the process input values (Table 2) assuming a triangular distribution. Therefore, it was assumed that the data whose DQI score is 5 has a permissible standard deviation of ±10%. As can be seen, a deviation from the most probable value occurs only at the level of ±10%, which is manifested by a narrow triangular distribution (strong concentration around the central value) and high (high probability) triangular distribution shape. The idea of triangular distribution is explained in Figure 2. This means that a person who performs an LCA assumes that the value of a given inventory (Table 3) (e.g., energy consumption, water consumption, CO_2 emissions) does not necessarily have to be x value (the result of a single measurement, estimation, collection from the company documentation, etc.), but that it can be a value from the range < x − 10% x; x + 10% x >.

Table 2. Aggregated data quality indicator (ADQI) values and deviation levels [54,60].

ADQI	Deviation (%)	ADQI	Deviation (%)	ADQI	Deviation (%)	ADQI	Deviation (%)
5.0	10	4.0	20	3.0	30	2.0	40
4.8	12	3.8	22	2.8	32	1.8	42
4.6	14	3.6	24	2.6	34	1.6	44
4.4	16	3.4	26	2.4	36	1.4	46
4.2	18	3.2	28	2.2	38	1.2	48
						1.0	50

Table 3. Sample electricity test results specified for the preform stretching and extending process.

Medium	Quantity	Distribution	Minimum	Maximum	DQI
Electricity	6.95 kWh	Triangular	6.26	7.64	5

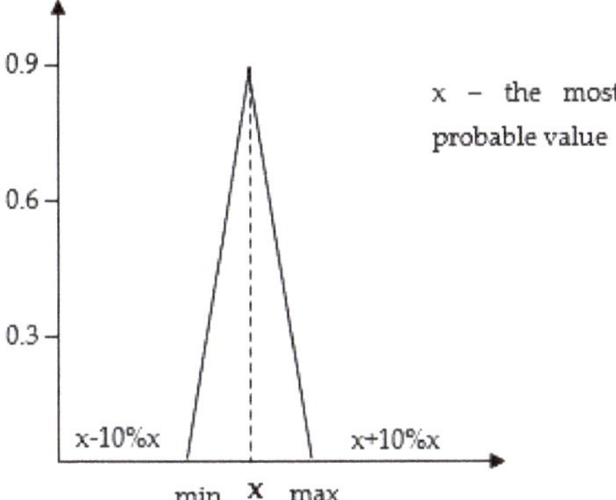

Figure 2. Dependence between the ADQI value and the shape of the triangular distribution curve in the process of polylactic acid bottle manufacturing.

The deviation ranges obtained on the basis of the ADQI values and the selected data distribution will be the input data in the next step of the stochastic uncertainty analysis approach using MC simulations, which will be described in detail in Section 2.1.2.

2.1.2. Stochastic Modeling Approach

The approach to stochastic modeling using MC simulation is a common method of quantitative uncertainty analysis in LCA. In this work, we used the MC simulation, because it allows one to quickly estimate the probability of results depending on the variability of input data, and thus identify the results with the highest sensitivity and uncertainty. The first assessment step is to correctly define the input parameters necessary to conduct the analysis properly. Triangular distributions based on two values—maximum and minimum—were used to estimate the impact category uncertainty. At this stage, the influence of the input parameters on the significant impact categories (impact categories which in total constituted 90% of damage area were considered as significant) was estimated. According to the adopted procedure, the values of the input parameter deviations depend on the ADQI result according to Table 2. For each input parameter, a value within the range of data variability was artificially generated and then values of selected impact categories were calculated. The procedure was repeated 1000 times to obtain the uncertainty distribution. This simulation was carried out using Sima Pro software.

2.1.3. Sensitivity Analysis

The correct interpretation of LCA results should assess the reliability of data and the uncertainty of results. For this purpose, among others, contribution analysis, perturbation analysis or sensitivity analysis are used [61]. Based on the sensitivity analysis, the usefulness of individual data can be inferred by indicating key variables that cannot be omitted in the analysis [62]. Performing a sensitivity analysis is considered to be one of the good practices in LCA [63,64]. There are three main approaches to sensitivity analysis mentioned in the literature: local, screening and global [62,65,66]. Local sensitivity analysis includes matrix perturbation and the once-at-a-time method. The method of elementary effects is a screening method designed by Morris and can be treated as an extension of the once-at-a-time method [66]. Global sensitivity analysis is mainly based on the analysis of variance of input variables. Among the methods of global sensitivity analysis, LCA uses the key issue analysis, the method of standardized regression coefficients, and Sobol sensitivity index [65].

In this work, the global sensitivity analysis method was used, namely the key issue analysis introduced by Heijungs [67,68]. This method is based on the analysis of the contribution of variables in variance and allows to determine the share of the input data uncertainty in the result uncertainty [68].

An analysis of data sensitivity was carried out for data which were in impact categories characterizing the overall impact of the PLA bottle shaping process in three areas: human health, ecosystem quality and resource avaliability. The procedure was carried out taking into account the variability of the analyzed parameters, using the MC simulation and the Crystal Ball (CB) software. The sensitivity analysis was presented in three formats: the grouped bar chart, the tornado type chart and the spider chart.

In the MC simulation the results obtained on the basis of LCA were used. The log-normal distribution, which is commonly used in analyzing data uncertainty, was used to estimate the value of the impact category in the Monte Carlo method [20,62]. Distribution parameters, i.e., population mean μ and standard deviation σ were adopted respectively: μ as the mean value of the impact category obtained as a result of data uncertainty analysis, σ as the standard deviation of the impact category obtained as a result of data uncertainty analysis. Tornado and spider charts were created based on the results of MC simulations. Relevant impact categories in LCA were used to generate charts.

2.2. Goal and Scope

The first step in LCA is to specify the objective and scope of the analysis, which can be determined based on the analysis and understanding of the product lifecycle. The research objective determines the degree of detail, thoroughness and scope of analyses, as well as the types of data needed to evaluate the lifecycle. To this end, the technological process of biodegradable polylactic acid (PLA) bottles shaping was subjected to evaluation. The process was broken down into six-unit operations, taking into account the demand for services and materials. The scope of the analysis covered the intake of pre-moulds into the heater, the heating of said moulds in the infrared heater, the stretching, extending, and pressure shaping of the pre-mould, as well as degassing and cooling the moulded bottles. The environmental impact of the bottle moulding process was performed using the ReCiPe 2016 method. The analysis covered 17 midpoint impact categories and three endpoint damage categories: human health, ecosystem quality and resource availability, which strictly correspond with three areas of protection: human health, ecosystem quality and resource scarcity [69,70]. Characterization factors from the endpoint level were obtained from the midpoint characterization factor using the constant midpoint to endpoint conversion factor [70,71]. The human health endpoint category includes impacts from the following midpoint level categories: particulate matter, trop. ozone formation (hum), ionizing radiation, ozone depletion, human toxicity (cancer), human toxicity (non-cancer), global warming, water use. The ecosystem quality endpoint category includes global warming, water use, freshwater ecotoxicity, freshwater eutrophication, trop. ozone (eco), terrestrial ecotoxicity, terrestrial acidification, land use/transformation, marine ecotoxicity. Finally, the resource availability category includes mineral resources and fossil resources. The LCA analysis was performed using the SimaPro software. Thereafter, relevant midpoint and endpoint categories were chosen. The categories with total contribution equal to 90% were considered as relevant. The aim of this analysis is to assess the uncertainty and accuracy of LCA results using the Monte Carlo method. The basis for looking for key factors was the uncertainty analysis of key results for which statistical and mathematical methods were used.

2.2.1. Functional Unit

The functional unit accepted for the analyses was determined based on data collected from the production plant. A 1 L PLA bottle was adopted as the functional unit.

2.2.2. System Boundaries

The analysis covers the entire cycle of the bottle shaping process, meaning that that all steps of this process are included, from the pre-mould delivery to the manufacturing company, up to the moment when beverage bottles are correctly shaped in the moulding process (Figure 3). Further processes, such as the filling of bottles with beverages, the labelling or storage/distribution thereof were not included in the system. Furthermore, the transportation of raw material and storage was also omitted.

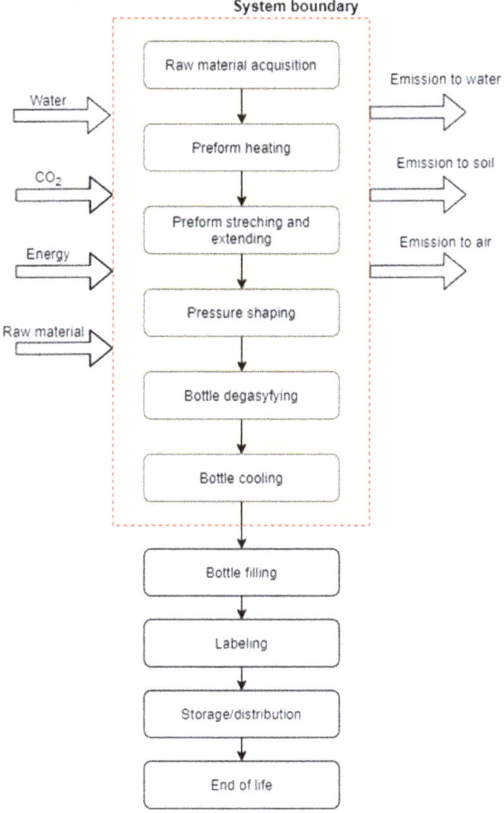

Figure 3. System boundaries for the environmental analysis of the polylactic acid (PLA) bottle production process.

2.2.3. Data Allocation

The allocation procedure is described in detail in ISO 14044 (clause 4.3.4. Allocation, Section 4.3.4.2 Allocation procedure) [2]. It is particularly important when considering multifunctional processes. For this reason, allocation, understood as the partitioning and attribution of environmental pressures to products/functions of the analysed system, is one of the most frequently applied multifunctional solutions. In the case of bottle moulding, the technological process was divided into smaller technological operations. Such a procedure does not require any partitioning; therefore allocation is not required.

2.3. Life Cycle Inventory

Designing a model for the analysis of the set of inputs and outputs is the second phase of LCA. The model reflects the whole product system, while its smaller elements represent technological operations. A technological operation should be understood as the smallest part of the system for which resource-related information is collected. Data collecting enables one to precisely specify the source of origin, geographic scope, representativeness and precision, all of these being indispensable elements of the uncertainty analysis [56]. The correct aggregation of input data helps to identify significant environmental points in the system.

Taking into account the confidential character of the LCI results presented in the study and company trade secrets, the values presented in Table 4 were changed by a coefficient ranging from 0.8 to 1.2. The data describing all of the process stages come from one company in Poland and pertain to the bottle moulding process taking place there. The figures apply to the year 2019. The modelling was performed using the Ecoinvent 3.2 database.

Table 4. Results LCI [55].

Technological Operations	Ecoinvent Activity	Amount
	Raw material acquisition	
PLA preform mass	Polylactide, granulate {GLO} market for/Alloc Def, S	18.24 g
Electrical energy	Electricity, medium voltage {PL} market for/Alloc Def, S	0.368 kWh
	Preform heating	
Electrical energy (infrared lamp 100 kW)		3.2 kWh
Electrical energy (infrared lamps 200 kW)	Electricity, medium voltage {PL} market for/Alloc Def, S	6.4 kWh
Electrical energy (supply chain)		0.16 kWh
	Preform stretching and extending	
Electrical energy	Electricity, medium voltage {PL} market for/Alloc Def, S	6.95 kWh
Compressed air	Compressed air, 1000 kPa gauge {RER} compressed air production/Alloc Def U	0.0016 kg/m^3
	Preform pressure shaping	
Electrical energy	Electricity, medium voltage {PL} market for/Alloc Def, S	5.66 kWh
	Bottle degasifying	
Electrical energy	Electricity, medium voltage {PL} market for/Alloc Def, S	1.01 kWh
	Bottle cooling	
Electrical energy	Electricity, medium voltage {PL} market for/Alloc Def, S	0.71 kWh
Water in a closed circulation	Tap water {Europe without Switzerland} market for/Alloc Def, S	2.4 m^3

3. Results

3.1. Contribution Analysis

Before commencing the uncertainty analysis, it is necessary to obtain LCA results. This paper analyses the quality, uncertainty and sensitivity of data in the LCA of the PLA bottle moulding process. [55] presents a complete set of the system characterisation results. The uncertainty analyses were performed using relevant impact categories only (with the summarised share of 90% at least). In the human health area, relevant impact categories included (Figure 4): fine particulate matter formation (50.68%), global warming human health (39.82%), water consumption human health (8.64%). As regards the ecosystem quality area, these were (Figure 5): global warming terrestrial ecosystems (35.31%), land use (30.77%) and water consumption, terrestrial acidification (15.44%), terrestrial acidification (13.59%). For the resource availability (Figure 6), the fossil resources scarcity category was significant (99.23%). The pre-form material and the electric energy consumption in the bottle moulding process were the most important determinants of the above-listed categories value.

In the damage category, the impact of pre-form material—PLA in this case—prevailed (Figure 7). In the category of human health, the impact of pre-mould (76.86%) was followed by the pre-form heating phase (7.72%) and the stretching and extending stages (5.93%), while in the ecosystem quality category, the role of pre-form heating (5.81) and bottle cooling (4.79%) can be noticed.

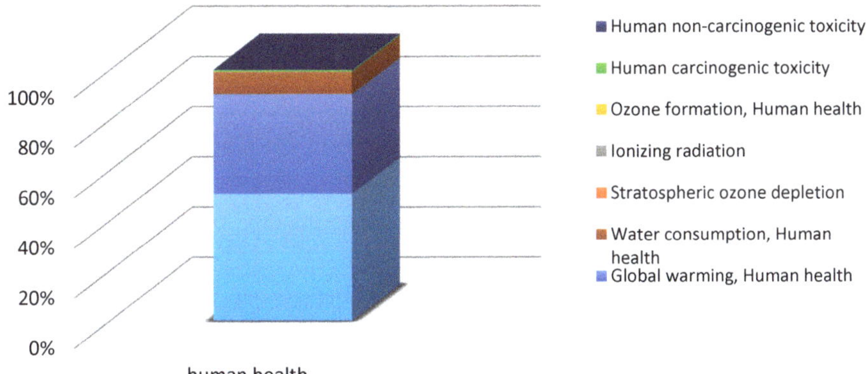

Figure 4. Contribution analysis in the human health damage category.

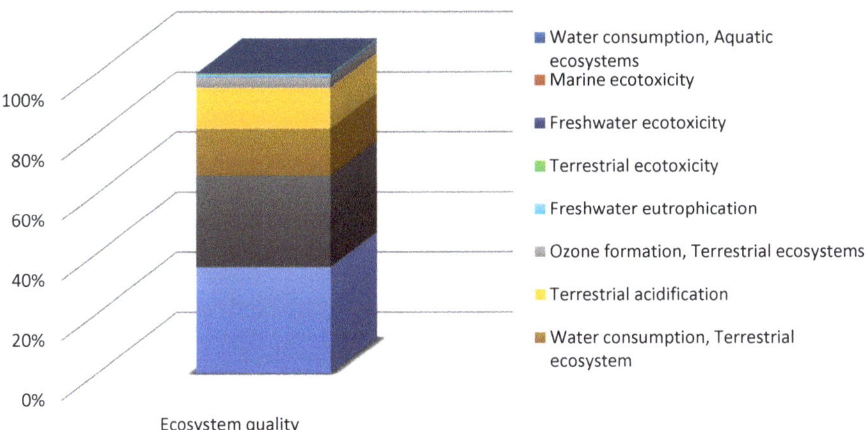

Figure 5. Contribution analysis in the ecosystem quality damage category.

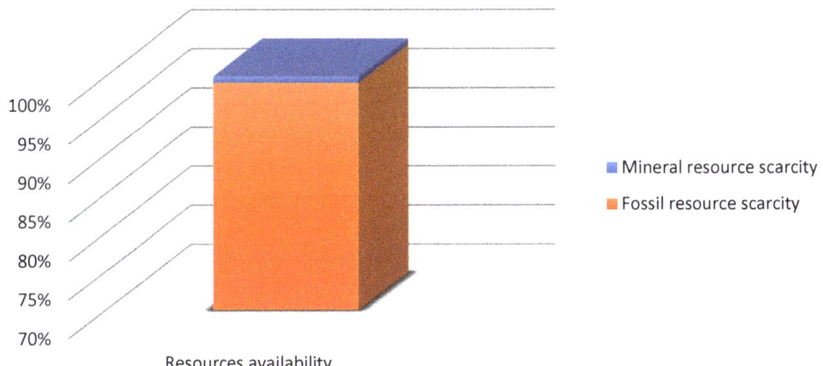

Figure 6. Contribution analysis in the resource availability damage category.

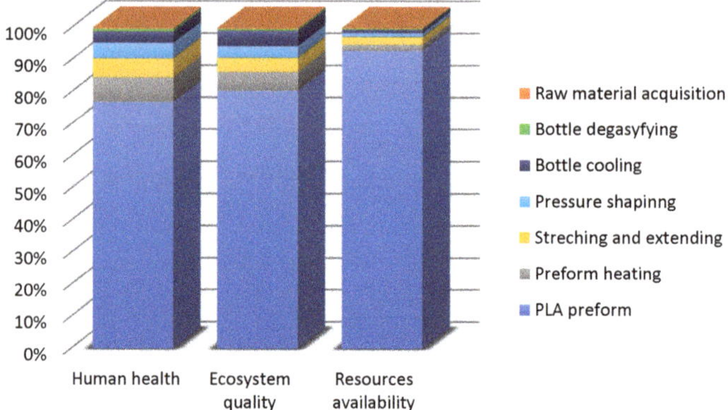

Figure 7. Contribution analysis of the operation of bottle shaping process in damage categories.

3.2. Semi-Quantitative DQI Aproach

For the first stage of the data quality and uncertainty assessment, the semi-quantitative DQI approach was adopted. The DQI and ADQI values were determined according to the procedure described in Section 2.1.1. A summary of results is presented in Table 5. The input data for each DQI were assigned 5 points, as they were obtained from measurements taken within one company and one process and were collected over a period of one year. Therefore, with such a high precision, data can be expected to provide reliable final results of the PLA bottle moulding process analysis. Following the multiplication of DQI by an appropriate weight coefficient, ADQI = 5 was obtained. Based on the ADQI value and Table 3, the deviation for each input value was determined. Using triangular distributions, the value provided in the inventory analysis is treated as the most probable value, and knowing its ADQI score, it is possible to define the minimum and maximum of input data.

Table 5. Values of the minimum and maximum for inventory elements of the process of polylactide bottle manufacturing.

Element LCI	Value		ADQI	Deviation (%)	Min.	Max.	Distribution
Raw material acquisition							
Electrical energy (three motors of the carousel)	0.368	kWh	5	10	0.33	0.404	Triangular
PLA preform mass	18.24	g	5	10	16.42	20.064	
Preform heating							
Electrical energy (infrared lamp 100 kW)	3.2	kWh	5	10	2.88	3.52	
Electrical energy (infrared lamps 200 kW)	6.4	kWh	5	10	5.76	7.04	Triangular
Electrical energy (supply chain)	0.16	kWh	5	10	0.14	0.176	
Preform stretching and extending							
Electrical energy	6.95	kWh	5	10	6.26	7.64	Triangular
Compressed air	0.0016	kg/m^3	5	10	0.00144	0.00176	
Preform pressure shaping							
Electrical energy	5.66	kWh	5	10	5.09	6.22	Triangular
Bottle degasifying							
Electrical energy	1.01	kWh	5	10	0.91	1.11	Triangular
Bottle cooling							
Electrical energy	0.71	kWh	5	10	0.64	0.78	Triangular
Water in closed circulation	2.4	m^3	5	10	2.16	2.64	

The data presented in Table 5 are the results of the estimated minimum and maximum values for the processes involved in PLA bottle shaping. According to the accepted minimum and maximum thresholds, it can be said that the value of energy consumption for the first operation ranges from +0.330 to +0.404.

3.3. Uncertainty Results

Using the function of uncertainty 1000 values of impact category on the basis of the MC analysis were generated for each entry and exit, according to the information about variability ranges of input parameters and their distribution (in this study triangular) presented in Section 3.2. During the simulation, values of input parameters such as: water, electrical energy use, material consumption and emissions are randomly selected for calculation of impact assessment. On this basis, distribution histograms of the selected relevant impact category results (Figures 3–8) were created for the randomly selected values of the input parameters. The analysis also provides basic parameters from the set of results obtained, such as standard deviation, mean, median and the coefficient of variation.

The unit point data (input) was used to analyze the relevant impact categories, and the results obtained are expressed in units corresponding to each endpoint damage category: human health, ecosystem quality and resources availability. In order to generate the diagrams presented in Figures 8–15, categories with a 90% share in damages were used.

Fine particle matter formation was the first significant category in the total impact of biodegradable bottle shaping process on human health. (Figure 8). For data generated in MC simulation, the total level of emissions was found to be 1.61×10^{-07} DALY. However, the value of the median was $= 1.59 \times 10^{-07}$ DALY, standard deviation $= 1.76 \times 10^{-08}$ DALY and coefficient of variation $= 10.96\%$.

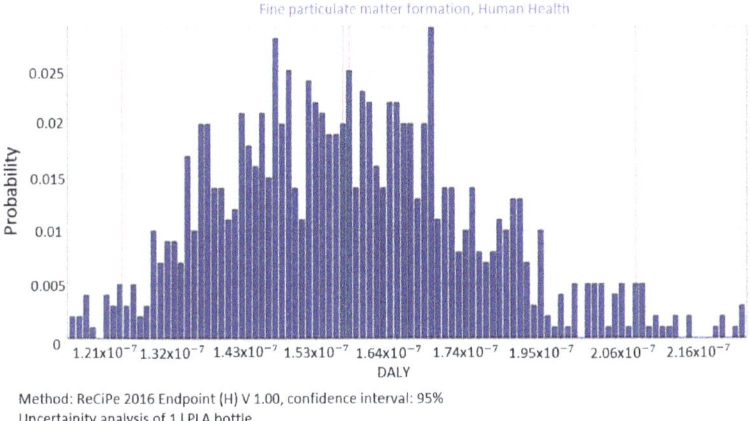

Figure 8. Results of the uncertainty analysis for the category of the formation of fine solid particles and their impact on human health.

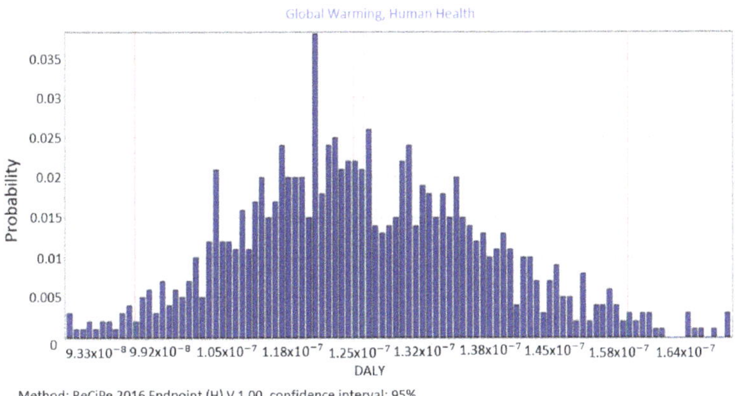

Figure 9. Results of the uncertainty analysis for the category of global warming and its impact on human health.

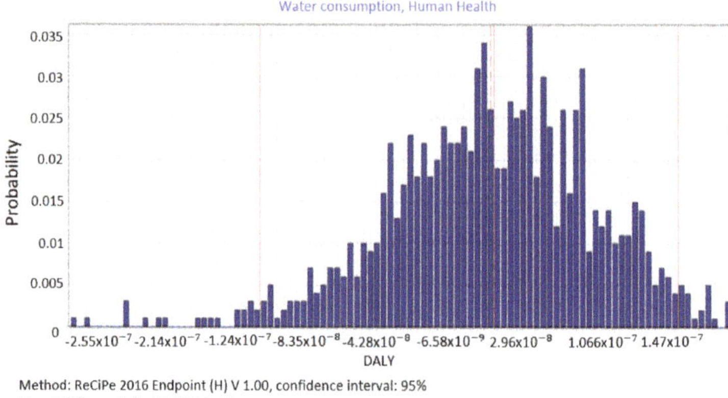

Figure 10. Results of the uncertainty analysis for the category of water consumption and its impact on human health.

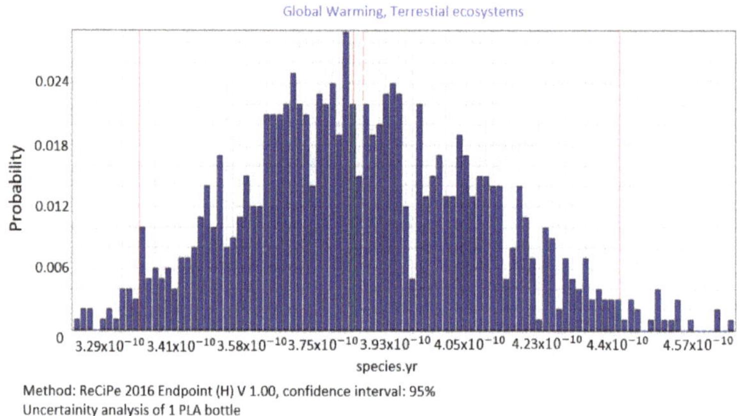

Figure 11. Results of the uncertainty analysis for the category of global warming and its influence on the ecosystem quality.

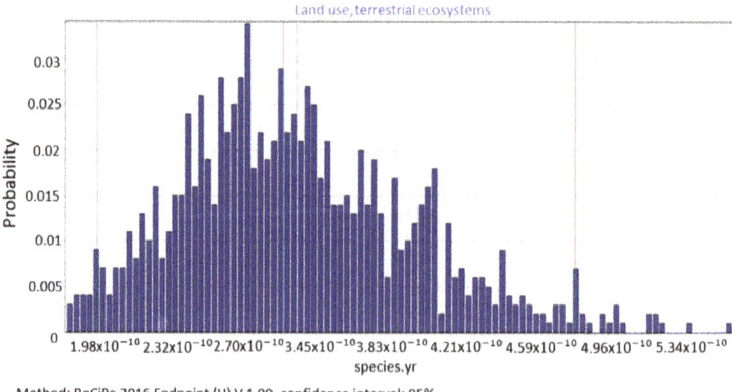

Figure 12. Results of the uncertainty analysis for the category of land use and its impact on the ecosystem quality.

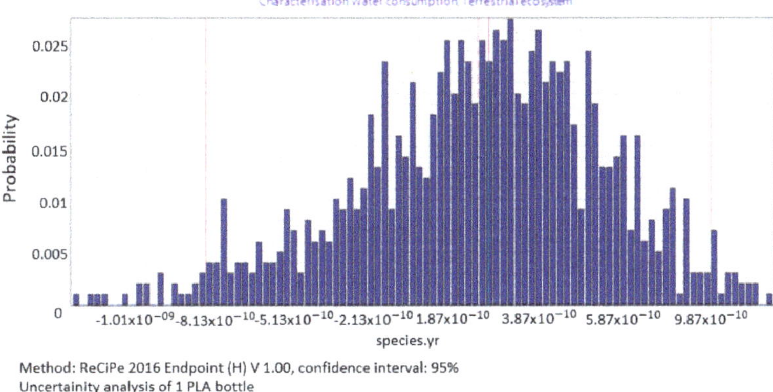

Figure 13. Results of the uncertainty analysis for the category of water consumption, terrestrial ecosystems and its impact on the ecosystem quality.

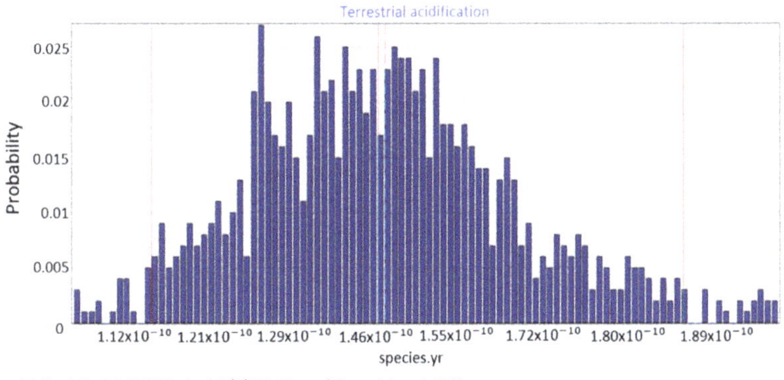

Figure 14. Results of the uncertainty analysis for the category of acidification and its impact on the ecosystem quality.

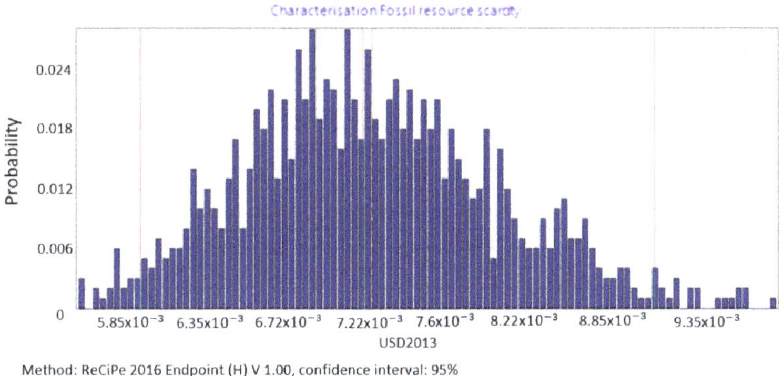

Figure 15. Results of the uncertainty analysis for the category of fossil resources scarcity and its impact on the resources availability.

The average impact of the manufacture of 1 L PLA bottles on the category of global warming corresponding to human health from 1000 scenarios of input data was 1.27×10^{-07} DALY (Figure 9), whereas, the remaining distribution parameters were as follows: median $= 1.26 \times 10^{-7}$ DALY, standard deviation $= 9.13 \times 10^{-09}$ DALY and the coefficient of variation $= 7.20\%$.

The analysis of all the stages involved in the process of PLA bottle shaping clearly shows that the accumulation of compounds has a high negative impact on the water consumption category (2.74×10^{-08} DALY), which adversely affects human health (Figure 10). The median for this is $= 3.05 \times 10^{-08}$ DALY, standard deviation $= 7.37 \times 10^{-09}$ DALY $= 26.8\%$.

Analyses of stochastic data for all the stages of PLA bottle shaping also covered the emissions of compounds that cause damage in the area of ecosystem quality by contributing to global warming (3.80×10^{-10} DALY) (Figure 11). The value of the median is $= 3.81 \times 10^{-10}$ species yr, standard deviation$= 2.75 \times 10^{-11}$ species yr and the coefficient of variation $= 7.20\%$.

The value of the median of the category of environmental impacts associated with land use (average: 3.32×10^{-10}) was found to be $= 3.17 \times 10^{-10}$ species yr, standard deviation $= 4.48 \times 10^{-11}$ species yr and coefficient of variation $= 25.1\%$ (Figure 12).

Categories showing significant impact in the ecosystem quality area include water consumption, terrestrial ecosystems, with a mean value of impacts equalling 1.66×10^{-10}, median $= 1.85 \times 10^{-10}$ species yr, a standard deviation of 4.62×10^{-11} species yr and a variation of 25.1% (Figure 13).

Figure 14 shows data for accumulated compounds that cause acidification (1.46×10^{-10})and their impact on the ecosystem quality. The values of the median $= 1.46 \times 10^{-10}$ species yr, standard deviation $= 1.32 \times 10^{-11}$ species yr and the coefficient of variation $= 9.01\%$.

Based on the data uncertainty analysis, a histogram of results for the fossil resources scarcity category was drawn up. The mean value of impact in the resource availability area equals 0.0072 USD2013, median $= 0.00715$ USD2013, standard deviation $= 0.00078$ USD2013, and variation $= 10.8\%$ (Figure 15).

3.4. Results of Sensitivity Analysis

Sensitivity analysis was performed for data represented by impact categories characterising the total impact of the biodegradable bottle preparation lifecycle in three damage areas human health, ecosystem quality and resources availability. Based on the uncertainty distribution of LCA results for relevant impact categories presented in Section 3.3, it was possible to perform the key issue analysis in order to assess the sensitivity of results in the given damage area. The procedure was performed using the MC simulation and the CB software. As Figure 16 and Table 6 show, the category of the formation of fine solid particles represents the greatest share of the variability of the total impact of bottle production on human health. Global warming is in second place. The lowest share in variability is observed in the case of water consumption.

Table 6. Results of the Monte Carlo (MC) simulation performed in Crystal Ball (CB) for the tornado-type sensitivity analysis of the significant categories of impact on human health.

	Human Health, 10^{-7} DALY				Input, 10^{-7} DALY		
Input Data	Lower Limit	Upper Limit	Range	Deviation Explained [1]	Lower Limit	Upper Limit	Base Case [2]
Fine particulate matter formation	2.94	3.39	0.45	70.28%	1.39	1.84	1.60
Global warming, human health	3.03	3.27	0.23	88.46%	1.15	1.38	1.26
Water consumption, human health	3.07	3.25	0.18	100.00%	0.19	0.37	0.26

[1] The explained deviation is cumulative, [2] The basic case for calculations in CB was the median value.

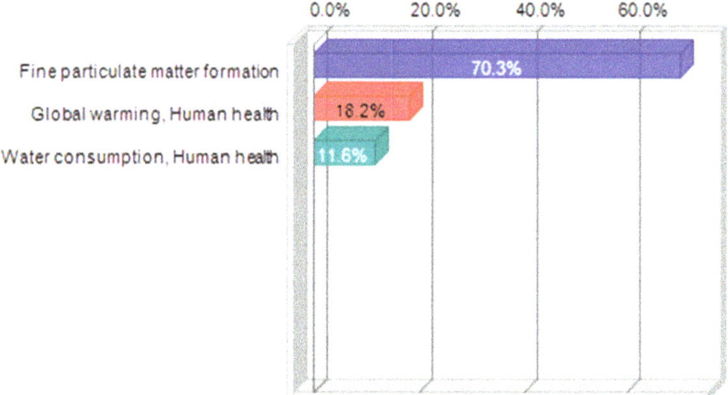

Figure 16. Shares of impact categories in the variance of the total impact of the bottle moulding process in the human health area.

The greatest deviations of the human health damage value are observed in the end values of impact categories (Figures 17 and 18). The increase in the fine particulate matter formation has the largest share in the increase of impacts on human health and at the same time a non-accurate estimation of this category will lead to the biggest errors in estimating the value of damages in the area of human health.

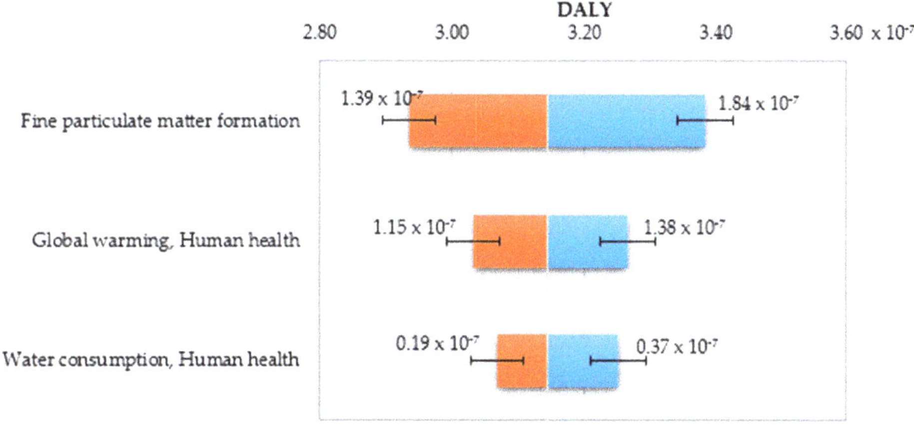

Figure 17. Tornado chart for the forecast of significant categories of impact on human health— error bars show the mean standard error value.

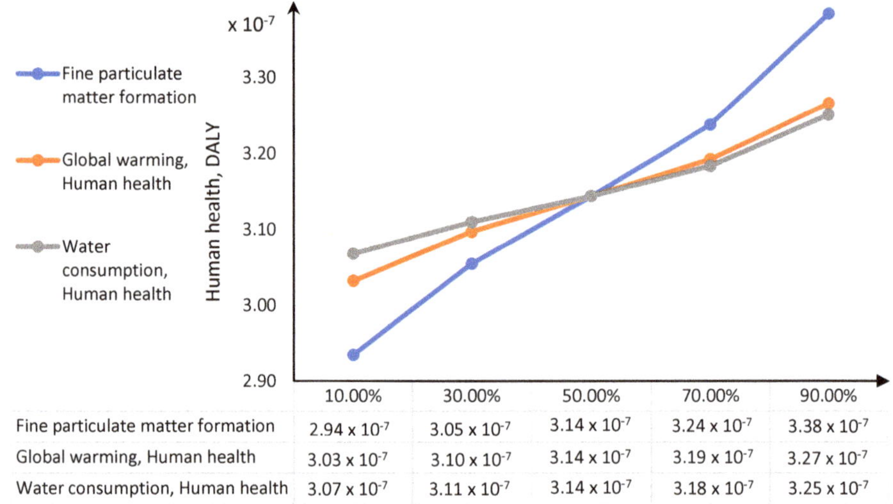

	10.00%	30.00%	50.00%	70.00%	90.00%
Fine particulate matter formation	2.94 x 10⁻⁷	3.05 x 10⁻⁷	3.14 x 10⁻⁷	3.24 x 10⁻⁷	3.38 x 10⁻⁷
Global warming, Human health	3.03 x 10⁻⁷	3.10 x 10⁻⁷	3.14 x 10⁻⁷	3.19 x 10⁻⁷	3.27 x 10⁻⁷
Water consumption, Human health	3.07 x 10⁻⁷	3.11 x 10⁻⁷	3.14 x 10⁻⁷	3.18 x 10⁻⁷	3.25 x 10⁻⁷

Figure 18. Spider chart of sensitivity for the human health damage category.

For damage category in area ecosystem quality the global water consumption, terrestrial ecosystem represents the greatest share and the next was land use (Table 7, Figure 19).

Table 7. Results of the MC simulation performed in CB for the tornado-type sensitivity analysis of the significant categories of impact on ecosystem quality.

Input Data	Human Health, 10^{-9} DALY				Input, 10^{-10} DALY		
	Lower Limit	Upper Limit	Range	Deviation Explained [1]	Lower Limit	Upper Limit	Base Case [2]
Water consumption, Terrestrial ecosystem	1.02	1.13	0.114	40.69%	1.13	2.27	1.60
Land use	1.01	1.12	0.113	81.06%	2.77	3.90	3.29
Global warming, Terrestrial ecosystems	1.03	1.10	0.070	96.46	3.46	4.16	3.79
Terrestrial acidification	1.05	1.08	0.033	100.00%	1.30	1.64	1.46

[1] The explained deviation is cumulative, [2] The basic case for calculations in CB was the median value.

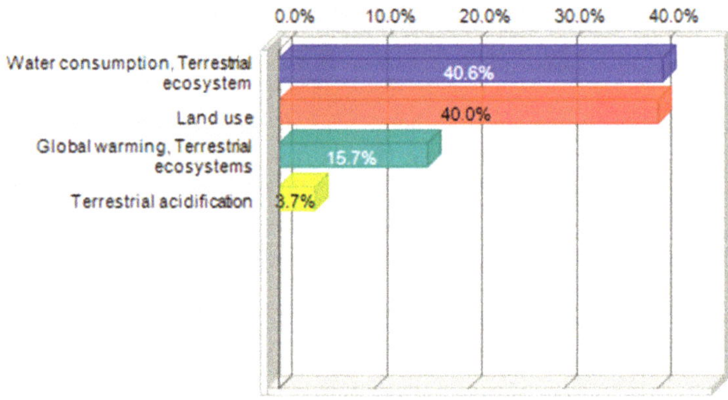

Figure 19. Shares of impact categories in the variance of the total impact of the bottle moulding process in the ecosystem quality area.

The greatest deviations of the ecosystem quality damage value are observed in the end values of impact categories (Figures 20 and 21). The increase in the water consumption terrestrial ecosystem category and land use values has the greatest share in the growth of impacts on the ecosystem quality and any imprecise estimation of these categories will lead to greatest errors in the estimation of damage values in the ecosystem quality area.

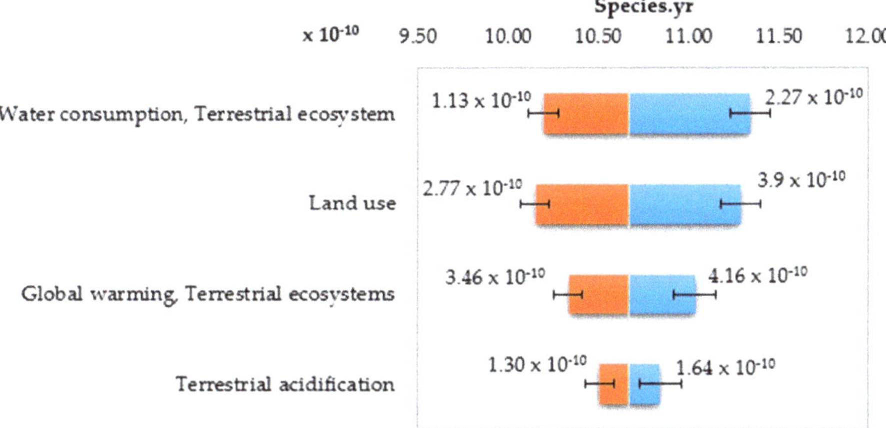

Figure 20. Tornado chart for the forecast of significant categories of impact on ecosystems—error bars show the mean standard error value.

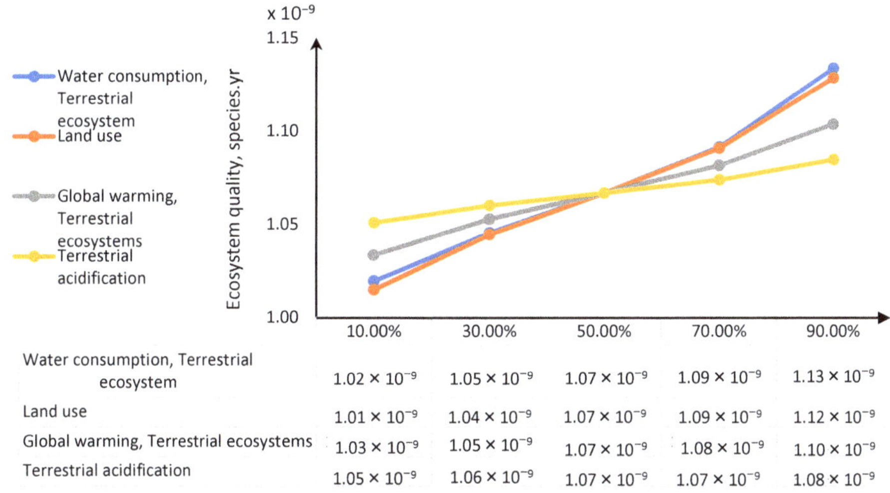

	10.00%	30.00%	50.00%	70.00%	90.00%
Water consumption, Terrestrial ecosystem	1.02×10^{-9}	1.05×10^{-9}	1.07×10^{-9}	1.09×10^{-9}	1.13×10^{-9}
Land use	1.01×10^{-9}	1.04×10^{-9}	1.07×10^{-9}	1.09×10^{-9}	1.12×10^{-9}
Global warming, Terrestrial ecosystems	1.03×10^{-9}	1.05×10^{-9}	1.07×10^{-9}	1.08×10^{-9}	1.10×10^{-9}
Terrestrial acidification	1.05×10^{-9}	1.06×10^{-9}	1.07×10^{-9}	1.07×10^{-9}	1.08×10^{-9}

Figure 21. Spider chart of sensitivity for the ecosystem quality damage category.

4. Discussion

The focus of this paper is on the evaluation and analysis of LCA results for the bottle moulding process. The results of contribution analysis presented in Section 3.1 show impact categories and stages of the bottle moulding process with the greatest effect on and share in damage categories for LCA results (Figures 4–7). When analysing shares of the process stages in midpoint impact categories, as well as in total impacts in damage areas, one has to conclude that pre-forms made of PLA take the first place and are followed by bottle degasifying and pressure shaping as far as negative environmental impacts are concerned (Figures 4–7). Stages of the bottle moulding process are closely related to input (inventory) data (Table 4), hence any solutions designed to improve the environmental impact of the bottle shaping process should be directed mainly towards reducing the PLA usage in the production of bottles and reducing the electric energy consumption in the pre-form heating processes (e.g., by means of replacing infrared lamps by less energy-intensive ones), pre-form stretching or cooling. The contribution analysis performed as the first step of the procedure enabled us to determine relevant impact categories to be subjected to the uncertainty and sensitivity analysis.

The procedure proposed in this study required us to perform data quality analysis in order to determine parameters of input data distribution for an uncertainty analysis to be conducted using the MC simulation. In this way, the process of stochastic search for the input parameters distribution was simplified by proposing a triangular distribution as PDF, with deviations from 10% to 50%, depending on the ADUI result. Determining the level of deviations is one of the elements of this search and affects the simulated values of expected results. The input data collected by us were characterised by high quality (ADQI = 5) and precision, as they were obtained directly form a manufacturing company. In the uncertainty analysis, a deviation of ±10% was used, according to the triangular distribution. The analysis of the diagrams presented in Figures 8–10 shows that the category associated with the use of water resources (Figure 10) has the largest variability range (deviation 26.8%), thus being the most critical input variable affecting human health throughout the life cycle of the product considered. Yet in the category of harmful impact on the ecosystem, the widest variability interval was found for the category of land use (deviation 25.1%, Figure 12) and water consumption connected, terrestrial ecosystems (deviation 25.1%, Figure 13). In the resource availability damage category, the variability of the prevailing impact category of fossil resources scarcity was 10.8% (Figure 15). The pre-form material was a key determinant of this category value. Input data deviations for the above-listed categories in the range of 10% translated into a significant variability of results, therefore the precision and accuracy of the input data will be of key importance to the values of these categories. With imprecise, low quality data, the result is least certain in these categories. As far as the final result is concerned, it is most important to measure the mass of the pre-form to be used in the blowing process and the consumption of water at the bottle cooling stage with greatest precision.

The results distribution characteristics obtained for significant impact categories as a result of the uncertainty analysis (i.e., mean, standard value) were used for performing the key issue analysis using the MC simulation and the CB software. The analysis enabled us to identify the impact categories with the greatest shares in the uncertainty of the final result in the damage category. The fine particulate matter formation category had the greatest share (69.5%) in the human health damage category variance. Depending on changes in this impact category, the total impact in the human health area will be subject to most serious changes. In the ecosystem quality area, water consumption had the greatest share as regards terrestrial ecosystem and land use. As regards these impact categories, the greatest changes should be also expected wherever even a slight decrease in the input parameters value takes place in order to improve the environmental impact of the PLA bottle shaping process. In the group of impact categories referred to above, low quality input data would introduce significant uncertainty of the final impact results in damage categories. The final impact result in the damage category is most sensitive to changes in the impact categories referred to above.

The method proposed for the evaluation of data quality and uncertainty permits to easily identify key parameters that affect the final result of LCA (without the need to calculate complex endpoint coefficients and parameters of the results distribution shape). A single procedure combines three types of analyses: the quality of data, uncertainty of results and sensitivity with respect to input (inventory) data and impact categories, thereby enabling one to determine qualitative and quantitative relationships between input data, impact categories and impact areas, as presented in Figure 22. Clearly, although some categories show a relatively high uncertainty, they will not have any significant effect on the final result, e.g., the water consumption category, despite its high uncertainty, will not have any serious impact on the human health damage category. But then, in the case of the ecosystem quality damage category, two impact categories will cause the greatest uncertainty of results: water consumption and land use, although they do not represent the greatest shares in this damage category.

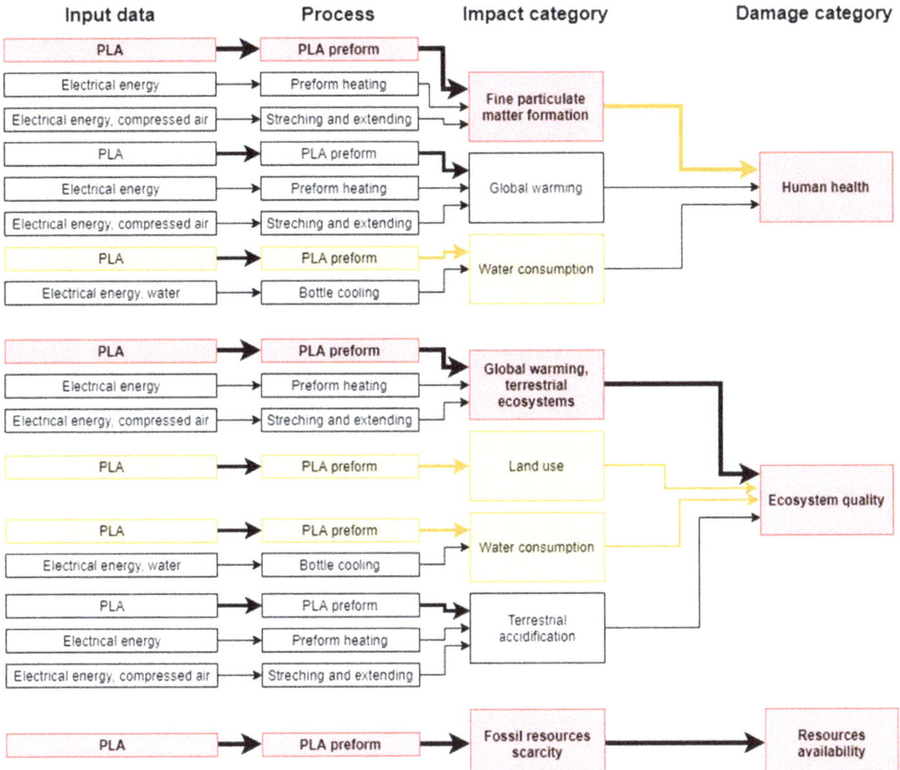

Figure 22. Paths of key links between input data and impact and damage categories. The figure presents input data for only significant stages of the PLA bottle shaping process and relevant impact categories. Bold arrows represent inputs with the greatest share in the process or category. Chains marked in red are those with the greatest share in the damage category, while yellow chains bring the greatest uncertainty with respect to the final result of the impact and damage category.

In practice, data uncertainty assessment procedures come down to one type of analysis, e.g., the semi-quantitative DQI approach [20], stochastic modelling with the use of the MC simulation [22,48], sensitivity analysis [10,65]. Combinations of several methods are not common, due to the significant input of labour and time required and a great number of data than need to be entered and collected [62,72]. However, the method proposed here, is based on inventory outcomes and each step produces a set of inputs for the next stage, thereby providing a network of links between input data and impact categories, as well as damage categories (Figure 22).

The types of variables distribution used in these studies (triangle distribution for input data and lognormal distribution for impact categories), as well as range endpoints, are not without effect on the value of results obtained either. Using other distributions might lead to small differences in the uncertainty results, as shown in [20,38,65,72]. It should be pointed out that the input data distribution is based on the semi-quantitative DQI approach result and it is largely the quality of data that will determine results of any further stages.

5. Conclusions

We have achieved the objective of the paper by proposing a method for the evaluation of uncertainty and precision of LCA results based on the DQI semi-quantitative approach, stochastic modelling and sensitivity analysis. The method permits to simply determine dependencies between the precision of input data and impact and damage categories, by identifying key elements that affect the LCA result. The strong point of the methodology used is that not much data need to be collected for the analyses.

The procedure identified input data of the PLA bottle shaping process, as well as sensitive impact categories. The material used for producing the pre-form (PLA) represents the input which has the greatest effect on the result of environmental impacts of the PLA bottle shaping process. The fact that the mass of material is reduced during the bottle shaping process is the greatest contributor to the reduction of environmental impacts. At the same time, the accuracy and precision of the PLA mass estimation will be the key element affecting the final result uncertainty, while the accuracy of water and electric energy consumption estimations will be less important. Impact categories with the greatest uncertainty include water consumption with respect to human health and land use and water consumption with respect to ecosystem quality. On the other hand, the uncertainty and development of the final result value in the human health damage category depend mostly on the fine particulate matter formation category and in the ecosystem quality damage category—on water consumption and land use. In the area of resources availability, the impact category of fossil resources scarcity is mainly responsible for both the uncertainty and value of results, its values being determined by the value of PLA pre-form material used in the process. In the context of improving the environmental balance of the bottle shaping process, the consumption of energy in the processes of heating, stretching and cooling should be brought down and water consumption should be reduced along with the pre-form material consumption.

Author Contributions: Conceptualization, P.B.-W.; J.F.; methodology, P.B.-W.; A.T.; software, P.B.-W., K.P.; validation, A.T., J.F., M.O., M.B. and R.K.; formal analysis, P.B.-W., A.T. and K.P.; investigation, A.T. and P.B.-W.; resources, P.B.-W.; data curation A.T., J.F., P.B.-W. and M.O.; writing—original draft preparation, P.B.-W.; writing—Review and editing, P.B.-W., K.P., A.T., and W.K.; visualization, R.K., M.B., P.B.-W., W.K. and A.T.; supervision, R.K., M.O., J.F., W.K. and K.P.; project administration, M.B., P.B.-W. All authors have read and agreed to the published version of the manuscript.

Funding: "The project/research was financed in the framework of the project Lublin University of Technology-Regional Excellence Initiative, funded by the Polish Ministry of Science and Higher Education (contract no. 030/RID/2018/19)". "Projekt/Badania zostały sfinansowane z Projektu Politechnika Lubelska-Regionalna Inicjatywa Doskonałości ze środków Ministerstwa Nauki i Szkolnictwa Wyższego na podstawie umowy nr 030/RID/2018/19".

Conflicts of Interest: The authors declare no conflict of interest.

References

1. International Organization for Standardization. *ISO 14040:2006—Environmental Management—Life Cycle Assessment—Principles and Framework*; International Organization for Standardization: Geneva, Switzerland, 2006.

2. International Organization for Standardization. *ISO 14044:2006—Environmental Management—Life Cycle Assessment—Requirements and Guidelines*; International Organization for Standardization: Geneva, Switzerland, 2006.

3. Mannheim, V.; Fehér, Z.S.; Siménfalvi, Z. Innovative solutions for the building industry to improve sustainability performance with Life Cycle Assessment modelling. In *Solutions for Sustainable Development*; CRC Press: Boca Raton, FL, USA; London, UK; New York, NY, USA, 2019; pp. 245–253.

4. Tomporowski, A.; Piasecka, I.; Flizikowski, J.; Kasner, R.; Kruszelnicka, W.; Mroziński, A.; Bieliński, K. Comparison Analysis of Blade Life Cycles of Land-Based and Offshore Wind Power Plants. *Pol. Marit. Res.* **2018**, *25*, 225–233. [CrossRef]

5. Huijbregts, M.A.J. Application of uncertainty and variability in LCA. *Int. J. Life Cycle Assess.* **1998**, *3*, 273. [CrossRef]

6. Heijungs, R.; Huijbregts, M.A.J. A Review of Approaches to Treat Uncertainty in LCA. In Proceedings of the International Congress on Environmental Modelling and Software, Osnabrück, Germany, 14–17 June 2004; p. 197.

7. Magrassi, F.; Del Borghi, A.; Gallo, M.; Strazza, C.; Robba, M. Optimal Planning of Sustainable Buildings: Integration of Life Cycle Assessment and Optimization in a Decision Support System (DSS). *Energies* **2016**, *9*, 490. [CrossRef]

8. Lloyd, S.M.; Ries, R. Characterizing, Propagating, and Analyzing Uncertainty in Life-Cycle Assessment: A Survey of Quantitative Approaches. *J. Ind. Ecol.* **2007**, *11*, 161–179. [CrossRef]

9. Pomponi, F.; D'Amico, B.; Moncaster, A.M. A Method to Facilitate Uncertainty Analysis in LCAs of Buildings. *Energies* **2017**, *10*, 524. [CrossRef]

10. Heijungs, R. Sensitivity coefficients for matrix-based LCA. *Int. J. Life Cycle Assess.* **2010**, *15*, 511–520. [CrossRef]

11. Heijungs, R.; Frischknecht, R. Representing Statistical Distributions for Uncertain Parameters in LCA. Relationships between mathematical forms, their representation in EcoSpold, and their representation in CMLCA (7 pp). *Int. J. Life Cycle Assess.* **2005**, *10*, 248–254. [CrossRef]

12. Maurice, B.; Frischknecht, R.; Coelho-Schwirtz, V.; Hungerbühler, K. Uncertainty analysis in life cycle inventory. Application to the production of electricity with French coal power plants. *J. Clean. Prod.* **2000**, *8*, 95–108. [CrossRef]

13. Heijungs, R. Identification of key issues for further investigation in improving the reliability of life-cycle assessments. *J. Clean. Prod.* **1996**, *4*, 159–166. [CrossRef]

14. Wang, E.; Shen, Z.; Neal, J.; Shi, J.; Berryman, C.; Schwer, A. An AHP-weighted aggregated data quality indicator (AWADQI) approach for estimating embodied energy of building materials. *Int. J. Life Cycle Assess.* **2012**, *17*, 764–773. [CrossRef]

15. Canter, K.G.; Kennedy, D.J.; Montgomery, D.C.; Keats, J.B.; Carlyle, W.M. Screening stochastic Life Cycle assessment inventory models. *Int. J. Life Cycle Assess.* **2002**, *7*, 18. [CrossRef]

16. Frischknecht, R.; Jungbluth, N.; Althaus, H.J.; Doka, G.; Dones, R.; Heck, T.; Hellweg, S.; Hischier, R.; Nemecek, T.; Rebitzer, G.; et al. The ecoinvent Database: Overview and Methodological Framework (7 pp). *Int. J. Life Cycle Assess.* **2005**, *10*, 3–9. [CrossRef]

17. Kennedy, D.J.; Montgomery, D.C.; Rollier, D.A.; Keats, J.B. Data Quality: Assessing Input Data Uncertainty in Life Cycle Assessment Inventory Models. *Int. J. Life Cycle Assess.* **1997**, *2*, 229–239. [CrossRef]

18. Weidema, B.P.; Wesnæs, M.S. Data quality management for life cycle inventories—An example of using data quality indicators. *J. Clean. Prod.* **1996**, *4*, 167–174. [CrossRef]

19. Weidema, B.P. Multi-user test of the data quality matrix for product life cycle inventory data. *Int. J. Life Cycle Assess.* **1998**, *3*, 259–265. [CrossRef]

20. Baek, C.Y.; Tahara, K.; Park, K.H. Parameter Uncertainty Analysis of the Life Cycle Inventory Database: Application to Greenhouse Gas Emissions from Brown Rice Production in IDEA. *Sustainability* **2018**, *10*, 922. [CrossRef]

21. Muller, S.; Lesage, P.; Ciroth, A.; Mutel, C.; Weidema, B.P.; Samson, R. The application of the pedigree approach to the distributions foreseen in ecoinvent v3. *Int. J. Life Cycle Assess.* **2016**, *21*, 1327–1337. [CrossRef]

22. Sonnemann, G.W.; Schuhmacher, M.; Castells, F. Uncertainty assessment by a Monte Carlo simulation in a life cycle inventory of electricity produced by a waste incinerator. *J. Clean. Prod.* **2003**, *11*, 279–292. [CrossRef]

23. International Organization for Standarization. *ISO/TR 14047:2003—Environmental Management—Life Cycle Impact Assessment—Examples of application of ISO 14042*; International Organization for Standardization: Geneva, Switzerland, 2003.

24. International Organization for Standarization. *ISO/TS 14048:2002—Environmental Management—Life Cycle Assessment—Data Documentation Format*; International Organization for Standardization: Geneva, Switzerland, 2002.

25. Hedbrant, J.; Sörme, L. Data Vagueness and Uncertainties in Urban Heavy-Metal Data Collection. *Water Air Soil Pollut. Focus* **2001**, *1*, 43–53. [CrossRef]

26. Lewandowska, A. Environmental life cycle assessment as a tool for identification and assessment of environmental aspects in environmental management systems (EMS) part 1: Methodology. *Int. J. Life Cycle Assess.* **2011**, *16*, 178–186. [CrossRef]

27. Shen, L.; Worrell, E.; Patel, M.K. Open-loop recycling: A LCA case study of PET bottle-to-fibre recycling. *Resour. Conserv. Recycl.* **2010**, *55*, 34–52. [CrossRef]

28. Lewandowska, A.; Matuszak-Flejszman, A. Eco-design as a normative element of Environmental Management Systems—The context of the revised ISO 14001:2015. *Int. J. Life Cycle Assess.* **2014**, *19*, 1794–1798. [CrossRef]

29. Steen, B. On uncertainty and sensitivity of LCA-based priority setting. *J. Clean. Prod.* **1997**, *5*, 255–262. [CrossRef]

30. Kowalski, Z.; Kulczycka, J. Ocena cyklu życia LCA jako podstawowy czynnik oceny czystszych produkcji. Odzysk odpadów—Technologie i możliwości. In *Materiały Konfrerencji, Waste Recycling*; Wydawnictwo IGSMiE PAN: Kraków, Poland, 2005; pp. 23–33.

31. Lewandowska, A.; Foltynowicz, Z. Eco-Design as a New Trend in Packing Materials Production. *Zesz. Nauk. Akad. Ekon. Pozn.* **2007**, *93*, 78–86.

32. Heijungs, R.; Suh, S. *The Computational Structure of Life Cycle Assessment*; Springer Science & Business Media: New York, NY, USA, 2002.

33. André, J.C.S.; Lopes, D.R. On the use of possibility theory in uncertainty analysis of life cycle inventory. *Int. J. Life Cycle Assess.* **2012**, *17*, 350–361. [CrossRef]

34. Benetto, E.; Dujet, C.; Rousseaux, P. Integrating fuzzy multicriteria analysis and uncertainty evaluation in life cycle assessment. *Environ. Model. Softw.* **2008**, *23*, 1461–1467. [CrossRef]

35. Egilmez, G.; Gumus, S.; Kucukvar, M.; Tatari, O. A fuzzy data envelopment analysis framework for dealing with uncertainty impacts of input–output life cycle assessment models on eco-efficiency assessment. *J. Clean. Prod.* **2016**, *129*, 622–636. [CrossRef]

36. Von Bahr, B.; Steen, B. Reducing epistemological uncertainty in life cycle inventory. *J. Clean. Prod.* **2004**, *12*, 369–388. [CrossRef]

37. Lasvaux, S.; Schiopu, N.; Habert, G.; Chevalier, J.; Peuportier, B. Influence of simplification of life cycle inventories on the accuracy of impact assessment: Application to construction products. *J. Clean. Prod.* **2014**, *79*, 142–151. [CrossRef]

38. Baek, C.Y.; Park, K.H.; Tahara, K.; Chun, Y.Y. Data Quality Assessment of the Uncertainty Analysis Applied to the Greenhouse Gas Emissions of a Dairy Cow System. *Sustainability* **2017**, *9*, 1676. [CrossRef]

39. Chou, J.S.; Yeh, K.C. Life cycle carbon dioxide emissions simulation and environmental cost analysis for building construction. *J. Clean. Prod.* **2015**, *101*, 137–147. [CrossRef]

40. Niero, M.; Pizzol, M.; Bruun, H.G.; Thomsen, M. Comparative life cycle assessment of wastewater treatment in Denmark including sensitivity and uncertainty analysis. *J. Clean. Prod.* **2014**, *68*, 25–35. [CrossRef]

41. Geisler, G.; Hellweg, S.; Hungerbühler, K. Uncertainty Analysis in Life Cycle Assessment (LCA): Case Study on Plant-Protection Products and Implications for Decision Making (9 pp + 3 pp). *Int. J. Life Cycle Assess.* **2005**, *10*, 184–192. [CrossRef]

42. Hong, J.; Shen, G.Q.; Peng, Y.; Feng, Y.; Mao, C. Uncertainty analysis for measuring greenhouse gas emissions in the building construction phase: A case study in China. *J. Clean. Prod.* **2016**, *129*, 183–195. [CrossRef]

43. Chevalier, J.L.; Téno, J.F.L. Life cycle analysis with ill-defined data and its application to building products. *Int. J. Life Cycle Assess.* **1996**, *1*, 90–96. [CrossRef]

44. Reap, J.; Roman, F.; Duncan, S.; Bras, B. A survey of unresolved problems in life cycle assessment. *Int. J. Life Cycle Assess.* **2008**, *13*, 290. [CrossRef]

45. Miller, S.A.; Moysey, S.; Sharp, B.; Alfaro, J. A Stochastic Approach to Model Dynamic Systems in Life Cycle Assessment. *J. Ind. Ecol.* **2013**, *17*, 352–362. [CrossRef]

46. Thiel, C.L.; Campion, N.; Landis, A.E.; Jones, A.K.; Schaefer, L.A.; Bilec, M.M. A Materials Life Cycle Assessment of a Net-Zero Energy Building. *Energies* **2013**, *6*, 1125–1141. [CrossRef]

47. Cambridge University Built Environment Sustainability (CUBES). Risk and Uncertainty in Embodied Carbon Assessment. In Proceedings of the Cambridge University Built Environment Sustainability (CUBES) Embodied Carbon Symposium 2016, Cambridge, UK, 11 April 2016.

48. Peters, G.P. Efficient algorithms for Life Cycle Assessment, Input-Output Analysis, and Monte-Carlo Analysis. *Int. J. Life Cycle Assess.* **2006**, *12*, 373. [CrossRef]

49. Heijungs, R.; Lenzen, M. Error propagation methods for LCA—A comparison. *Int. J. Life Cycle Assess.* **2014**, *19*, 1445–1461. [CrossRef]

50. Heijungs, R.; Tan, R.R. Rigorous proof of fuzzy error propagation with matrix-based LCI. *Int. J. Life Cycle Assess.* **2010**, *15*, 1014–1019. [CrossRef]

51. Imbeault-Tétreault, H.; Jolliet, O.; Deschênes, L.; Rosenbaum, R.K. Analytical Propagation of Uncertainty in Life Cycle Assessment Using Matrix Formulation. *J. Ind. Ecol.* **2013**, *17*, 485–492. [CrossRef]

52. Ciroth, A.; Fleischer, G.; Steinbach, J. Uncertainty calculation in life cycle assessments. *Int. J. Life Cycle Assess.* **2004**, *9*, 216. [CrossRef]

53. Piotrowska, K.; Kruszelnicka, W.; Baldowska-Witos, P.; Kasner, R.; Rudnicki, J.; Tomporowski, A.; Flizikowski, J.; Opielak, M. Assessment of the Environmental Impact of a Car Tire throughout Its Lifecycle Using the LCA Method. *Materials* **2019**, *12*, 4177. [CrossRef] [PubMed]

54. Bałdowska-Witos, P.; Kruszelnicka, W.; Kasner, R.; Tomporowski, A.; Flizikowski, J.; Mrozinski, A. Impact of the plastic bottle production on the natural environment. Part 2. Analysis of data uncertainty in the assessment of the life cycle of plastic beverage bottles using the Monte Carlo technique. *Przem. Chem.* **2019**, *98*, 1668–1672.

55. Bałdowska-Witos, P.; Kruszelnicka, W.; Kasner, R.; Tomporowski, A.; Flizikowski, J.; Kłos, Z.; Piotrowska, K.; Markowska, K. Application of LCA Method for Assessment of Environmental Impacts of a Polylactide (PLA) Bottle Shaping. *Polymers* **2020**, *12*, 388. [CrossRef] [PubMed]

56. Edelen, A.; Ingwersen, W. *Guidance on Data Quality Assessment for Life Cycle Inventory Data*; United States Environmental Protection Agency: Washington, DC, USA, 2016.

57. Bulle, C.; Margni, M.; Patouillard, L.; Boulay, A.M.; Bourgault, G.; De Bruille, V.; Cao, V.; Hauschild, M.; Henderson, A.; Humbert, S.; et al. IMPACT World+: A globally regionalized life cycle impact assessment method. *Int. J. Life Cycle Assess.* **2019**, *24*, 1653–1674. [CrossRef]

58. Fregonara, E.; Ferrando, D.G.; Pattono, S. Economic–Environmental Sustainability in Building Projects: Introducing Risk and Uncertainty in LCCE and LCCA. *Sustainability* **2018**, *10*, 1901. [CrossRef]

59. Ogilvie, S.; Collins, M.; Aumônier, S. *Life Cycle Assessment of the Management Options for Waste Tyres*; Environment Agency: Bristol, UK, 2004; ISBN 978-1-84432-289-3.

60. Lewandowska, A. *Środowiskowa Ocena Cyklu Życia Produktu na Przykładzie Wybranych Typów Pomp Przemysłowych*; Wydawnictwo Akademii Ekonomicznej w Poznaniu: Poznań, Poland, 2006.

61. International Organization for Standardization. *ISO 14042:2002—Environmental Management—Life Cycle Assessment—Life Cycle Impact Assessment*; International Organization for Standardization: Geneva, Switzerland, 2002.

62. Bieda, B. *Metoda Monte Carlo w Ocenie Niepewności w Stochastycznej Analizie w Przemyśle Stalowniczym i Inżynierii Środowiska*; Wydawnictwa AGH: Kraków, Poland, 2014.

63. Kennedy, P. *A Guide to Econometrics*; MIT Press: Cambridge, MA, USA, 2003; ISBN 978-0-262-61183-1.

64. Saltelli, A.; Annoni, P. How to avoid a perfunctory sensitivity analysis. *Environ. Model. Softw.* **2010**, *25*, 1508–1517. [CrossRef]

65. Groen, E.A.; Bokkers, E.A.M.; Heijungs, R.; de Boer, I.J.M. Methods for global sensitivity analysis in life cycle assessment. *Int. J. Life Cycle Assess.* **2017**, *22*, 1125–1137. [CrossRef]

66. Groen, E.A.; Heijungs, R.; Bokkers, E.A.M.; de Boer, I.J.M. Sensitivity analysis in life cycle assessment. In Proceedings of the 9th International Conference on Life Cycle Assessment in the Agri-Food Sector, San Francisco, CA, USA, 8–10 October 2014; ACLCA: Vashon, WA, USA, 2014.

67. Heijungs, R. The use of matrix perturbation theory for addressing sensitivity and uncertainty issues in LCA. In Proceedings of the Fifth International Conference on EcoBalance—Practical Tools and Thoughtful Principles for Sustainability, Tsukuba, Japan, 6–8 November 2002; p. 4.

68. Heijungs, R.; Suh, S.; Kleijn, R. Numerical Approaches to Life Cycle Interpretation—The case of the Ecoinvent'96 database (10 pp). *Int. J. Life Cycle Assess.* **2005**, *10*, 103–112. [CrossRef]

69. Huibregts, M.A.J.; Steinmann, Z.J.N.; Elshout, P.M.F.; Stam, G.; Verones, F.; Vieira, M.D.M.; Hollander, A.; Zijp, M.; Van, Z. *ReCiPe 2016 A Harmonized Life Cycle Impact Assessment Method at Midpoint and Endpoint Level Report I: Characterization*; RIVM Report 2016-0104; National Institute for Public Health and the Environment: Bilthoven, The Netherlands, 2016.

70. Huijbregts, M.A.J.; Steinmann, Z.J.N.; Elshout, P.M.F.; Stam, G.; Verones, F.; Vieira, M.; Zijp, M.; Hollander, A.; van Zelm, R. ReCiPe2016: A harmonised life cycle impact assessment method at midpoint and endpoint level. *Int. J. Life Cycle Assess.* **2017**, *22*, 138–147. [CrossRef]
71. Bakshi, B.R. *Sustainable Engineering: Principles and Practice*; Cambridge University Press: Cambridge, UK, 2019; ISBN 978-1-108-42045-7.
72. Mutel, C.L.; de Baan, L.; Hellweg, S. Two-Step Sensitivity Testing of Parametrized and Regionalized Life Cycle Assessments: Methodology and Case Study. *Environ. Sci. Technol.* **2013**, *47*, 5660–5667. [CrossRef] [PubMed]

Article

Biodegradable Polylactide–Poly(3-Hydroxybutyrate) Compositions Obtained via Blending under Shear Deformations and Electrospinning: Characterization and Environmental Application

Svetlana Rogovina [1,*], Lubov Zhorina [1], Andrey Gatin [1], Eduard Prut [1], Olga Kuznetsova [1], Anastasia Yakhina [1], Anatoliy Olkhov [1,2], Naum Samoylov [3], Maxim Grishin [1], Alexey Iordanskii [1] and Alexandr Berlin [1]

[1] Semenov Institute of Chemical Physics, Russian Academy of Sciences, 119991 Moscow, Russia; 301119481@bk.ru (L.Z.); akgatin@yandex.ru (A.G.); evprut@mail.ru (E.P.); 123zzz321@inbox.ru (O.K.); nastya_0496@mail.ru (A.Y.); aolkhov72@yandex.ru (A.O.); mvgrishin68@yandex.ru (M.G.); aljordan08@gmail.com (A.I.); berlin@chph.ras.ru (A.B.)

[2] Department of Chemistry and Physics, Plekhanov Russian University of Economics, 117997 Moscow, Russia

[3] Department of Petrochemistry and Chemical Technology, Ufa State Petroleum Technological University, 450062 Ufa, Russia; naum.samoilow@yandex.ru

* Correspondence: s.rogovina@mail.ru

Received: 21 April 2020; Accepted: 7 May 2020; Published: 10 May 2020

Abstract: Compositions of polylactide (PLA) and poly(3-hydroxybutyrate) (PHB) thermoplastic polyesters originated from the nature raw have been obtained by blending under shear deformations and electrospinning methods in the form of films and nanofibers as well as unwoven nanofibrous materials, respectively. The degrees of crystallinity calculated on the base of melting enthalpies and thermal transition temperatures for glassy state, cold crystallization, and melting point for individual biopolymers and ternary polymer blends PLA-PHB- poly(ethyleneglycol) (PEG) have been evaluated. It has been shown that the mechanical properties of compositions depend on the presence of plasticizers PEG with different molar masses in interval of 400–1000. The experiments on the action of mold fungi on the films have shown that PHB is a fully biodegradable polymer unlike PLA, whereas the biodegradability of the obtained composites is determined by their composition. The sorption activity of PLA–PHB nanofibers and unwoven nanofibrous PLA–PHB composites relative to water and oil has been studied and the possibility of their use as absorbents in wastewater treatment from petroleum products has been demonstrated.

Keywords: polylactide; poly(3-hydroxybutyrate); poly(ethyleneglycol); blending under shear deformations; electrospinning; biodegradability; oil absorption

---

1. Introduction

In recent years the creation of biodegradable ecofriendly polymer materials capable to degradation under the environmental action forming substances safe for nature and thereby favoring the solution of ecological problems acquires a growing significance. These materials may be used in packaging as the biodegradable barriers and in responding to environmental challenging as the innovative pollution absorbents and separation membranes [1–6].

The biopolyesters, such as poly(α-hydroxyacides), namely polylactide (PLA), and poly(β-hydroxyacides) (PHA), namely poly(3-hydroxybutyrate) (PHB) as a basic homolog of the PHA family, are credible alternatives to the petrol-based plastics.

PLA is one of the most promising biopolymers obtaining on polymerization of D-, L- lactic acids originated through enzyme fermentation of renewable natural products such as corn, potato, sugar beet, etc. By a range of physical characteristics commercial-grade PLA resembles the biostable synthetic polymers polystyrene and poly(ethylene terephthalate) [7], but its biodegradability can occur only in hydrolytic and enzymatic media [8–10]. Being a good thermoplastic, it can be extruded into films and pellets, molded into goods of diverse shape and via electrospinning formed into ultrafine fibers [11,12]. Microbial PHB, which is more costly than PLA, is a fully biodegradable and renewable thermoplastic with the high crystallinity up to 70% brittle and high melting temperature. Owing to biocompatibility, biodegradability, and nontoxicity [13,14] PHB is used in different medical, packaging, and environmental applications, although its wide application is limited by current high cost. In fact, PLA, PHB, and their blends can be considered favorably for biomedical applications [15,16] especially in the tissue engineering, regenerative medicine and as the biodegradable therapeutic systems for drug delivery [17–19].

Both PLA and PHB application is limited by their mechanical properties, including brittleness, poor ductility, caused by high crystallinity, and relatively low thermostability due to a small difference between temperatures of melting and decomposition. According to the previous comprehensive studies [20–23], blending of these two biopolyesters, being quite similar in chemical structure, is very promising for elaboration of novel ecocompatible materials with improved transport and mechanical characteristics. For example, in the work [24] the authors showed that PHB incorporation into PLA films strongly increased oxygen barrier and simultaneously decreased wettability that is quite reasonable for food packaging application. Moreover, blending of PLA and PHB has considerable promise, since the addition of PHB to PLA endows the compositions with biodegradability, and PLA allows one to reduce the cost of the corresponding products as compared with the cost of PHB based materials [25]. The studied compositions can be prepared by different methods determining their structure, properties, and hence the application areas. The innovation methods for obtaining of these compositions such as blending under shear deformations and electrospinning are believed to be of great prospect.

It is noteworthy, that combining PLA and PHB as the binary blend without a structure-modifying additive has rarely led to miscibility of the polymer components. To resolve the above drawbacks including their rigidity some of authors have produced PHB-PLA compositions by using plasticizers such as limonene [26], trybutyrine [27], oligomeric lactic acid [28], polyethylene glycol (PEG) [24], and others. Due to strict requirements to food packaging demands, a series of conventional plasticizers such as t-butylphenyl phosphate and dioctyl phthalate should be excluded. In this context, the application of a series of low molar mass PEGs as the nontoxic plasticizing agents open the promising prospects for the more effective design of PLA-PHB systems. Given the PEG benefits as an ecofriendly and nontoxic plasticizer, it should be noted that its effect on thermal, mechanical and diffusion characteristics of the triple system have been explored in very few cases. According to Scopus® search engine data, the corresponding results of low molar mass PEG in the region of 400–1000 impact have not been published yet besides some rare papers, where PEG was used as the additive with a single molar mass, e.g., 300 [29].

The blending of PLA with PHB is often carried out in solution (regularly in chloroform) that requires essential volumes of solvent and the process becomes durable and costly. However, the blending under high-intensity shear deformations in the absent of solvents [30] allows one to obtain PLA–PHB compositions by ecologically safe procedure and in this case their properties are determined by the mixture composition.

At the same time electrospinning presents a promising method for obtaining fibers in the micro- and nano-range. Nanosized fibrous structures formed during electrospinning are characteristic of high specific surface area and controlled porosity that makes promising their application as efficient filters, high-sensitivity sensors, and absorbents with a high sorption capacity [31,32]. Recently, Liu, Cao, Iordanskii et al. [33] have reported the production and characterization of PHA-PHB spun fibers produced by melt electrospinning. In the following paper of the same authors [34], the comparison of melt electrospun PHB-PLA fibers and molded films with analogous compositions has been presented to elucidate the diffusivity impact on drug release features. In addition to the usage of electrospun fibers in biomedicine as well as at development of innovative packaging materials, they are starting to be actively used for the development of innovative absorbents for pollutants removing [35]. Therefore, a current tendency to production of ecologically safe high-selectivity nanofibrillar porous sorbents on the basis of PLA–PHB compositions obtained by electrospinning for recovery of oil pollutants from wastewater is of great practical interest.

Considering the arguments presented above, this work is devoted to preparation of PLA-PHB compositions by two different methods: via blending with low-molecular plasticizer PEG under shear deformations and electrospinning as well as subsequent examination of their reasonable characteristics. This study involves primarily the food packaging elaboration and ecofriendly fibrillar absorbent application for oil pollutions removing.

2. Materials and Methods

2.1. Materials

PLA (the trade mark "Ingeo™ 4043D") is a product from Nature Works (Minnetonka, MN, USA) as pellets with a diameter of 3 mm was used for blending with PHB and PEG. According to the manufacturer, the weight average (Mw) and number average (Mn) molar masses of PLA are found to be 2.20×10^5 g/mol and 1.65×10^5 g/mol respectively with a polydispersity index Đ = Mw/Mn = 1.35. Preliminary the pellets were dried in a vacuum oven for 8 h at 70 °C. The PLA density was 1.245 g/cm^3 and transparency 2.1%. This trade mark of PLA designated for food packaging was proved in the list of FDA effective [36].

PHB with weight average molar mass (M_w) being equal to 2.05×10^5 was kindly provided by Biomer company (Krailling, Germany), personally by Dr. U. Haenggi. The biopolymer was originally in the form of a white powder with particle size of 3–7 μm. It has density of 1.290 g/cm^3; 0, 43 mas % humidity, and therefore the material was dried overnight at 70 °C before mixing. In 2007 the FDA had cleared its marketing in the USA, indicating a bright future for a practical application of PHB in biomedical areas, e.g., as medical devices [37].

Poly(ethylene glycol) (PEG), CAS Number: 25322-68-3, with M_w equals to 400, 600, and 1000 g/mol was supplied by Sigma-Aldrich S.A. PEG has been found to be nontoxic and is approved by the FDA for use as excipients or as a carrier in different pharmaceutical formulations, foods, and cosmetics.

Chloroform, CAS Number 67-66-3, was also purchased from Sigma-Aldrich (St. Louis, MO, USA).

All materials were used as received, without any further purification.

2.2. Production of Compositions

The polymer compositions were prepared under action of shear deformations in a Brabender type mixer (Plastograph Brabender® GmbH and Co. KG, Duisburg, Germany) at 180 °C for 10 min at the rate of 100 rpm. For further investigation, films of 0.3 mm thick were pressed on a Carver press (Carver Inc., Wabach, IN, USA) at 180 °C and pressure of 10 MPa for 10 min followed by cooling with the rate of 15 °C/min. As a result, the films of 0.2–0.25 mm thick were formed.

2.3. Investigation of Thermophysical Properties

Thermophysical characteristics and thermal stability of the initial polymers and their blends were studied by the differential scanning calorimetry (DSC) method on a DSC-204 F1 calorimeter (Erich Netzch GmbH and Co. Holding KG, Selb, Germany) at the heating rate of 10 K/min in the temperature range of 30–200 °C. Sample weight was approximately 10 mg. In the air, the experiment involved the first heating, cooling, and the second heating with the same rate.

2.4. Mechanical Tests

Mechanical characteristics of samples were determined on an Instron-3365 tensile machine (High Wycombe, UK) under uniaxial stretching, at the rate of the upper traverse motion of 50 mm/min at room temperature. From the obtained diagrams, elastic modulus E, tensile strength σ_p, and elongation at break ε_p were calculated. The results were averaged by five samples.

2.5. Biodegradability

The biodegradation was estimated by carrying out microbiological tests of the compositions with different component ratios for stability to the effect of mold fungi. The principle of the method is in the holding of samples infected with an aqueous suspension of mold fungi spores under the optimum conditions of their growth and the estimation of fungi resistance (the index of fungi growth) of samples by variations in their color, shape, and texture according to the six-point scale.

The biodegradability of compositions was studied also by modeling the processes occurring under the natural conditions. For this purpose, the samples were placed into a container with wet soil (pH = 7.5) meant for place growing. The containers were kept in a thermostat at 30 °C during several months. The rate of biodegradation in soil was evaluated by the weight loss of samples at regular time intervals.

2.6. Atomic Force Microscopy (AFM)

The fiber surfaces were studied on a Solver HV (NT-MDT) atomic force microscope. The tests were carried out in the semicontact mode with the use of a HA_NC probe sensor (chip size 3.6 mm × 1.6 mm × 0.45 mm, cantilever size 87 μm × 32 μm × 1.75 μm, curvature radius of a tip 10 nm, force constant 5.8 N/m).

2.7. Production of Fibers and Unwoven Materials by Electrospinning

The 7% solution for electrospinning was prepared by mixing for all PLA/PHB ratios (1/0, 0.5/0.5, 0.1/0.9, and 0/1) in cosolvent, chloroform, at magnetic stirring for 8 h at room temperature. The electrical conductivity of polymer blend solutions, as one of key factors being responsible for fiber perfection, were 10 μS/cm. After mixing the solutions were placed in a 20 mL syringe fitted with a stainless-steel needle with inner diameter of 0.1 mm. The fibers were produced using the laboratory installation described earlier [31] as a single-capillary vertical setup with the following operative parameters such as flow rate of 0.5 mL/h, voltage of 12 kV, and the distance between the needle tip and the electrode collector was 18 cm.

2.8. Oil Absorption Measurements

As sorbate was used, West Siberian crude oil with a density of 848 kg/m^3 (according to RF standard: GOST 51069-97) and a sulfur content of 10 mas. % (according to standard RF: GOST 32139-2013) was also used. Oil absorption properties for electrospun fibrils forming the corresponding mats with different PLA–PHB compositions (100:0, 50:50, 10:90, and 0:100 wt %) were evaluated by a weight method with the use of an HR-200 analytical balance (A&D Weighing Co, Milpitas, CA, USA) with a readability of 10^{-4} g.

2.9. Statistics and Data Availability Statement

The mat thicknesses were equal to 86 ± 4.1 μm with accuracy 0.0434 for all the polymer ratios. The averaged diameter of electrospun fibers was calculated by seven independent measurements of AFM microphotographs and was depended on the PLA–PHB ratio in the range from 8.3 ± 0.9 to 11.9 ± 1.4 μm. The measurement error of absorbed oil after triple weighing was ± 4.6%. Temperature was measured to an accuracy of 0.1 °C. Relative standard deviation of mechanical testing for E, σ_p, and ε_p with 5-fold averaged values was spanned in a 5–10% interval.

3. Results and Discussion

3.1. Thermophysical Properties of PHB, PLA, and their Compositions

PLA–PHB blends with different component ratios were obtained in a Brabender mixer under conditions of shear deformation. Since PLA and PHB are rigid and brittle polymers, their blending was carried out in the presence of PEG of various molar mass (400, 600, and 1000). For further thermophysical and mechanical measurements as well as the biodegradability evaluation, films from PLA, PHB, and their blends were formed.

Figure 1 shows DSC curves of PLA (*1,1′*), PHB (*2,2′*), and their blend with PEG (PLA–PHB (80: 20 wt %) + 5 wt % PEG$_{400}$) (*3,3′*) at the first (without prime) and second (primed) heating. It follows from the figure that the DSC curve of the first heating displays four peaks belonging correspondingly to glass transition (T_g = 65 °C), cold crystallization (T_{cc} = 124 °C), and melting doublet (T_{m1} = 159.4 °C, T_{m2} = 161.8 °C). At the second heating, because of change in PLA crystalline structure, the temperatures of transitions decreased on several degrees (Table 1).

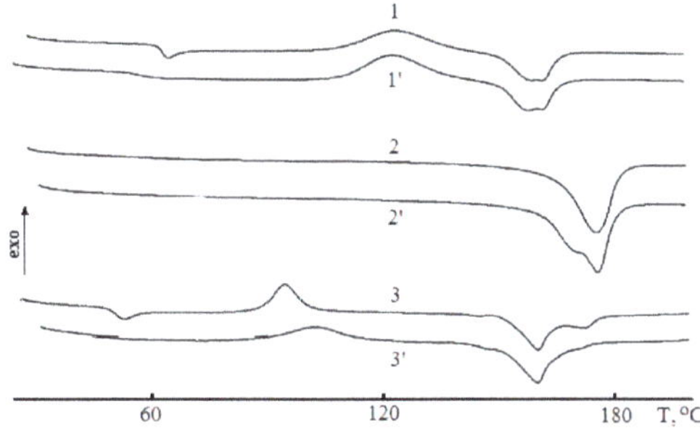

Figure 1. Differential scanning calorimetry (DSC) curves of PLA (*1,1′*), PHB (*2,2′*), and their blend with PEG (PLA–PHB (80: 20 wt %) + 5 wt % PEG$_{400}$) (*3,3′*) at the first (without prime) and second (primed) heating.

As it is shown in Figure 1, the DSC curve for high crystalline PHB demonstrated a single melting peak, and the peak reflecting glass transition was absent due to minor content of amorphous phase. The determined PHB crystallinity at the first and second heating was 49% and 52% correspondingly. For the neat PHB after the first heating the T_m slight reduction was also associated with crystalline structure reorganization. In the blends, owing to a sharp decrease in PHB concentration by 4 times, the evaluation of its crystallinity was practically impossible therefore corresponding values are not given in Table 1.

In Figure 2 the DSC thermograms for the ternary blends are presented. In DSC curves of PLA–PHB blend with 5% PEG$_{400}$, at the first heating, five peaks were observed due to glass transition, cold crystallization, melting temperature doublet for PLA, and melting temperature of PHB (Table 1). Depression of PLA transition temperatures, T_g and T_{cc}, in the blends as compared to the analogous temperature transitions of the neat PLA was due to the presence of two other components, PHB and especially PEG, which affect the mobility of PLA molecules. At the second heating of the triple blends, all temperatures of the thermal effects (melting and cold crystallization) remain the same while the glass transition peaks of PLA were absent due to the crystallinity increase (Table 1).

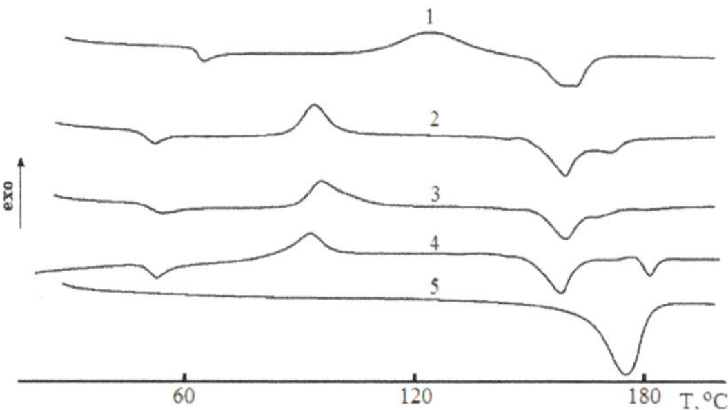

Figure 2. DSC curves of PLA (1), PHB (5) and PLA–PHB blends (80: 20 wt %) with 5 wt % PEG of different molar mass 400 (2), 600 (3), and 1000 (4).

Table 1. Thermophysical characteristics of initial PLA and in its blends of various compositions at the first and second heating.

Blend Composition PLA-PGB, wt %	Heating	T_g, °C	T_m, °C	ΔH_m, J/g	T_{cc}, °C	ΔT_{ccg}, °C	χ_{cr},% *
			Polylactide				
PLA = 100	first	65.0	doublet 159.4; 162.0	25.6	124.0	59.0	27.3
	second	63.0	doublet 158.4; 161.7	27.7	123.3	60.3	29.6
(80:20) + 5% PEG$_{400}$	first	52.0	doublet 144.4; 159.3	31.3	94.0	42.0	42.0
	second	–	doublet 146.4; 159.3	36.6	102.0		49.0
(80:20) + 10% PEG$_{400}$	first	–	157.6	41.3	83.0		55.0
	second	–	156.8	42.7	doublet 81.5; 127.0		57.0
(80:20) + 5% PEG$_{600}$	first	54.2	doublet 145.6; 159.3	22.8	96.0	42.0	30.0
	second	–	doublet 148.0; 159.4	32.4	106.0		43.0
(80:20) +5% PEG$_{1000}$	first	53.5	159.0	29.6	107.0	53.5	34.0
PHB = 100	first	-	175.0	71.0	-		49.0
	second	-	170.0	75.5	-		52.0

* Crystallinity of individual PLA and PHB was calculated by formula $\chi_{cr} = \frac{\Delta H_{exp}}{\Delta H_m^0}$, where ΔH_m^0 are melting enthalpies of PLA and PHB at 100% crystallinity equal to 93.7 [38] and 146 J/g [39], respectively. Crystallinity of PLA in blends χ_{cr} was calculated by formula $\chi_{cr} = \frac{\Delta H_{exp}}{\Delta H_m^0 \cdot W_{PLA}}$, where W_{PLA} is weight fraction of PLA in blend.

Exploring the PEG impact on thermal blend behavior, it is possible to find that the general patterns of the DSC curves changed. In particularly, the PEG$_{400}$ content increment up to 10% led to vanishing the PLA glassy transition and simultaneously to increasing to maximal crystallinity of 57%. By acting as the plasticizer, PEG declined the positions of T_g and T_{cc} on the temperature scale. Moreover, the most pronounced effect of plasticization, expressed as the low temperature shift, the growth of crystallinity, and the decrease of T_{cc} values was found for the blends containing PEG$_{400}$ that is the plasticizer with the lowest molecular weight.

Generally, the glass transition temperatures (T_g) of the blends indicate polymer miscibility degree. If blending leads to the composition having a lower single T_g than the brittle component, namely PLA, that implies the emergence of flexible chains, thereby lowering brittleness and, hence, improving applicability. Otherwise, if the binary composition has a higher single T_g, the brittleness could arise because of macromolecular stiffness increasing [40]. Two or even more glass transition points could testify immiscibility of the blend [41] especially when both polymer components are crystallizable [42]. In our case, the DSC curves and data in Table 1 show the single T_g that is more typical for glassy transition of PLA. From the same data, it follows that T_g in all ternary blends containing 5% PEG was essentially decreased. This decrease was especially remarkably observed for PEG of the lowest M_w (400) that testified the plasticizing impact of PEG in the range of M_w from 400 to 1000.

It is well established that the low molecular PEG was characterized by low molar volume and high concentration of terminal hydroxide groups. These characteristics promoted its diffusivity into the biopolymer blends and facilitated the interaction with the functional groups of the biopolyesters (PHB and PLA) that could lead to the plasticizing effect [43,44]. In contrast, high molecular PEG is denied such an opportunity and in that case the plasticization does not occur at all [45]. For example, recent research [46] has shown that relatively high molecular PEG (M_w > 2000) was not only a poor plasticizer, but its loading has caused well-observed phase separation as well.

A pronounced effect of PEG addition was observed for PLA-PHB blends as T_g depression for all plasticized samples in the informative works [24,46]. Following the concept of the work [24], we reckon that in the framework of free volume theory [47,48], the plasticizing effect of PEG increases the segmental mobility of the biopolymers and, ultimately, enhancing molecular contact between PLA and PHB molecules. For all the ternary blends PLA-PHB-PEG, the shift of T_{cc} values to the field of lower temperatures is evidence of PEG plasticizing effect and PLA segmental mobility enhancing. This effect can be demonstrated more clearly if instead of T_{cc} values, the difference between two characteristic temperatures, namely $T_{cc} - T_g = \Delta T_{ccg}$, should be used. The results of the calculation are presented in Table 1.

The feature ΔT_{ccg} reflects the temperature interval between the glassy-rubber transition and the development of cold crystallization, e.g., the situation when the segmental motion becomes high enough and a low temperature crystallization as the dynamic process can happen. It should be noted that the narrower this interval, the more effective the PEG action as the plasticizer. With this consideration, in contrast to PEG_{1000}, PEG_{400} along with PEG_{600} could be treated as the most efficient plasticizer. Enhancing in PLA segmental mobility as the expression of plasticizing effect is confirmed also by the doubled growth of PLA crystallinity degree in the ternary blends (Table 1), namely from 27% to 42% and to 55 % for 5% and 10% of PEG_{400} content correspondingly.

3.2. Mechanical Testing

By the data of mechanical testing for PLA, PHB, and their blends containing PEG of different molar masses, elastic modulus E, tensile strength σ_b, and elongation at break ε_b of the initial polymers and their blends were evaluated. The mechanical parameters of PLA and PHB are characteristic of glassy polymers with a low breaking strain. Characterization of PLA with E = 2700 MPa, σ_b = 45.4 MPa, and ε_b = 4% demonstrated its high rigidity, while PHB is an even more rigid polymer with higher elastic modulus and lower elongation at break, such as E = 2900 MPa, σ_b = 19.5 MPa, and ε_b = 1%. In the literature the brittleness and the poor ductility of both PLA and PHB have been broadly discussed to avoid the basic drawbacks that prevent their widespread applications, see e.g., [49–51]. As mentioned above in the Introduction, simple blending of these two biopolymers has rarely led to their mechanical improvement. On the contrary, their characteristics even have worsened as it was reported in the paper [52], where σ_p and ε_p values of PLA–PHB blends decreased especially at a higher PHB content in the blends.

Figure 3 shows the dependences of E, σ_b, and ε_b values for PLA–PHB–PEG compositions (5 wt % PEG) on the PHB content in blend. As can be seen from the data in Figure 3a, an increase in the content of more rigid polymer PHB in the ternary compositions practically does not influence on elastic modulus of the blends. However, as it follows from the curve in Figure 3b, the gain in the PHB content leads to some decrease of σ_b value monotonically and the lowest value of σ_b observed for a composition involving 30 wt % PHB. Additionally, finally, an increase of the PHB content in compositions resulted also in a drop of elongation at break (Figure 3c). The observed scatter in the experimental values was supposedly connected with structural inhomogeneity of PLA-PHB compositions.

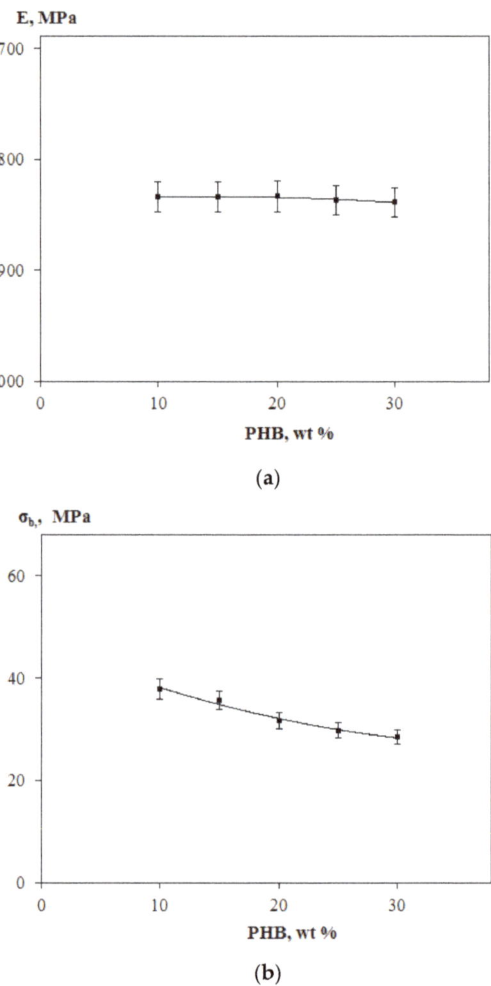

Figure 3. *Cont.*

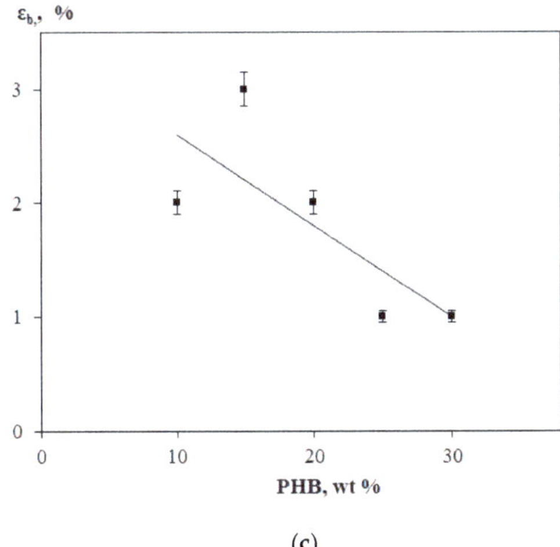

(c)

Figure 3. Elastic modulus E (**a**), tensile strength σ_b (**b**), and elongation at break ε_b (**c**) of PLA–PHB–PEG$_{1000}$ compositions vs. PHB content.

Figure 4 shows dependences of mechanical parameters of PLA–PHB–PEG$_{400}$ compositions on the PEG content (the PLA–PHB ratio is equal to 80:20 wt % for all compositions). As can be seen from Figure 4a,b elastic modulus and tensile strength of compositions decreased as the PEG content in compositions grew. Elongation at break for the above mentioned blends at small PEG contents remains very low (from 1% to 2%) and an increase of its value begins with further gain in the plasticizer content (Figure 4c). For compositions with 10% PEG$_{400}$, supposedly the character of deformation changes from brittle fracture to plastic flow; as a result, elongation at break attains 13%, elastic modulus decreases to 800 MPa, and tensile strength decreases from 28.2 to 11. 6 MPa. So, the elastic modulus was dropped approximately three times, tensile strength decreased by 2.5 times, and elongation at break increased by more than seven times reaching the value of 14%. Hence, as in the case of thermal findings (Section 3.1), the ternary composition revealed itself as essentially more flexible than binary PHB–PLA system.

For PLA-PHB compositions the loading of different plasticizers is a quite common way to improve both mechanical and thermal behavior of these binary blends. In the paper [53] Averous and Martin have considered the influence of low molecular PEG and some other plasticizers on thermal and mechanical characteristics of PLA. For the PLA-PEG system they have shown the remarkable decrease of T_g, and as in the case of our findings, PEG$_{400}$ showed the higher thermal effect comparing to PEG$_{1000}$. Simultaneously under storage and loss modules measurement the presence of 20% PEG$_{400}$ had demonstrated the better effect comparing to other plasticizers. The ternary PLA-PHB-PEG blends remain scantily investigated, except of two relevant publications, namely of the paper of Wang et al. [54] considering the similar composition PLA-PHBHV-PEG and the work of Arietta and coworkers [29] that has described the analogous ternary blend PLA-PHB-PEG. In the first of the mentioned publications the PEG addition led to an essential improvement of tensile and impact characteristics of PLA-PHB blends. Particularly, impact strength of the ternary composition was 2–4 times higher than that of the initial binary PLA-PHBV blend, as well as the elongation at break.

It should be pointed out, that the change of M_w of PEG in the range 400–1000 practically did not affect the mechanical characteristics of the ternary system (Table 2). However and in this case the slight decline in the basic characteristics, namely E иσ$_b$, was observed in the presence of the PEG$_{400}$, which had the lowest value of M_w.

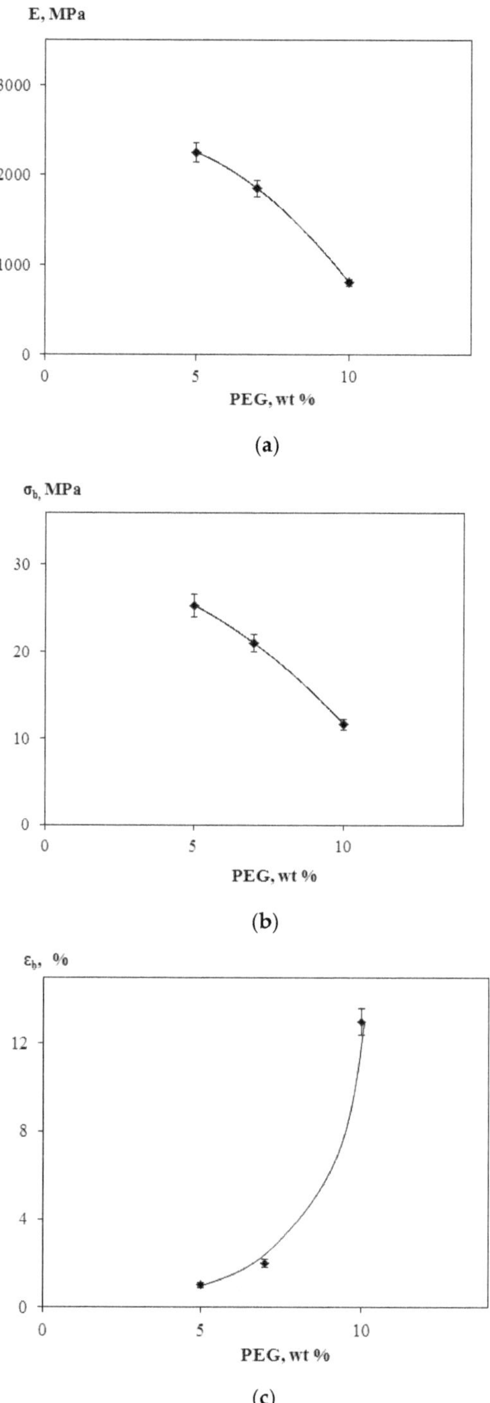

Figure 4. Elastic modulus E (**a**), tensile strength σ_b (**b**), and elongation at break ε_b (**c**) of PLA–PHB–PEG$_{400}$ compositions vs. PEG content.

Table 2. Influence of PEG with different molar mass on mechanical properties of PLA-PHB blends.

Blend Composition, PLA-PHB, wt %	E, MPa	σ_b, MPa	ε_b, %
80:20 + 5% PEG$_{1000}$	2500 ± 125	31.6 ± 1.6	2 ± 0.1
80:20 + 5% PEG$_{600}$	2700 ± 130	21.0 ± 1.1	1 ± 0.1
80:20 + 5% PEG$_{400}$	2250 ± 110	25.3 ± 1.2	2 ± 0.1

3.3. Biodegradation of Composites

In the previous paper the authors investigated the hydrolysis of PLA, PHB, and PHB-PLA blends in phosphate buffer [55]. In this work, Bonartsev et al. showed that the rate of hydrolytic degradation of PLA-PHB blend differed from the corresponding homopolymers. Most recently, in the paper of Bonartsev, Berlin, Iordanskii et al. [56] it was demonstrated the impact of the PHB+HV (hydroxyvalereat) copolymer content on the hydrolytic behavior of the membranes. Additionally to the results obtained in aquatic medium, it would be interesting to explore the PLA-PHB compositions under fungi contact as the reasonable situation during contact the films with a microbial medium.

As the capacity of decomposition under the environmental action is one of the most important properties of the biodegradable composites the biodegradation of PLA, PHB, and their composites was investigated by two independent methods: by the estimation of the degradation under the action of the set of fungi as the biodegradation agents and by the exposure in soil. The tests on the fungus resistance performed with PLA, PHB, and their blends with PEG showed the difference in the intensity of mold fungi growth.

The data on the fungi resistance of samples depending on the testing time with determination of the fungi species are given in Table 3.

As can be seen from the data given in Table 3, PLA is virtually not subject to degradation under the fungi action, whereas, with PHB, the maximum intensity of mold fungi growth (5 points) was found. In PLA–PHB–PEG composition (70: 30 wt % + 5 wt % PEG$_{1000}$), an insignificant growth of mycelium was revealed in tenth day of testing, in this case as well as for the initial PHB, the spore formation was provided by *Aspergillus brasiliensis, Trichoderma virens,* and *Paecilomyces variotii* species. A decline in PHB concentration to15 wt % is accompanied by the decrease of fungi growth intensity from 4 to 2 points.

Table 3. Fungi resistance of samples in points depending on testing time.

Testing Time, Days	Fungi Growth Intensity, Points			
	PLA	PHB	PLA-PHB (70:30 wt %) + 5 wt % PEG$_{1000}$	PLA-PHB (85:15 wt %) + 5 wt % PEG$_{1000}$
10	0	2	1	1
15	0	2	1	1
21	0	2	2	1
28	0	5	2	1
50	0	5	3	1-2
84	0-1	5	3-4	2

The photographs of PLA–PHB–PEG compositions before (a) and after (b) exposure in soil during 3 months are present in Figure 5. As is seen from the data obtained, after exposure in soil the films demonstrated spots, which are the first stages of the biodegradation process.

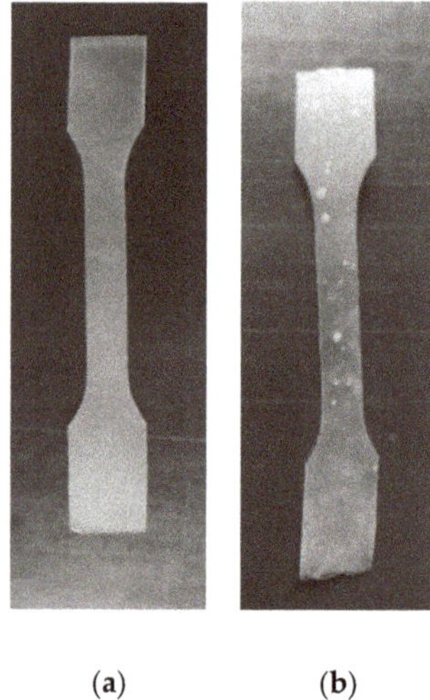

(a)   (b)

Figure 5. Photographs of PLA–PHB composition films (80: 20 wt % + 10 wt % PEG$_{400}$) before (**a**) and after (**b**) exposure in soil within 3 months.

These data coincided with our previous works, where by using the SEM method it was shown that on the compositions PLA-cellulose [57], as well as the on compositions PLA-starch [58] after exposure in soil the deep cracks and the open-end holes on the surface of the films are formed. These cracks and holes are the initial stages of the sample fragmentation leading to their following destruction due to the biodegradation.

The estimating the capacity of the studied compositions of biodegradation on exposure in soil the weight loss was measured at intervals. It occurs that the weight loss of the investigated composite film after exposure in soil within 12 months was approximately 10%.

In the same time, for a pure PLA film, during 12 months exposure in soil weight loss was absent, whereas, for PHB, the weight loss was equal to 36% within 6 months, and further exposure in soil resulted in the complete disintegration of sample film.

Thus, the study of fungi resistance of PLA, PHB, and their compositions showed that unlike PLA PHB was a fully biodegradable polymer. These data coincided also with the results on weight loss obtained at the exposure of these polymers in soil. Generally, the biodegradability of the investigated PLA-PHB films depends on the composition and increases with the PHB content.

3.4. Oil Sorption by Electrospun Fibrous Absorbents

For estimating the potential application of PLA/PHB blends particularly for oil recovery from aqueous media, the unwoven fibrous materials in the shape of plane fibrillar mats were prepared by solution electrospinning. As a result, the 2D structures were obtained with the developed interfibrillar pores system and the high specific surface. The water permeability and the absorption selectivity of the above membranes relative to organic components make them promising products used for separation of oil–water systems.

The AFM study of the fiber surfaces demonstrated the distinction in their morphology that could clearly be observed only at the large magnification (Figure 6). Whereas the PLA-PHB blend fibers (1:1) had a relatively uniform surface without the noticeable roughness, the homopolymers PHB and PLA demonstrate surface heterogeneity. The cause of heterogeneity may be determined by the rapid solvent evaporation and the crystallinity of the polyesters, in which the size of crystals (spherulites and lamellae) is comparable with the fiber diameter.

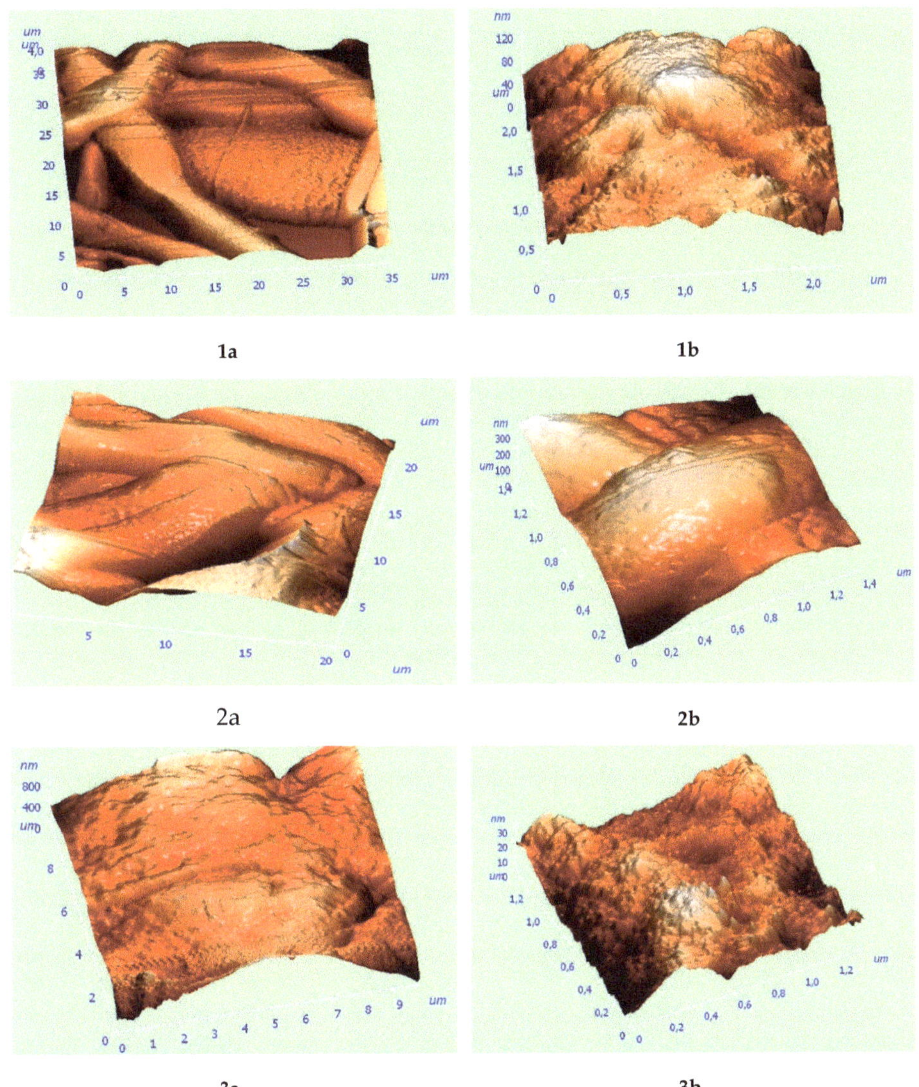

Figure 6. 3D Atomic Force Microscopy AMF images of fibers surfaces for PLA (1), (2) PLA–PHB (50: 50 wt %), and (3) PHB fibers at different magnification (**a,b**).

The sorption capacity of unwoven fibrous materials (mats) of PLA, PHB, and their compositions with different component ratio (Table 4) was determined by measuring the sorption of oil from aqueous medium at the oil layer thickness no less than 10 mm. The homogeneous polyesters PLA and PHB proved to be characteristic of high oil absorption (30 and 45 *g/g*, respectively), whereas their blends demonstrated a decrease in this fiber characterization namely 16 and 15 *g/g* for 50: 50 and 90: 10 wt % respectively. Thus, the mechanism of oil absorption by the unwoven materials under consideration is a relevant example of oil sorption on the surface of oleophilic fibers as well as the formation of films between fibers through surface tension forces [59].

Table 4. Oil absorption for the productive polymer absorbents.

Absorbent	Oil Capacity, *g/g*	Absorbent Condition	References
PDOS	3.4	Peat Dust Oil Sorbent	[60]
Lessorb^{TM}	5.6	Dispersed powder	[60]
PET	14	Nonwoven Fabric	[61]
PHB-PLA (90:10)	15	Nonwoven mat	[61]
PHB-PLA (50:50)	16	Nonwoven mat	[61]
Sintapeks	24	Cotton-processing product	[62]
PLA	30	Nonwoven mat	[61]
PU	37	Foam pellets (5–8 mm)	[61]
PHB	45	Nonwoven mat	[61]
EGS	50	Exfoliated Graphite Sorbent	[60]

As was mentioned above, the biodegradability of PHB significantly differed from PLA biodegradability, besides, its cost exceeded the cost of PLA about twice. To decrease the PHB-PLA blend absorbent expenses and to vary the rate of their biodegradation, it is advisable to use these electrospun fibers for oil recovery via absorption and following utilization in soil. Absorption capacity comparison for several promising absorbents designated for oil recovery from aqueous media shows that the maximal productivity demonstrated exfoliated graphite (EGS). From the Table 4 follows, PHB and PLA fibers were somewhat inferior to it and their blend fibers had absorbed this pollutant even less. However, the high hydrophobicity of the biopolyesters' mats, its continually falling prices due to cheaper raw materials for their large-scale production, and being natural renewable resources, as well as their ecofriendly character of biodegradation creates for them a promising prospect for making novel materials being able to replace a nonbiodegradable absorbent.

4. Conclusions

PLA-PHB compositions were prepared by environmentally safe methods, namely by blending under shear deformations in the absent of solvents and electrospinning of biodegradable blends. Preliminary literature analysis partly presented in the article has shown that PLA-PHB compositions were both semicrystalline polymers with poor miscibility. This circumstance encouraged the authors to explore thermal and mechanical characteristics of PLA-PHB blends modified by low molecular PEG as the plasticizer. The crystallinity and the specific thermal transitions (T_g, T_{cc}, and T_m) for the individual PLA, PHB, and their compositions with PEG were calculated. Thermal and mechanical features of the ternary blends testified the pronounced behavior of PEG as the plasticizer with the noticeable increase in the elastic modulus, tensile strength, and elongation at break and with the remarkable decrease of T_g that was particularly evident for the PEG$_{400}$ with the lowest Mw. It was shown that the second heating resulted in a positive increment of the crystallinity degree demonstrating a significant importance of the thermal prehistory of the blend samples.

On estimation of the fungi resistance of polyesters and their compositions, it was found that, in contrast to PLA, PHB was a completely biodegradable polymer, and the biodegradability of related compositions depended on the component ratio and increased with the PHB content.

Unwoven fibrous materials on the basis of electrospun fibers of PLA, PHB, and PLA-PHB compositions with different component ratios were characteristic of a high absorption capacity relative to oil and moderate water absorption that allowed one to consider these materials as efficient absorbents of oil products on ecological accidents and water pollution. The biodegradability of the developed fibrillar systems makes possible their utilization after service completion presented also their important advantage to successfully solve environmental challenges.

Author Contributions: Conceptualization, S.R.; Data curation, S.R., M.G.; Funding acquisition, A.I.; Investigation, L.Z., A.G., E.P., O.K., A.Y., A.O. and N.S.; Methodology, E.P., A.I., M.G.; Project administration, A.B.; Supervision, A.B.; Validation, L.Z.; Visualization, L.Z.; Writing—original draft, O.K.; Writing—review and editing, S.R. All authors have read and agreed to the published version of the manuscript.

Funding: This work was supported by the Russian Foundation for Basic Research, grant № 18-29-05017.

Conflicts of Interest: The authors declare no conflict of interest.

References

1. Yadav, B.; Pandey, A.; Kumar, L.R.; Tyagi, R.D. Bioconversion of waste (water)/residues to bioplastics-A circular bioeconomy approach. *Bioresour. Technol.* **2020**, *298*, 122584. [CrossRef] [PubMed]

2. Rameshkumar, R.; Shaiju, P.; O'Connor, K.E.; Babu, P.R. Bio-based and biodegradable polymers-State-of-the-art, challenges and emerging trends. *Curr. Opin. Green Sustain. Chem.* **2020**, *21*, 75–81. [CrossRef]

3. Murariu, M.; Dubois, P. PLA composites. From production to properties. *Adv. Drug Deliv. Rev.* **2016**, *107*, 17–46. [CrossRef] [PubMed]

4. Nakajima, H.; Dijkstra, P.; Loos, K. Recent Developments in Biobased Polymers toward General and Engineering Applications: Polymers that are Upgraded from Biodegradable Polymers, Analogous to Petroleum-Derived Polymers, and Newly Developed. *Polymers* **2017**, *9*, 523. [CrossRef]

5. Soares, R.M.D.; Siqueira, N.M.; Prabhakaram, M.P.; Ramakrishna, S. Electrospinning and electrospray of bio-based and natural polymers for biomaterials development. *Mater. Sci. Eng. C* **2018**, *921*, 969–982. [CrossRef]

6. Rogovina, S.Z. Biodegradable polymer Composites Based on Synthetic and Natural Polymers of Various Polymers of Various Classes. *Polym. Sci. Ser. C* **2016**, *58*, 62–73. [CrossRef]

7. Dorgan, J.R.; Lehermeier, H.; Mang, M. Thermal and Rheological Properties of Commercial-Grade Poly(Lactic Acid)s. *J. Polym. Environ.* **2000**, *8*, 1–9. [CrossRef]

8. Rodriguez, E.J.; Marcos, B.; Huneault, M.A. Hydrolysis of polylactide in aqueous media. *J. Appl. Polym. Sci.* **2016**, *133*, 44152. [CrossRef]

9. Valentina, I.; Haroutioun, A.; Fabrice, L.; Vincent, V.; Roberto, P. Poly(Lactic Acid)-Based Nanobiocomposites with Modulated Degradation Rates. *Materials* **2018**, *11*, 1943. [CrossRef]

10. Tsuji, H.; Miyauchi, S. Enzymatic Hydrolysis of Poly(lactide)s: Effects of Molecular Weight, l-Lactide Content, and Enantiomeric and Diastereoisomeric Polymer Blending. *Biomacromolecules* **2001**, *2*, 597–604. [CrossRef]

11. Lim, L.-T.; Auras, R.; Rubino, M. Processing technologies for poly(lactic acid). *Prog. Polym. Sci.* **2008**, *33*, 820–852. [CrossRef]

12. Gupta, B.; Revagade, N.; Hilborn, J. Poly(lactic acid) fiber: An overview. *Prog. Polym. Sci.* **2007**, *32*, 455–482. [CrossRef]

13. Follain, N.; Chappey, C.; Dargent, E.; Chivrac, F.; Crétois, R.; Marais, S. Structure and Barrier Properties of Biodegradable Polyhydroxyalkanoate Films. *J. Phys. Chem.* **2014**, *118*, 6165–6177. [CrossRef]

14. Iordanskii, A.L.; Ol'khov, A.A.; Karpova, S.G.; Kucherenko, E.L.; Kosenko, R.Y.; Rogovina, S.Z.; Chalykh, A.E.; Berlin, A.A. Influence of the structure and morphology of ultrathin poly(3-hydroxybutyrate) fibers on the diffusion kinetics and transport of drugs. *Polym. Sci. Ser. A* **2017**, *59*, 343–353. [CrossRef]

15. Saini, P.; Arora, M.; Kumar, M.N.V.R. Poly(lactic acid) blends in biomedical applications. *Adv. Drug Deliv. Rev.* **2016**, *107*, 47–59. [CrossRef]

16. Butt, F.I.; Muhammad, N.; Hamid, A.; Moniruzzaman, M.; Sharif, F. Recent progress in the utilization of biosynthesized polyhydroxyalkanoates for biomedical applications–Review. *Int. J. Biol. Macromol. Part A* **2018**, *120*, 1294–1305. [CrossRef]

17. Da Silva, L.P.; Kundu, S.C.; Reis, R.L.; Correlo, V.M. Electric Phenomenon: A Disregarded Tool in Tissue Engineering and Regenerative Medicine. *Trends Biotechnol.* **2020**, *38*, 24–49. [CrossRef]

18. Cheng, H.; Yang, X.; Che, X.; Yang, M.; Zhai, G. Biomedical application and controlled drug release of electrospun fibrous materials. *Mater. Sci. Eng. C* **2018**, *90*, 750–763. [CrossRef]

19. Tyler, B.; Gullotti, D.; Mangraviti, A.; Utsuki, T.; Brem, H. Polylactic acid (PLA) controlled delivery carriers for biomedical applications. *Adv. Drug Deliv. Rev.* **2016**, *107*, 163–175. [CrossRef]

20. Arrieta, M.P.; Samper, M.D.; Aldas, M.; López, J. On the Use of PLA-PHB Blends for Sustainable Food Packaging Applications. *Materials* **2017**, *10*, 1008. [CrossRef]

21. Yu, L.; Dean, K.; Li, L. Polymer blends and composites from renewable resources. *Prog. Polym. Sci.* **2006**, *31*, 576–602. [CrossRef]

22. Armentano, I.; Fortunati, E.; Burgos, N.; Dominici, F.; Luzi, F.; Fiori, S.; Jiménez, A.; Yoon, K.; Ahn, J.; Kang, S.; et al. Processing and characterization of plasticized PLA/PHB blends for biodegradable multiphase systems. *eXPRESS Polym. Lett.* **2015**, *9*, 583–596. [CrossRef]

23. Abdelwahab, M.A.; Flynn, A.; Chiou, B.-S.; Imam, S.; Orts, W.; Chiellini, E. Thermal, mechanical and morphological characterization of plasticized PLA-PHB blends. *Polym. Degrad. Stab.* **2012**, *97*, 1822–1828. [CrossRef]

24. Balogová, A.F.; Tóth, R.H.T.; Schnitzer, M.; Feranc, J.; Bakoš, D.; Živcak, J. Determination of geometrical and viscoelastic properties of PLA/PHB samples made by additive manufacturing for urethral substitution. *J. Biotechnol.* **2018**, *284*, 123–130. [CrossRef] [PubMed]

25. Pavan, F.A.; Junqueira, T.L.; Watanabe, M.D.B.; Bonomi, A.; Quines, L.K.M.; Schmidell, W.; Aragao, G.M.F. Economic analysis of polyhydroxybutyrate production by Cupriavidus necator using different routes for product recovery. *Biochem. Eng. J.* **2019**, *146*, 97–104. [CrossRef]

26. Arrieta, M.P.; López, J.; Hernández, A.; Rayón, E. Ternary PLA–PHB–Limonene blends intended for biodegradable food packaging applications. *Eur. Polym. J.* **2014**, *50*, 255–270. [CrossRef]

27. D'Amico, D.A.; Montes, M.L.I.; Manfredi, L.B.; Cyras, V.P. Fully bio-based and biodegradable polylactic acid/poly(3-hydroxybutirate) blends: Use of a common plasticizer as performance improvement strategy. *Polym. Test.* **2016**, *49*, 22–28. [CrossRef]

28. Armentano, I.; Fortunati, E.; Burgos, N.; Dominici, F.; Luzi, F.; Fiori, S.; Jiménez, A.; Yoon, K.; Ahn, J.; Kang, S.; et al. Biobased PLA-PHB plasticized blend films. Part I: Processing and structural characterization. *LWT-Food Sci. Technol.* **2015**, *64*, 980–988. [CrossRef]

29. Arrieta, M.P.; Samper, M.D.; Lo'pez, J.; Jime´nez, A. Combined Effect of Poly(hydroxybutyrate) and Plasticizers on Polylactic acid Properties for Film Intended for Food Packaging. *J. Polym. Environ.* **2014**, *22*, 460–470. [CrossRef]

30. Enikolopian, N.S. Some aspects of chemistry and physics of plastic flow. *Pure Appl. Chem.* **1985**, *57*, 1707–1711. [CrossRef]

31. Filatov, Y.; Budyka, A.; Kirichenko, V. *Electrospinning of Micro-and Nanofibers: Fundamentals in Separation and Filtration Processes*; Begell House, Inc.: Redding, CT, USA, 2007; p. 488.

32. Saleem, H.; Trabzon, L.; Kilic, A.; Zaidi, S.J. Recent advances in nanofibrous membranes: Production and applications in water treatment and desalination. *Desalination* **2020**, *478*, 114178. [CrossRef]

33. Cao, K.; Liu, Y.; Olkhov, A.A.; Siracusa, V.; Iordanskii, A.L. PLLA-PHB fiber membranes obtained by solvent-free electrospinning for short-time drug delivery. *Drug Deliv. Translat. Res.* **2018**, *8*, 291–302. [CrossRef] [PubMed]

34. Liu, Y.; Cao, K.; Karpova, S.; Olkhov, A.A.; Iordanskii, A.L. Comparative Characterization of Melt Electrospun Fibers and Films Based on PLA-PHB Blends: Diffusion, Drug Release, and Structural Features. *Macromol. Symp.* **2018**, *381*, 1800130. [CrossRef]

35. Bagbi, Y.; Pandey, A.; Solank, P.R. Electrospun Nanofibrous Filtration Membranes for Heavy Metals and Dye Removal. In *Nanoscale Materials in Water Purification*; Elsevier: Amsterdam, The Netherlands, 2019; pp. 275–288.

36. Wang, Y.; Qu, W.; Choi, S.H. FDA's Regulatory Science Program for Generic PLA-PLGA-Based Drug Products. *Am. Pharm. Rev.* **2016**, *19*, 5–9.

37. Absorbable Poly(hydroxybutyrate) Surgical Suture Produced by Recombinant DNA Technology-Class II Special Controls Guidance for Industry and FDA Staff. Available online: https://www.fda.gov/medical-devices/guidance-documents-medical-devices-and-radiation-emitting-products/absorbable-polyhydroxybutyrate-surgicalutureproduced-recombinant-dna-technology-class-ii-special (accessed on 28 June 2018).

38. Cheng, S.Z.D. (Ed.) *Handbook of Thermal Analysis and Calorimetry*, 2nd ed.; Elsevier: Amsterdam, The Netherlands, 2002; Volume 3, p. 810.

39. Lin, K.-W.; Lan, C.-H.; Sun, Y.-M. Poly[(R)3-hydroxybutyrate] (PHB)/poly(L-lactic acid) (PLLA) blends with poly(PHB/PLLA urethane) as a compatibilizer. *Polym. Degrad. Stab.* **2016**, *134*, 30–40. [CrossRef]

40. Mousavioun, P.; Halley, P.J.; Doherty, W.O.S. Thermophysical properties and rheology of PHB/lignin blends. *Ind. Crops Prod.* **2013**, *50*, 270–275. [CrossRef]

41. Qiu, J.; Xing, C.; Cao, X.; Wang, H.; Wang, L.; Zhao, L.; Li, Y. Miscibility and Double Glass Transition Temperature Depression of PLA (PLLA-POM) Blends. *Macromolecules* **2013**, *46*, 5806–5814. [CrossRef]

42. Florez, J.P.; Fazeli, M.; Simão, R.A. Preparation and characterization of thermoplastic starch composite reinforced by plasma-treated poly (hydroxybutyrate) PHB. *Int. J. Biol. Macromol.* **2019**, *123*, 609–621. [CrossRef]

43. Faradilla, R.F.; Lee, G.; Sivakumar, P.; Stenzel, M.; Arcot, J. Effect of PEG molecular weight and nanofillers on properties of banana nanocellulose films. *Carbohydr. Polym.* **2019**, *205*, 330–339. [CrossRef]

44. Qussi, B.; Suess, W.G. The Influence of Different Plasticizers and Polymers on the Mechanical and Thermal Properties, Porosity and Drug Permeability of Free Shellac Films. *Drug Dev. Ind. Pharm.* **2006**, *32*, 403–412. [CrossRef]

45. Yu, Y.; Cheng, Y.; Ren, J.; Cao, E.; Fu, X.; Guo, W. Plasticizing effect of poly(ethylene glycol)s with different molecular weights in poly(lactic acid)/starch blends. *J. Appl. Polym. Sci.* **2015**, *132*, 41808. [CrossRef]

46. Zhang, M.; Thomas, N.L. Blending polylactic acid with polyhydroxybutyrate: The effect on thermal, mechanical, biodegradation properties. *Adv. Polym. Technol.* **2011**, *30*, 67–79. [CrossRef]

47. Fujita, H. Notes on free volume theories. *Polym. J.* **1991**, *23*, 1499–1506. [CrossRef]

48. Sharma, J.; Tewari, K.; Arya, R.K. Diffusion in polymeric systems–A review on free volume theory. *Prog. Org. Coat.* **2017**, *111*, 83–92. [CrossRef]

49. Modi, S.; Koelling, K.; Vodovotz, Y. Assessing the mechanical, phase inversion, and rheological properties of poly-[(R)-3-hydroxybutyrate-co-(R)-3-hydroxyvalerate] (PHBV) blended with poly-(L-lactic acid) (PLA). *Eur. Polym. J.* **2013**, *49*, 3681–3690. [CrossRef]

50. Arrieta, M.P.; López, J.; López, D.; Kenny, J.M.; Peponi, L. Development of flexible materials based on plasticized electrospun PLA–PHB blends: Structural, thermal, mechanical and disintegration properties. *Eur. Polym. J.* **2015**, *73*, 433–446. [CrossRef]

51. Montes, M.L.I.; Cyras, V.P.; Manfredi, L.B.; Pettarín, V.; Fasce, L.A. Fracture evaluation of plasticized polylactic acid/poly (3-HYDROXYBUTYRATE) blends for commodities replacement in packaging applications. *Polym. Test.* **2020**, *84*, 106375–106386. [CrossRef]

52. Zhang, L.; Xiong, C.; Deng, X. Miscibility, crystallization and morphology of poly(b-hydroxybutyrate)-poly(d,l-lactide) blends. *Polymer* **1996**, *37*, 235–241. [CrossRef]

53. Martin, O.; Averous, L. Polylactic acid: Plasticization and properties of biodegradable multiphase systems. *Polymer* **2001**, *42*, 6209–6219. [CrossRef]

54. Wang, S.; Ma, P.; Wang, R.; Wang, S.; Zhang, Y.; Zhang, Y. Mechanical, thermal and degradation properties of poly(d,l-lactide)/poly(hydroxybutyrate-co-hydroxyvalerate)/poly(ethylene glycol) blend. *Polym. Degrad. Stab.* **2008**, *93*, 1364–1369. [CrossRef]

55. Bonartsev, A.P.; Boskhomodgiev, A.P.; Iordanskii, A.L.; Bonartseva, G.A.; Rebrov, A.V.; Makhina, T.K.; Myshkina, V.L.; Yakovlev, S.A.; Filatova, E.A.; Ivanov, E.A.; et al. Hydrolytic Degradation of Poly(3-hydroxybutyrate), Polylactide and their Derivatives: Kinetics, Crystallinity, and Surface Morphology. *Mol. Cryst. Liq. Cryst.* **2012**, *556*, 288–300. [CrossRef]

56. Zhuikov, V.A.; Berlin, A.A.; Bonartsev, A.P.; Iordanskii, A.L. Comparative Structure-Property Characterization of Poly(3-Hydroxybutyrate-Co-3-Hydroxyvalerate)s Films Under Hydrolytic And Enzymatic Degradation: Finding a Transition Point in 3-Hydroxyvalerate Content. *Polymers* **2020**, *12*, 728. [CrossRef] [PubMed]

57. Rogovina, S.Z.; Aleksanyan, K.V.; Kosarev, A.A.; Ivanushkina, N.E.; Prut, E.A.; Berlin, A.A. Biodegradable Polymer Composites Based on Polylactide and Cellulose. *Polym. Sci. Ser. B* **2016**, *58*, 38–46. [CrossRef]

58. Rogovina, S.Z.; Aleksanyan, K.V.; Loginova, A.A.; Ivanushkina, N.E.; Vladimirov, L.V.; Prut, E.A.; Berlin, A.A. Influence of PEG on Mechanical Properties and Biodegradability of Composites Based on PLA and Starch. *Starch* **2018**, *70*, 1700268. [CrossRef]

59. Samoilov, N.A.; Khlestkin, R.N.; Shemetov, A.V.; Shammazov, A.A. *Sorption Method of Liquidation of Emergency Spills of Oil AND Oil Products*; Khimiya: Moscow, Russia, 2001; p. 189.

60. Blinovskaya, Y.Y. Efficiency Sorbents Comparative Analysis for Heavy Oil Products in the Conditions of Low Temperatures. *IOP Conf. Ser. Earth Environ. Sci.* **2019**, *272*, 032116. [CrossRef]

61. Iordanskii, A.L.; Samoilov, N.A.; Olkhov, A.A.; Markin, V.S.; Rogovina, S.Z.; Kildeeva, N.R.; Berlin, A.A. New Fibrillar Composites Based on Biodegradable Poly(3-hydroxybutyrate) and Polylactide Polyesters with High Selective Absorption of Oil from Water Medium. *Dokl. Phys. Chem.* **2019**, *487*, 106–108. [CrossRef]
62. Sirotkina, E.E.; Novoselova, L.Y. Materials for Adsorption Purification of Water from Petroleum and Oil Products. *Chem. Sustain. Dev.* **2005**, *13*, 359–375.

Article

Influence of Genipin Crosslinking on the Properties of Chitosan-Based Films

Nataliya Kildeeva [1,*], Anatoliy Chalykh [2], Mariya Belokon [1,3], Tatyana Petrova [2], Vladimir Matveev [2], Evgeniya Svidchenko [4], Nikolay Surin [4] and Nikita Sazhnev [1]

1   Department of Chemistry and Technology of Polymer Materials and Nanocomposites, The Kosygin State University of Russia, 119071 Moscow, Russia; m.solyankina@mail.ru (M.B.); nsazhnev@mail.ru (N.S.)
2   Frumkin Institute of Physical Chemistry and Electrochemistry, Russian Academy of Sciences, 119071 Moscow, Russia; chalykh@mail.ru (A.C.); petrttt@mail.ru (T.P.); matveev46@yandex.ru (V.M.)
3   Engelhardt Institute of Molecular Biology, Russian Academy of Sciences, 119071 Moscow, Russia
4   Enikolopov Institute of Synthetic Polymeric Materials, Russian Academy of Sciences, 119071 Moscow, Russia; evgensv@yandex.ru (E.S.); niksurin@yandex.ru (N.S.)
*   Correspondence: kildeeva@mail.ru

Received: 20 March 2020; Accepted: 6 May 2020; Published: 10 May 2020

Abstract: Chitosan is a promising environment friendly active polymer packaging material due to its biodegradability, exceptional film forming capacity, great mechanical strength, appropriate barrier property along with intrinsic antioxidant and antimicrobial features. Bifunctional reagent was used for producing water insoluble chitosan films. Biopolymeric films crosslinked by Genipin (Gp), which is a reagent of natural origin, should have high potential in food packaging. The influence of the ratio of functional groups in the chitosan-Gp system on film absorption in the visible and ultraviolet regions of the spectrum, sorption, physical, and mechanical properties of the films has been studied. The degree of chitosan crosslinking in the films obtained from solutions containing Gp was estimated using the experimental data on film swelling and water vapor sorption isotherms. It is demonstrated that crosslinking with genipin improves swelling, water resistance, and mechanical properties of the films.

Keywords: chitosan; polymeric films; crosslinking; genipin; mechanical properties; sorption isotherm; degree of crosslinking

---

1. Introduction

Chitosan is one of the most promising biodegradable biopolymers, which is an amorphous-crystalline polysaccharide consisting mainly of D-glucosamine residues and it is produced via chitin processing. It is a potentially biocompatible, chemically versatile material due to the presence of amino groups and various possible M_w. Fungi (micromycetes) and a number of enzymes are the effective destructors of chitosan [1–3].

Chitosan is a promising environment friendly polymer packaging material due to its biodegradability exceptional film forming capacity, great mechanical strength, appropriate barrier property along with intrinsic antioxidant and antimicrobial features [4,5]. These properties are used in the development of materials for drug delivery and tissue engineering [6–8]. Chitosan is approved as a bioactive additive in many countries and it has been adopted by the Food and Drug Administration for use in wound dressings [9,10].

The presence of a primary amino group makes chitosan suitable for chemical modification by covalent attachment of different functional groups or for chemical crosslinking. Because of the large amount of hydrogen bonds, chitosan degrades before melting, so it is usually recycled in solution form. Chitosan has rather low pKa value—6.4 [11], which determines its water solubility, depending on pH.

Chitosan is soluble in diluted aqueous acid solutions, which, along with fiber- and film-forming ability, ensures its processing into polymeric products. Traditionally, chitosan films are obtained by casting from diluted carboxylic acid solutions. The solution is poured onto a flat surface and the solvent is allowed to evaporate.

Films and fibers obtained from chitosan solutions are water-soluble. To convert them into insoluble form, the treatment of samples with ammonia or alkaline solutions is traditionally used, which provides deprotonation of the chitosan amino groups [12]. However, as a result of such treatment the samples not only lose water solubility, but also sharply reduce the sorption capacity. In this case, bifunctional reagents are used to obtain insoluble chitosan-based films, fibers, and other materials with a high rate of water swelling. In the presence of such reagents, a three-dimensional spatial grid of chemical bonds is formed in chitosan solutions. The presence of such a grid provides the material with high and regulated sorption capacity of water vapor, controlled release of biologically active and medicinal compounds, if any are incorporated in the material [13,14].

Covalent or ionic type crosslinking reagents are used to form such grids of spatial bonds in films. Reagents of different structure and functionality are used for covalent crosslinking of chitosan: formaldehyde, toluene-2,4-diisocyanate, epichlorohydrine, and diglicidyl ether of ethylene glycol [15–17]. Most of the works carried out in this direction describe the technology of synthesis of polymeric materials based on chitosan modified with dialdehyde, most often with glutaraldehyde [13]. However, there are studies that indicate the toxicity of products of interaction between chitosan and aldehyde, which is an obstacle to their use [14,18], although it has been shown [19] that the use of low, but sufficient for the loss of film water solubility, amounts of glutaraldehyde does not lead to an increase in the cytotoxicity of the polysaccharide. Thus, for the production of consumable films or food packaging materials, the above-mentioned crosslinking reagents cannot be used and it is reasonable to use less toxic reagents.

Particularly, crosslinking with genipin (Gp) (Figure 1), which is a reagent of natural origin, improves the swelling, water resistance, and mechanical properties of the films [14,20–22]. Genipin has recently been used in biomedical applications and for controlled drug release due to its biocompatibility and low toxicity. The low toxicity of the chitosan materials crosslinked by Gp is demonstrated by a number of studies [15,20,23,24]. Genipin crosslinked biopolymeric films should have high potential in food packaging due to these facts. However, the reversibility of the reaction of genipin with chitosan amino groups [25] and possibility of cross-linking reagent polymerization, which leads to a decrease in the number of molecules available for reaction with chitosan amino groups, require special attention to the conditions of chitosan cross-linking by genipin upon film formation.

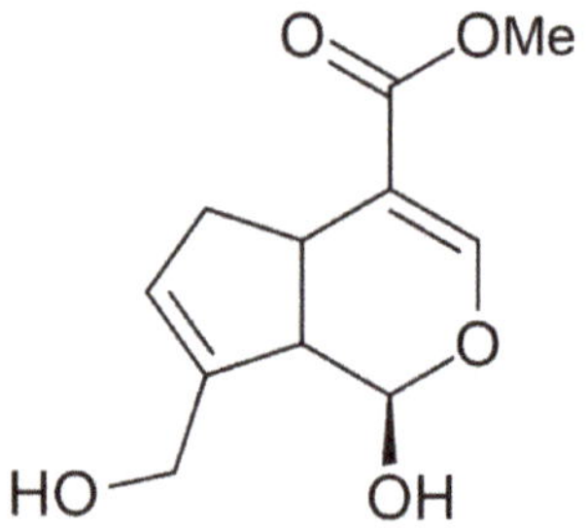

Figure 1. Chemical structure of genipin (Gp).

The literature provides contradictory data regarding the impact of crosslinking on the strength and elastic properties of chitosan films. In [26], it was shown that, in comparison with non-crosslinked films derived from a mixture of chitosan/poly(ethylene oxide), a film that was based on chitosan crosslinked with genipin had greater strength and elasticity. There are reports that, when chitosan films are crosslinked with genipin, the strength increases, but the breaking elongation decreases [27,28].

In other studies on the influence of different types of crosslinking reagents on the properties of the films, including their use as food packaging [29–31], it is noted that, at a high degree of crosslinking, the films become more rigid, and their strength decreases. Several authors have used special techniques in order to improve the mechanical properties of the cross-linked films: introduction of nanocrystalline cellulose [32] or additional flexible-chain polymers [33]. We assume that such contradictions may be related to different initial conditions of sample formation: temperature and the solvent removal rate, the use of alkaline treatment, the amount of added crosslinking reagent, pH, etc.

In the review [31] devoted to the integrity of natural biopolymer films used in food packaging, which were obtained by crosslinking approach, the cases of different influence on the physical and mechanical properties of films of low and high cross-linking reagent concentrations are noted. Two different behaviors of soy protein cross-linked by genipin have been observed in [34]: increase in the elasticity of protein films at 1% concentration of genipin and decrease in elastic modulus at 10% of cross-linking reagent.

The aim of the present work was to study the properties of Gp cross-linked chitosan films with significantly different cross-linking reagent amounts, which might be useful in the determination of the tuning direction of biopolymer films properties for food packaging, drug delivery, or wound dressings.

The content of the cross-linking reagent affects the degree of cross-linking, the reaction rate, and time, during which the gelation process is completed in the chitosan solution. On the basis of preliminary experiments and previous studies [13], two extreme values of molar ratios of genipin-chitosan amino group (Gp/NH_2) were chosen: the minimum, at which gelation in the used chitosan solution is possible—0.003 mol/mol, and the maximum—0.02 mol/mol, at which cross-linking is much faster, but still allows for completing the operations on preparation of the molding solution and film casting before gelation. For the films differing (almost seven times) in cross-linking reagent content, comparative studies of the form of electronic absorption spectra, sorption, structural and physical-mechanical properties of samples have been conducted. In a number of experiments, the films containing genipin in an amount exceeding the established limits, or not at all containing cross-linking reagent, were studied for a more complete interpretation of the obtained results. With the use of experimental data on film swelling and isotherms of water vapor sorption, the theoretical evaluation of the degree of cross-linking of chitosan in the films that were obtained from solutions containing Gp was carried out.

2. Materials and Methods

2.1. Materials

Chitosan with a molecular weight of 320 kDa and deacetylation degree of 88% was purchased from Bioprogress, Voronezh Oblast, Russia. Genipin (Gp) and acetic acid ware obtained from Aldrich Chemicals, Gillingham, UK. All other reagents and solvents used were of reagent grade and they were used without further purification.

2.2. Experimental Method

2.2.1. Preparation of Chitosan Films

The chitosan films used in the study were fabricated by means of casting/solvent evaporation technique. Chitosan was dissolved in 2 wt % aqueous acetic acid at room temperature overnight in order to obtain a 2.1 wt/v% solution. The viscous chitosan solution was filtered through filter paper to remove any undissolved gel. Chitosan solutions with a concentration of approximately 2 wt % and pH 4.1 were obtained. The solutions were kept at room temperature for 3 h for degassing. The degassed solutions were then cast onto glass plates and dried to a constant weight at ambient temperature. The thickness of the dried films was 50 ± 5 μm. The resulting films were water-soluble.

2.2.2. Preparation of Genipin-Crosslinked Films (Chitosan-Gp)

Chitosan solution was prepared, as described in Section 2.2.2. The concentration of chitosan solution was specified after solution filtration and weighing of the paper filter mass dried up until constant. Chitosan solution containing 0.2 g of polymer was mixed with 1.25 mL of aqueous solution of genipin obtained by dilution of its 0.5% aqueous solution. Table 1 shows concentrations of genipin solution. The obtained solutions were kept at room temperature for 30 min, and then poured onto a glass disk and left for crosslinking and solvent evaporation. The resulting films were dried at an ambient temperature to a constant weight (after 3 weighing every 2–3 h). The films considered ready were obtained after 70 ± 5 h. The thickness of the films was determined while using the EP-10–60 Ms thickness indicator. The scale interval was 0.01 mm. The thickness of the dried films was 52 ± 5 μm.

Table 1. Concentrations of genipin solution used for genipin cross-linked chitosan film production.

Ratio Gp/NH$_2$, mol/mole	Genipin Solution Concentration, %	Genipin Content of the Molding Solution, % of Chitosan Mass
0.02	0.24	1.5
0.003	0.036	0.225
0.0025	0.03	0.187
0.002	0.024	0.15

2.2.3. Spectrophotometric Studies

The absorption spectra of the pure chitosan and genipin crosslinked chitosan (chitosan-Gp) films were recorded on a Shimadzu UV-2501PC spectrophotometer (Kyoto, Japan) in the spectral region from 190 to 800 nm.

2.2.4. Swelling Studies

The swelling kinetics of chitosan and genipin crosslinked chitosan films with the size of 2 × 2 cm^2 in water were gravimetrically studied. Measurements were made on HR-200 A&D Co. LTD scales (Tokyo, Japan). Before the sorption experiments, the films were vacuumized to a constant weight at a residual pressure of ~10^{-2} Pa and temperature of 120 °C. Preliminary thermogravimetry measurements showed that the residual water content in chitosan and genipin crosslinked chitosan films before the experiments did not exceed 0.2 wt %. All of the samples were conditioned in a dry desiccator at zero humidity before measurements. The swelling ratio of each studied chitosan membrane was determined by immersing the membrane in water (pH = 7.0) at room temperature. The swelling ratio of the chitosan membrane was calculated, as follows:

$$M(\text{wt \%}) = \frac{m - m_0}{m_0} \times 100 \tag{1}$$

where M (%) is the percentage water absorption of the film; and, m_0 and m are the weights of the samples in the dry and swollen states, respectively. Each swelling experiment was repeated three times, and the average values are reported. A cross plot of swelling degree on time $M = f(t)$ was constructed according to the obtained data.

2.2.5. Sorption Measurements

McBen vacuum scale was used to study the processes of interaction of water vapor with chitosan and genipin crosslinked chitosan membranes (2 × 5 cm^2). The quantity of sorbed (desorbed) water vapors was determined by stretching of the calibrated quartz spiral with the help of optical registration system. The accuracy of measurement was ±0.00001 g. The sorption measurements were performed under isobaric-isothermal conditions in desiccators at a temperature of 20 ± 1 °C. The methods of integral and interval sorption were used [35]. All of the samples were conditioned in a dry desiccator at zero humidity before taking measurements. The range of relative changes in water vapor pressure (p/ps) was changed from 0.20 to 0.95. The sorption of the chitosan membrane was calculated as (1).

2.2.6. Mechanical Property Measurements

Rectangular specimens of 5×1 cm^2 of chitosan and genipin crosslinked chitosan membranes were used to determine the limiting strength characteristics. Single axis tensile testing at a deformation rate of 10 mm/s was carried out on a universal testing machine BM-50 (Russia) with the fixation of tensile strength and elongation at break. The experiment was carried out five times (tensile strength at break point was determined on Russian state standard (GOST) 11262-80 type 5 blades).

2.2.7. Structural-Morphological Studies

Structural-morphological studies were carried out by the method of transmission electron microscopy. The samples were etched in oxygen discharge plasma for 40 min. to reveal the fine structure (etch depth of 150–200 nm). Electron energy in the etching zone was 5–6 eV, oxygen pressure was 0.1 mm Hg, etching time was 20 min. Imaging of the etched surface was carried out by transmission electron microscopy while using single-stage carbon-platinum replicas using the EM 301 (Philips, Dutch, The Netherlands) electron microscope.

3. Results and Discussion

3.1. Optical Properties of the Gp Crosslinked Chitosan Films

In contrast to practically colorless films of pure chitosan, the films with introduced Gp are characterized by pronounced blue color, the intensity of which depends on the Gp/NH$_2$ ratio (Figure 2). Such coloration is typical for various products of genipin and chitosan interaction: hydrogels, microparticles, etc. [20].

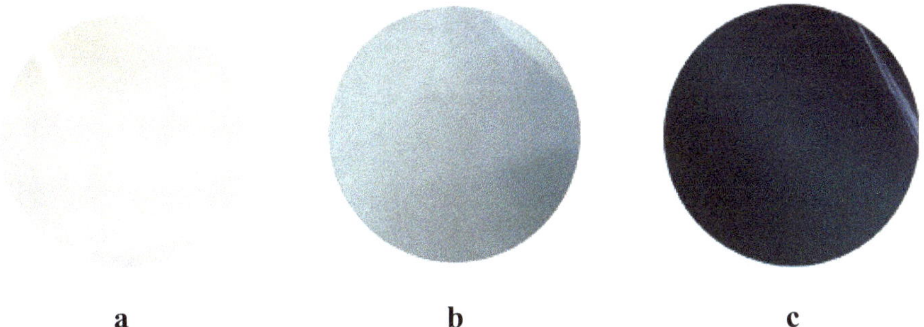

a   **b**   **c**

Figure 2. Photographs of the studied films. Pure chitosan film (**a**); chtosan films, crosslinked in Gp/NH$_2$ molar ratio of 0.003 (**b**) and 0.02 (**c**).

Figure 3a shows the electronic absorption spectra of films that are based on chitosan crosslinked with genipin at different Gp/NH$_2$ ratios. The absorption spectrum of pure chitosan exhibits weak bands in the area below 400 nm. Intensive absorption bands in the range of 260–700 nm appear in films with Gp (Figure 3a).

For the analysis, the pure chitosan absorbance spectrum was subtracted from the spectra of crosslinked films. The obtained spectra are well approximated by a sum of four Gaussian bands: A$_{608\,nm}$ = 0.78; A$_{460\,nm}$ = 0.4; A$_{348\,nm}$ = 1.18; and, A$_{283\,nm}$ = 4.1, which is illustrated in Figure 3b for the film with the Gp/NH$_2$ ratio of 0.02.

The form of the film absorption spectrum with a Gp/NH$_2$ ratio of 0.003 is similar to the one that is discussed above, which testifies to the identical set of chromophore groups in the samples under consideration and indicates the same nature of chemical reactions occurring during crosslinking with different genipin/chitosan ratio. The ratio of absorption bands intensity in the studied films is close to the ratio of genipin concentration in them and is about 6.

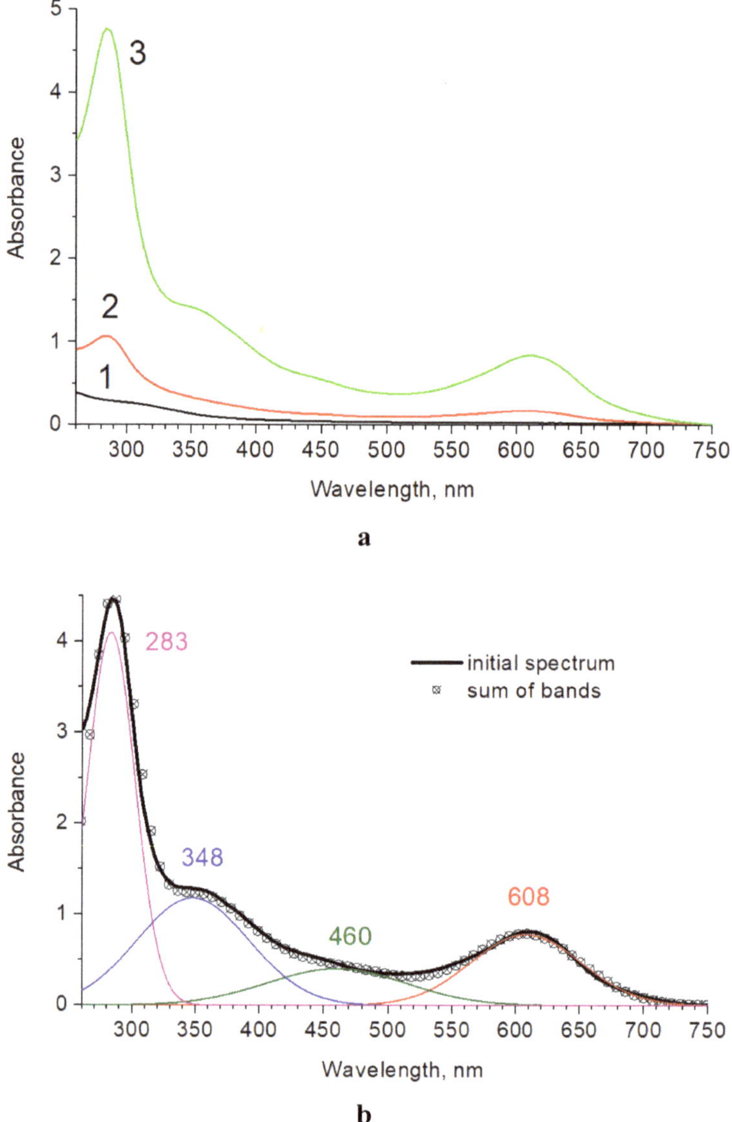

Figure 3. (a) UV-vis absorbance of chitosan film (1), and the films obtained at Gp/NH$_2$ molar ratios of 0.003 (2) and 0.02 (3). (b) Deconvolution of the absorbance spectrum for the film with Gp/NH$_2$ molar ratio of 0.02.

A similar form of the absorption spectra was previously observed upon crosslinking performed in solutions [36]. In [21], the absorption band at 280–290 nm was associated with the inclusion of chitosan amino group in the heterocycle of genipin, which leads to the formation of heterocyclic amine, and the absorption band at 610 nm was assigned to the reaction of radical polymerization of genipin, induced by oxygen in air, occurring as one of the stages in the process of the crosslinking of chitosan. The last absorption band determines the intense blue color of the genipin crosslinked chitosan films. The increase of the peak intensity at 280–290 nm indicates an increase in the degree of binding of Gp by chitosan at a high Gp/NH_2 ratio.

3.2. Kinetics of the Swelling and Water Vapor Sorption by the Films

The nature of the swelling kinetic curves of Gp cross-linked chitosan films in water (Figure 4) shows that films containing Gp at molar ratios of 0.003 and 0.02 mol/mol exhibit limited swelling in water, and the degree of their swelling at the initial stage of the process is linearly dependent on $t^{1/2}$, which indicates a Fick diffusion mechanism of water absorption. The equilibrium swelling decreases with an increasing concentration of cross-linking reagent: 1020% wt. for film with Gp/NH_2 ratio of 0.003 and 340% wt. for Gp/NH_2 of 0.02 mol/mol. This clearly indicates an increase in grid density in cross-linked Gp films. Upon a decrease of the genipin content to 0.002 mole/mole, the film dissolves in water, its kinetic curve is similar to that of an unrestricted swelling of a non-crosslinked film (curves 3 and 4, Figure 4).

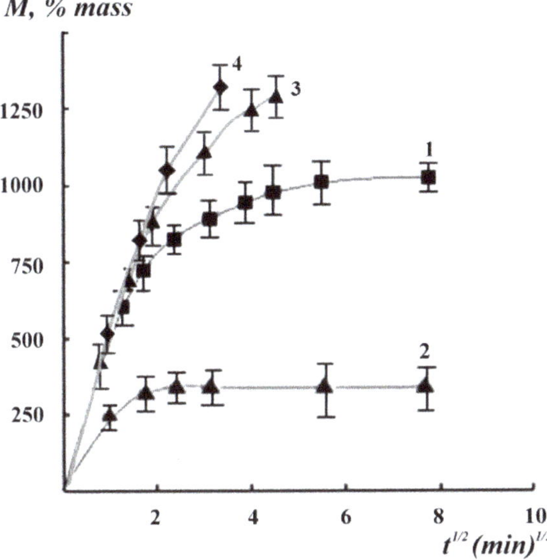

Figure 4. The kinetics of swelling of genepin crosslinked chitosan films with Gp/NH_2 molar ratios of 0.003 (1), 0.02 (2), 0.002 (3), and chitosan film without genipin (4) in water.

Figure 5 shows the kinetic curves of interval sorption of water vapor for chitosan films crosslinked with genipin at different ambient humidity. It can also be seen that the Fick kinetics of the sorption equilibrium establishment is observed for all types of crosslinked and non-crosslinked samples in these conditions. The difference in sorption kinetics and swelling kinetics is only in the speed of sorption equilibrium establishment. It is most vividly manifested in the values of translational coefficients of water diffusion, which we estimated from experimental data M–$t^{1/2}$, while using Equation (2)

$$D = 1.96 \cdot L^2 / \pi^2 t_{0.5} \tag{2}$$

where L is the membrane thickness and $t_{0.5}$ is the time it takes to reach half saturation of the membrane sorption capacity.

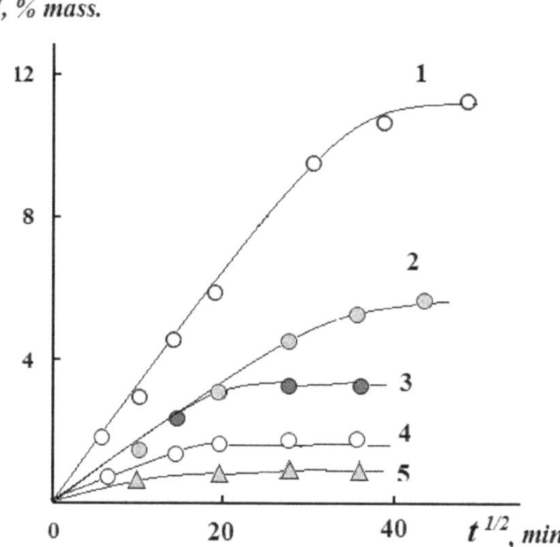

Figure 5. Kinetics of water vapor sorption of genipin crosslinked chitosan films crosslinked: 1—Gp/NH$_2$ ratio 0.003 mol/mol; 2—0.02 mol/mol. 1, 4—Gp/NH$_2$ ratio of 0.02 mole/mol; 2,3,5—0.003 mole/mole at 293 K and P/P$_0$ 0.2(5), 0.3(4), 0.6(3), 0.7(1), 0.8(2).

Figure 6 shows the coefficients of water diffusion in chitosan films, depending on the relative humidity of the environment. It is seen that, with increasing humidity, the translational mobility of sorbed water molecules increases, which indicates the plasticization of chitosan films crosslinked with genipin. Obviously, this result indicates that it is possible to apply the Flory–Riener Equation (3) to calculate the chemical bonds grid density of chitosan films crosslinked with genipin.

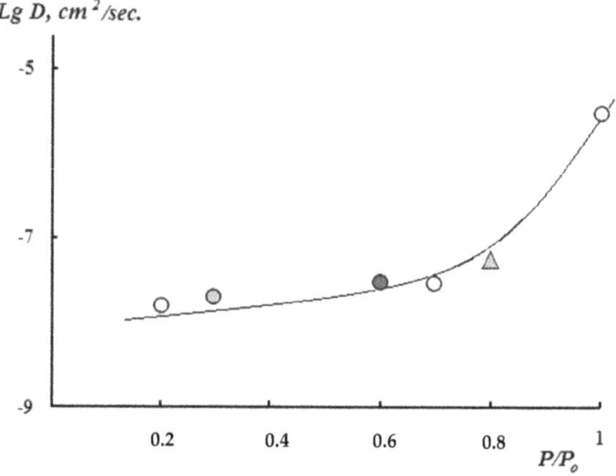

Figure 6. Dependence of the water molecule diffusion coefficients in the genipin crosslinked chitosan films on the ambient humidity.

Figure 7 shows the isotherms of water vapor sorption by the films obtained from 2% solutions of chitosan with different content of cross-linking reagent: Gp/NH$_2$ of 0.003 and 0.02 mol/mol. It can be seen that, for all samples of genipin crosslinked chitosan films, the isotherms of water vapor sorption are S-shaped. In this case, their sorption capacity decreases with increasing amount of genipin in the whole interval of water vapor pressure. Since, genipin crosslinked chitosan films are non-porous sorbents with homogeneous nanoscale domain structure (Figure 8), according to electron microscopic studies, we did not consider it expedient to use the BET model to analyze the sorption isotherms.

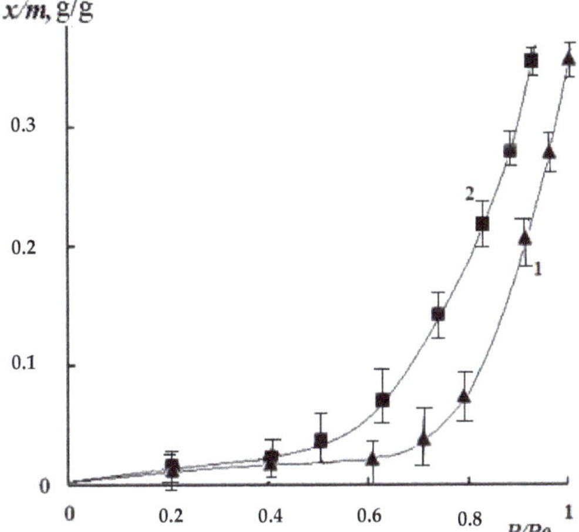

Figure 7. Isotherms of water vapor sorption of chitosan films, crosslinked by genipin. 1—Gp/NH$_2$ ratio 0.02 mol/mol; 2—0.003 mol/mol.

Figure 8. Microphotograph of the supramolecular structure of the same formed with Gp/NH$_2$ molar ratio of 0.003.

Following the model of double sorption and the method proposed earlier [35], decomposition of the sorption isotherms in the TWOFL package into Langmuir and Flory–Haggins components was performed (Figure 9). It can be seen that the filling of accessible (free) functional groups of chitosan with water molecules is realized in the region of low vapor activity, whereas, in the region of high humidity, the main fraction is made up of free and mobile mode molecules. The numerical values of paired interaction parameters χ obtained for the Flory–Haggins mode amounted to 0.85 for Gp cross-linked chitosan at 0.003 mol/mol and 0.99 for 0.02 mol/mol ratios. It is important to note that the critical value for aqueous solutions of chitosan amounts to 0.5 [35]. It can be assumed that this difference in the numerical values of the paired interaction parameter is due to the contribution of genipin molecules to the intermolecular interactions of the system components.

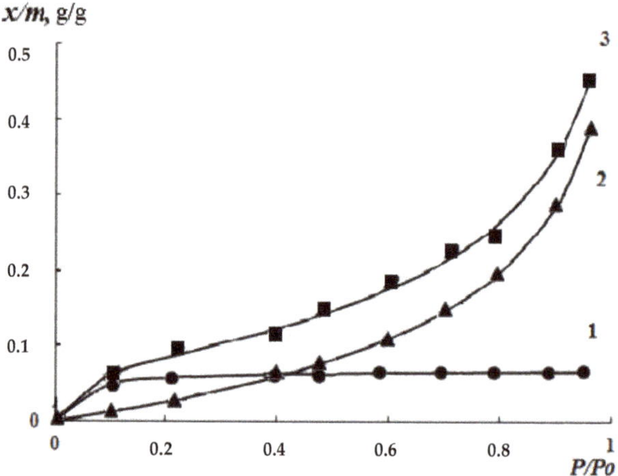

Figure 9. Decomposition of the film water vapor sorption isotherm (3) in the framework of the Flory–Haggins model: 1—Langmuir isotherm, 2—Flory-Haggins isotherm.

We calculated the molecular mass of the chemical bond grid between the nodes M_c using the values of the equilibrium degree of swelling (Figure 4) and the paired interaction parameter χ according to Flory–Riener Equation (3) [37].

$$\ln(1 - \phi_2) + \phi_2 + \chi_1\phi_2^2 + \frac{d_2\overline{V}_1}{M_c}\left(\phi_2^{1/3} - \frac{2\phi_2}{f}\right) = 0 \tag{3}$$

where φ_2 is the volume fraction of the polymer in the swollen sample; d_2 is the polymer density; $\overline{V}_1$ is partial molar volume of the solvent; M_c—molecular weight of a chain segment between nodes of the grid; and, $\chi_1; f$ is the grid functionality. The density of the chitosan films (~1.021 g/cm³), crosslinked with genipin (~1.018 g/cm³) were determined by hydrostatic weighing; the partial mole volume of sorbate was taken equal to mole volume of water in the calculations. The volume fraction of the polymer was calculated within the limits of additive model of density of swollen sorbent. Table 2 provides the obtained results.

Table 2. The results of the degree of chitosan crosslinking by genipin calculation.

Molar Ratio Gp/NH$_2$	Equilibrium Swelling Degree, %	Paired Interaction Parameter χ	Molecular Weight of the Chitosan Chains between the Mesh Nodes M$_c$, g/mole	Number of Elementary Links of Chitosan between Mesh Nodes n$_c$	Number of Meshes per 100 Elementary Links of Chitosan
0.003	1025	0.85	6934	42	2.7
0.02	340	0.99	1015	6	18.8

The values of the molecular weight of the grid between the mesh nodes M_c can be calculated while using the values of the equilibrium degree of swelling (Figure 4) and the paired interaction parameter χ obtained from sorption isotherms according to the Flory–Riener equation [37].

The change in the degree of modification of amino groups is proportional to the changes in the content of crosslinking reagent in chitosan solution; thus, the molar ratio Gp/NH$_2$ can be used to assess the degree of crosslinking, as can be seen from Table 2. However, the absolute values of the number of meshes 2.7 and 18.8 exceed the maximum possible degree of crosslinking of chitosan macromolecules, which amounts to 6 and 42 in calculation per 100 elementary links for films obtained while using different Gp/NH_2 molar ratios. This indicates that intermolecular contacts, more precisely hydrogen bonds, make an additional contribution to the formation of the grid of meshes. Chemical crosslinking only fixes the structure of the film formed by the rapprochement of the macromolecules due to evaporation of the solvent.

The non-equilibrium domain structure of the polymer would be fixed if the interaction with Gp that leads to the chemical crosslinking of macromolecules occurs before the equilibrium arrangement of the macromolecular chains (Figure 8). If the crosslinking is completed with low residual solvent content, when the mesh grid is already mostly formed, the more perfect equilibrium structure of the film is fixed, and one can expect that such material will have a higher level of physical and mechanical properties. This fact is important for the technology of obtaining polymer materials from solutions while using crosslinking reagents and it provides a tool for programming the properties of polymer materials.

3.3. Mechanical Properties of Gp-Crosslinked Chitosan Films

To obtain water-insoluble films, except for the Gp/NH_2 ratio of 0.02, the minimum Gp/NH_2 molar ratio of 0.003, which provides crosslinking of chitosan in the solution, and a ratio of 0.0025, the use of which did not lead to gelation in the solution. The concentration of a solution increases upon film formation in the process of the solvent evaporation, and smaller quantity of cross-linking reagent, than for gel formation in 2–5% solutions of chitosan, is sufficient for the completion of cross-linking reaction. After the evaporation of the solvent, the film obtained at this Gp/NH_2 ratio remained colorless during 24 h, but it later acquired blue coloring and water resistance. The films, obtained at lower genipin content (0.002 mol/mol ratio), are water soluble (Figure 4).

Obviously, the crosslinking of chitosan by bifunctional reagent in solution fixes mutual arrangement of polymer chains, not allowing them to take equilibrium conformations in the process of solvent evaporation. This fact is confirmed by the revealed sharp decrease in the level of physical and mechanical properties of films: the strength of films obtained using Gp/NH_2 molar ratios of 0.003 and 0.02 is much lower than the strength of films that were obtained in conditions of polymer chains relaxation in the absence of crosslinking reagent and slow evaporation of solvent at 20 °C. However, reducing the content of genipin in chitosan solution to the concentration at which the evaporation of the solvent occurs faster than the completion of crosslinking of macromolecular chains of chitosan not only does not lead to a decrease in strength, but even increases the strength of the film, probably due to the fixation of macromolecular chains in an equilibrium position, in which the intermolecular interactions are realized to a higher degree (p. 4, Table 3). The ratio of 0.0025 makes it possible, on the one hand, to obtain a film insoluble in water and, on the other hand, to increase the breaking load under certain temperature (20–22 °C) and humidity conditions. Similar data were obtained in [34] in the study of soy protein films that were obtained in the presence of genipin: the improvement of mechanical properties only occurred when the content of the cross-linking reagent was less than 1%. With changing conditions of chitosan Gp crosslinking reaction, the optimal ratio of functional groups can be different.

Table 3. Mechanical properties of the chitosan films formed at different Gp/NH_2 ratios.

Gp/NH_2 Molar Ratio	Breaking Load, MPa	Breaking Elongation, %
0	73	33
0.02	65	33
0.0030	55	25
0.0025	90	15

4. Conclusions

The influence of the ratio of functional groups in the chitosan-cross-linking reagent Gp system on the film absorption in the visible and ultraviolet spectral regions, sorption, physical, and mechanical properties of the films was studied at significantly different concentrations of cross-linking reagent 0.02 and ≤ 0.003 mol/mol ratios. The analysis of electronic absorption spectra showed the formation of an identical set of chromophore groups in the studied samples, which indicates the same nature of chemical reactions occurring during cross-linking in systems, with a genipin/chitosan molar ratio that varies by a factor of 6.7. Kinetic and thermodynamic parameters of swelling and water vapor sorption processes (diffusion coefficients, equilibrium sorption values, and paired interaction parameters χ) change in proportion to the cross-linking reagent content. The decomposition of sorption isotherms into Langmuir and Flory–Haggins components within the double sorption model according to the proposed technique has shown that the filling of chitosan functional groups with water molecules is realized in the region of low vapour activity, while the main fraction is made up of molecules of free and mobile mode in the region of high humidity. At the same time, genipin cross-linking contributes to the intermolecular interactions, as evidenced by the difference in numerical values of the paired interaction parameter χ for different genipin contents in the films (0.85 at Gp/NH_2 ratio of 0.003 mol/mol, and 0.99 for 0.02 mol/mol). The cross-linking of the chitosan by a bifunctional reagent in the solution fixes the mutual location of polymer chains, not allowing them to take equilibrium conformations during solvent evaporation, which leads to a deterioration in the physical and mechanical properties of the resulting films. Only at a very low content of the cross-linking reagent (0.0025 mol/mol), at which there is no gel formation in the molding solution, the fixation of macromolecular chains occurs in the position where intermolecular interactions are realized to the greatest extent, which, in its turn, leads to a significant increase in the strength of chitosan films. The results of research into the properties of chitosan films with significantly different cross-linking reagent contents can be used to determine the direction of programming of cross-linked biopolymer film properties for food packaging, edible coatings, or for medical use to increase strength and regulate moisture absorption.

Author Contributions: Writing—View and Edit, N.K.; Conceptualization, A.C.; Writing-review, N.S. (Nikita Sazhnev); Writing—original draft, M.B.; Data curation, T.P.; Research, V.M.; Methodology, N.S. (Nikolay Surin); Visualization, E.S. All authors have read and agreed to the published version of the manuscript.

Funding: This work was supported by Russian Foundation for Basic Research (project no. 18-29-17059). UV/Vis spectroscopy was performed with financial support from the Ministry of Science and Higher Education of the Russian Federation, using the equipment of the Collaborative Access Center "Center for Polymer Research" of ISPM RAS.

Conflicts of Interest: The authors declare no conflict of interest.

References

1. Jung, W.-J.; Park, R.-D. Bioproduction of chitooligosaccharides. Review. *Mur. Drugs* **2014**, *12*, 5328–5356. [CrossRef]

2. Thadathil, N.; Velappan, S.P. Recent developments in chitosanase research and its biotechnological applications: A review. *Food Chem.* **2014**, *150*, 392–399. [CrossRef] [PubMed]

3. Kean, T.; Thanou, M. Biodegradation, biodistribution and toxicity of chitosan. *Adv. Drug Deliv. Rev.* **2010**, *62*, 3–11. [CrossRef] [PubMed]

4. Aider, M. Chitosan application for active bio-based films production and potential in the food industry: Review. *Lwt-Food Sci. Technol.* **2010**, *43*, 837–842. [CrossRef]

5. Souza, V.G.L.; Pires, R.A.J.; Rodrigues, C.; Coelhoso, I.M.; Fernando, A.L. Chitosan Composites in Packaging Industry—Current Trends and Future Challenges. *Polymers* **2020**, *12*, 417. [CrossRef] [PubMed]

6. Mukhopadhyay, P.; Bhattacharya, S.; Nandy, A.; Bhattacharyya, A.; Mishra, R.; Kundu, P.P. Assessment of in vivo chronic toxicity of chitosan and its derivates used as oral insulin carriers. *Toxicol. Res.* **2015**, *4*, 281–290. [CrossRef]

7. Lu, B.; Lv, X.; Le, Y. Chitosan-Modified PLGA Nanoparticles for Control-Released Drug Delivery. *Polymers* **2019**, *11*, 304. [CrossRef]

8. Tiwari, S.; Patil, R.; Bahadur, P. Polysaccharide Based Scaffolds for Soft Tissue Engineering Applications. *Polymers* **2019**, *11*, 1. [CrossRef]

9. Thanou, M.; Verhoef, J.C.; Junginger, H.E. Oral drug absorption enhancement by chitosan and its derivatives. *Adv. Drug Delivery Rev.* **2001**, *52*, 117–126. [CrossRef]

10. Wedmore, I.; McManus, J.G.; Pusateri, A.E.; Holcomb, J.B. A special report on the chitosan-based hemostatic dressing: experience in current combat operations. *J. Trauma* **2006**, *60*, 655–658. [CrossRef]

11. Strand, S.P.; Tømmeraas, K.; Vårum, K.M.; Østgaard, K. Electrophoretic light scattering studies of chitosans with different degrees of N-acetylation. *Biomacromolecules* **2001**, *2*, 1310–1314. [CrossRef] [PubMed]

12. Wang, X.; Yan, Y.; Xiong, Z.; Lin, F.; Wu, R.; Zhang, R.; Lu, Q. Preparation and evaluation of ammonia-treated collagen/chitosan matrices for liver tissue engineer-ing. *J. Biomed. Mater. Res. Part B Appl. Biomater.* **2005**, *75*, 91–98. [CrossRef] [PubMed]

13. Kil'deeva, N.R.; Kasatkina, M.A.; Mikhailov, S.N. Peculiarities of obtaining biocompatible films based on chitosan cross linked by genipin. *Polym. Sci. Ser. D* **2017**, *10*, 189–193. [CrossRef]

14. Kildeeva, N.R.; Perminov, P.A.; Vladimirov, L.V.; Novikov, V.V.; Mikhailov, S.N. About mechanism of chitosan crosslinking with glutaraldehyde. *Russ. J. Bioorganic Chem.* **2009**, *35*, 360–369. [CrossRef]

15. Devi, D.A.; Smitha, B.; Sridhar, S.; Aminabhavi, T.M. Pervaporation separation of isopropanol/water mixtures through crosslinked chitosan membranes. *J. Membr. Sci.* **2005**, *262*, 91–99. [CrossRef]

16. Zeng, X.; Ruckenstein, E. Crosslinked macroporous chitosan anion-exchange membranes for protein separations. *J. Membr. Sci.* **1998**, *148*, 195–205. [CrossRef]

17. Vieira, R.S.; Beppu, M.M. Interaction of natural and crosslinked chitosan membranes with Hg (II) ions. *Colloids Surf. A Physicochem. Eng. Asp.* **2006**, *279*, 196–207. [CrossRef]

18. Lai, J.Y.; Li, Y.T.; Wang, T.P. In vitro response of retinal pigment epithelial cells exposed to chitosan materials prepared with different crosslinkers. *Interna-Tional J. Mol. Sci.* **2010**, *11*, 5256–5272. [CrossRef]

19. Kil'deeva, N.R.; Kasatkina, M.A.; Drozdova, M.G.; Demina, T.S.; Uspenskii, S.A.; Mikhailov, S.N.; Markvicheva, E.A. Biodegradable scaffolds based on chitosan: Preparation, properties, and use for the cultivation of animal cells. *Appl. Biochem. Microbiol.* **2016**, *52*, 515–524. [CrossRef]

20. Muzzarelli, R.A.A. Genipin-crosslinked chitosan hydrogels as biomedical and pharmaceutical aids. *Carbohydr. Polym.* **2009**, *77*, 1–9. [CrossRef]

21. Butler, M.F.; Pudney, P.D.A. Mechanism and kinetics of the crosslinking reaction between biopolymers containing primary amine groups and genipin. *J. Polym. Sci. Part A Polym. Chem.* **2003**, *41*, 3941–3953. [CrossRef]

22. Shah, R.; Stodulka, P.; Skopalova, K.; Saha, P. Dual Crosslinked Collagen/Chitosan Film for Potential Biomedical Applications. *Polymers* **2019**, *11*, 2094. [CrossRef] [PubMed]

23. Berger, J.; Reist, M.; Mayer, J.M.; Felt, O.; Peppa, N.A. Gurny. Structure and interactions in covalently and ionically crosslinked chitosan hydrogels for biomedical applications. *Eur. J. Pharm. Biopharm.* **2004**, *57*, 19–34. [CrossRef]

24. Sazhnev, N.A.; Drozdova, M.G.; Rodionov, I.A.; Kil'deeva, N.R.; Balabanova, T.V.; Markvicheva, E.A.; Lozinsky, V.I. Preparation of chitosan cryostructurates with controlled porous morphology and their use as 3D-scaffolds for the cultivation of animal cells. *Appl. Biochem. Microbiol.* **2018**, *54*, 459–467. [CrossRef]

25. Matcham, K. Novakovic. Fluorescence Imaging in Genipin Crosslinked Chitosan–Poly(vinyl pyrrolidone) Hydrogels. *Polymers* **2016**, *8*, 385. [CrossRef]

26. Jin, J.J.; Song, M.; Hourston, D.J. Novel Chitosan-Based Films Crosslinked by Genipin with Improved Physical Properties. *Biomacromolecules* **2004**, *5*, 162–168. [CrossRef]

27. Nunes, C.; Maricato, É.; Cunha, Â.; Nunes, A.; da Silva, J.A.L.; Coimbra, M.A. Chitosan–caffeic acid–genipin films presenting enhanced antioxidant activity and stability in acidic media. *Carbohydr. Polym.* **2013**, *91*, 236–243. [CrossRef]

28. Mi, F.-L.; Tan, Y.-C.; Liang, H.-C.; Huang, R.-N.; Sung, H.-W. In vitro evaluation of a chitosan membrane crosslinked with genipin. *J. Biomater. Sci. Polym. Ed.* **2001**, *12*, 835–850. [CrossRef]

29. Frick, J.M.; Ambrosi, A.; Pollo, L.D.; Tessaro, I.C. Influence of Glutaraldehyde Crosslinking and Alkaline Post-treatment on the Properties of Chitosan-Based Films. *J. Polym. Environ.* **2018**, *26*, 2748–2757. [CrossRef]

30. Silva, R.M.; Silva, G.A.; Coutinho, O.P.; Mano, J.F.; Reis, R.L. Preparation and characterisation in simulated body conditions of glutaraldehyde crosslinked chitosan membranes. *J. Mater. Sci. Mater. Med.* **2004**, *15*, 1105–1112. [CrossRef]

31. Garavand, F.; Rouhi, M.; Razavi, S.H.; Cacciotti, I.; Mohammadi, R. Improving the integrity of natural biopolymer films used in food packaging by crosslinking approach: A review. *Int. J. Biol. Macromol.* **2017**, *104*, 687–707. [CrossRef] [PubMed]

32. Mirzaei, E.; Faridi-Majidi, R.; Shokrgozar, M.A.; Paskiabi, F.A. Genipin cross-linked electrospun chitosan-based nanofibrous mat as tissue engineering scaffold. *Nanomed. J.* **2014**, *1*, 137–146.

33. Pomari, A.A.d.N.; Montanheiro, T.L.d.A.; de Siqueira, C.P.; Silva, R.S.; Tada, D.B.; Lemes, A.P. Chitosan Hydrogels Crosslinked by Genipin and Reinforced with Cellulose Nanocrystals: Production and Characterization. *J. Compos. Sci.* **2019**, *3*, 84. [CrossRef]

34. González, A.; Strumia, M.C.; Igarzabal, C.I.A. Cross-linked soy protein as material for biodegradable films: synthesis, characterization and biodegradation. *J. Food Eng.* **2011**, *106*, 331–338. [CrossRef]

35. Chalykh, A.E.; Petrova, T.F.; Khasbiullin, R.R.; Ozerin, A.N. Water sorption on and water diffusion in chitin and chitosan. *Polym. Sci. Ser. A* **2014**, *56*, 614–622. [CrossRef]

36. Mi, F.L.; Shyu, S.S.; Peng, C.K. Characterization of ring-opening polymerization of genipin and pH-dependent crosslinking reactions between chitosan and genipin. *J. Polym. Sci. Part A* **2005**, *43*, 1985–2000. [CrossRef]

37. Flory, P.J. *Principles of Polymer Chemistry*; Cornell University Press: Ithaca, NY, USA, 1953; p. 688.

Article

Comparative Structure-Property Characterization of Poly(3-Hydroxybutyrate-Co-3-Hydroxyvalerate)s Films under Hydrolytic and Enzymatic Degradation: Finding a Transition Point in 3-Hydroxyvalerate Content

Vsevolod A. Zhuikov [1,*], Yuliya V. Zhuikova [1], Tatiana K. Makhina [1], Vera L. Myshkina [1], Alexey Rusakov [2], Alexey Useinov [2], Vera V. Voinova [3], Garina A. Bonartseva [1], Alexandr A. Berlin [4], Anton P. Bonartsev [1,3] and Alexey L. Iordanskii [4]

[1] Research Center of Biotechnology of the Russian Academy of Sciences 33, Bld. 2 Leninsky Ave, 119071 Moscow, Russia; zhuikova.uv@gmail.com (Y.V.Z.); tat.makhina@gmail.com (T.K.M.); v.l.myshkina@gmail.com (V.L.M.); bonar@inbi.ras.ru (G.A.B.); ant_bonar@mail.ru (A.P.B.)

[2] Federal State Budgetary Institution "Technological Institute for Superhard and Novel Carbon Materials", 7a Tsentralnaya Street, Troitsk, 108840 Moscow, Russia; rusakov.alexey@gmail.com (A.R.); useinov@mail.ru (A.U.)

[3] Faculty of Biology, M.V. Lomonosov Moscow State University, Leninskie Gory 1-12, 119234 Moscow, Russia; veravoinova@mail.ru

[4] Research Center of Chemical Physics the Russian Academy of Sciences, Kosygin str. 4, 119991 Moscow, Russia; berlin@chph.ras.ru (A.A.B.); aljordan08@gmail.com (A.L.I.)

* Correspondence: vsevolod1905@yandex.ru; Tel.: +79-153-207-380

Received: 20 February 2020; Accepted: 20 March 2020; Published: 24 March 2020

Abstract: The hydrolytic and enzymatic degradation of polymer films of poly(3-hydroxybutyrate) (PHB) of different molecular mass and its copolymers with 3-hydroxyvalerate (PHBV) of different 3-hydroxyvalerate (3-HV) content and molecular mass, 3-hydroxy-4-methylvalerate (PHB4MV), and polyethylene glycol (PHBV-PEG) produced by the *Azotobacter chroococcum 7B* by controlled biosynthesis technique were studied under in vitro model conditions. The changes in the physicochemical properties of the polymers during their in vitro degradation in the pancreatic lipase solution and in phosphate-buffered saline for a long time (183 days) were investigated using different analytical techniques. A mathematical model was used to analyze the kinetics of hydrolytic degradation of poly(3-hydroxyaklannoate)s by not autocatalytic and autocatalytic hydrolysis mechanisms. It was also shown that the degree of crystallinity of some polymers changes differently during degradation in vitro. The total mass of the films decreased slightly up to 8–9% (for the high-molecular weight PHB with the 3-HV content 17.6% and 9%), in contrast to the copolymer molecular mass, the decrease of which reached 80%. The contact angle for all copolymers after the enzymatic degradation decreased by an average value of 23% compared to 17% after the hydrolytic degradation. Young's modulus increased up to 2-fold. It was shown that the effect of autocatalysis was observed during enzymatic degradation, while autocatalysis was not available during hydrolytic degradation. During hydrolytic and enzymatic degradation in vitro, it was found that PHBV, containing 5.7–5.9 mol.% 3-HV and having about 50% crystallinity degree, presents critical content, beyond which the structural and mechanical properties of the copolymer have essentially changed. The obtained results could be applicable to biomedical polymer systems and food packaging materials.

Keywords: poly(3-hydroxybutyrate); poly(3-hydroxybutyrate-co-3-hydroxyvalerate); poly(3-hydroxybutyrate-co-4-methyl-3-hydroxyvalerate); biodegradation; hydrolysis; pancreatic lipase; mechanical behavior

1. Introduction

Poly (3-hydroxybutyrate) (PHB), the main polymer of the polyhydroxyalkanoates family (PHA), is the most well-known microbiological polyester, which is a promising alternative to biodegradable synthetic thermoplastics [1–3] and other biocompatible polymers [4–11]. PHA are obtained microbiologically. This production method allows varying the physicochemical properties of polymers of this type over a wide range [12]. Because PHB has the ability to biodegrade and has high biocompatibility, it is widely used for biomedical applications in regenerative medicine and tissue engineering [13–22] and medicine forms [23–26]. It is able to create composites with synthetic polymers, inorganic materials [19,27–29] and also as a new environmentally friendly material, including application as a material for the packaging industry [15,24,30,31].

However, PHB has a number of disadvantages: fragility, lack of hydrophilicity, etc. In order to improve the physicochemical properties, copolymers of PHB with other polyhydroxyalkanoates are made. The copolymer poly(3-hydroxybutyrate-co-3-hydroxyvalerate) is made to compensate disadvantages of PHB. Incorporation of HV into the PHB homopolymer chain considerably improves the physicochemical properties, such as the melting temperature. The copolymer PHBV is more plastic, extensible, and resilient due to a decrease in the value of Young's modulus with an increase in the HV molar fraction in PHB–HV polymer chain Reference [32] The copolymerization of PHB with polyethylene glycol (PEG) increases water permeability and solubility due to the hygroscopicity of PEG. In addition, biocompatibility and hydrophilicity are improved compared to a homopolymer [15].

It is well known that the biodegradation of PHB both in living systems and in the environment occurs through enzymatic and non-enzymatic processes that occur simultaneously in natural conditions, compared to other biodegradable polymers (e.g., poly(lactic-co-glycolic) acid [33]). PHB is considered to be moderately resistant to in vitro degradation, as well as to biodegradation in animal tissues. The degradation rate is influenced by polymer characteristics, such as chemical composition, crystallinity, morphology, and molecular mass [34,35].

Therefore, to develop novel medical devices and packing materials based on PHA and its copolymers, it is necessary to know how the physicochemical properties of these polymers change during their degradation. In order to understand what changes, occur with polymers in the human body during degradation, it is necessary to study the kinetics of the change in the basic physicochemical properties when they degrade in vitro under conditions simulating the internal environment of the organism. Studies of degradation of PHA, especially on long periods, are rare [36]. Thus, the purpose of this study is to obtain and compare the kinetic curves of the long-term degradation of PHB and its copolymers. Considerable attention is paid to the change in the molecular mass and degree of crystallinity of polymer films. In addition, at present in the literature, there are no exact data on the effect of the monomer composition of the copolymer on the polymer decomposition process. There is fragmentary evidence that the inclusion of > 10% 3-hydroxyvalerate (3-HV) in the composition of the copolymer leads to a change in the structure of the crystalline component of the polymer [37].

An understanding of the processes of hydrolytic and enzymatic degradation of PHAs is very important for the development of a new biodegradable and harmless packaging for meat, fish, dairy, and vegetable food products [38–41].

Therefore, the goals of this work are to trace the changes in the physicochemical properties of the copolymers and to search for the value of the included hydroxyvalerate in the PHV chain, which has a key effect on the degradation mechanisms of the product.

2. Materials and Methods

2.1. Production of Films from PHAs

To study the in vitro biodegradation, a series of films 50 ± 10 µm in thickness and 90 mm in diameter was made from PHB and its copolymers (Table 1) obtained by controlled biosynthesis using producing strain Azotobacter chroococcum (Table S1, Figures S1–S3). The polymer (~400 mg) was dissolved in chloroform (35 mL) overnight at room temperature. Films were prepared by casting from a chloroform solution to the bottom of Petri dishes previously degreased. The films were dried until the solvent was completely removed for at least 72 h at room temperature. Then, the plates were cut from the resulting films with dimensions of 30 mm × 10 mm.

Table 1. The list of polymers used in the work. HV = hydroxyvalerate; PEG = polyethylene glycol; PHB = poly(3-hydroxybutyrate); PHBV = poly(3-hydroxybutyrate-co-3-hydroxyvalerate); MV = methylvalerate.

Substrate	Molecular Mass, kDa	The Content of 3-HV/(3-H4MV)/(PEG) in the Copolymer, %
PHB 1095	1095	0
PHBV 2.5% 768	768	2.5
PHBV 5.9% 819	819	9.0
PHBV 9% 1010	1010	17.6
PHBV 17.6% 1190	1190	5.9
PHBV-PEG 290	290	4.69% 0.15% (PEG)
PHB-4MV 1340	1340	0.60 (3-H4MV)

2.2. Determination of MM

The molecular mass (MM) of PHB and its copolymers were determined by gel filtration chromatography (GPC). Chloroform with the addition of 3% v/v methanol was used as a solvent. The elution rate was 1 mL/min. Used as the detector was a Waters 2414 differential refractive index detector and a UV detector and a waters 1525 pump. The sample was 100 µL with a concentration of 5 mg/mL. Four Waters styragel columns (Waters, Milford, MA, USA) (Styragel HT 6E, 4.6 mm × 300 mm) were used. Calibration was carried out using polystyrene reference samples having a narrow distribution [31]. The data obtained by GPC were correlated with viscometric data estimated by viscosimetry [40]. The viscosity was measured in a 30 °C solution of chloroform in a Ubbelohde viscometer. Molecular mass was calculated using the Mark-Hauwink-Kuhn equation. The 6 specimens per sample were analyzed (Table S2).

A mathematical description for the not autocatalytic and autocatalytic degradation of aliphatic polyesters mechanisms was proposed in Reference [42]. Assuming that the degree of degradation is low, the authors proposed the following kinetic dependence based on the mean molecular mass of polymers (1):

$$1/MM = 1/MM_0 + kt, \tag{1}$$

where MM and MM_0 are the mean molecular mass of the polymer component at time t and at the initial time, respectively, and k is the rate constant. An equation that took autocatalysis into account, which is the consequence of the appearance of the terminal groups of carboxylic acids, was also proposed. The following equation can describe this process (2):

$$\ln(MM) = -kt + \ln MM_0. \tag{2}$$

2.3. Determination of Crystallinity of a Polymer

The crystallinity of PHA were measured by DSC (DSC 204 F1 Phoenix, Netzsch, Germany). The samples were heated from 25 to 220 °C. at a heating rate of 10 K/min under an argon atmosphere. The crystallinity of the PHA structure (Xc) can be calculated as follows (3):

$$Xc = (\Delta Hm/\Delta H_0 m(PHB)) \times 100\%, \tag{3}$$

where ΔHm is the enthalpy change caused by the melting of the test sample, respectively, and $\Delta H_0 m$ (PHB) is the theoretical value for the thermodynamic enthalpy of melting that would have been obtained for 100%-crystalline samples of the PHB (146.6 J/g) [43]. All calculations were carried out for the second heating cycles (Figure S4). The 6 specimens per polymer sample were analyzed.

2.4. Exploration of Mechanical Properties

The mechanical properties of the films were studied using the nanoindentation method in accordance with ISO 14577. The measurements were carried out using a NanoScan-4D scanning nano-hardness tester (TISNCM, Troitsk, Moscow, Russia). Nanoindentation was performed on the smooth side of the films. Films with dimensions of 2 mm × 2 mm were fixed with phenyl salicylate. The load was carried out in a linear mode, the peak load on the sample was 5 mN. The load time was equal to the unloading time and was 30 s. The peak load was maintained for 5 s. The average penetration depth into the sample was not more than 10% of the film thickness (Figure S5). The 6 specimens per polymer sample were analyzed.

2.5. Contact Angle Measurement

The hydrophilicity of the polymer surfaces was evaluated by measuring the contact angle between the drops of water and the "smooth" surface of the samples using the Contact Angle Meter 110 VAC (Cole-Parmer, Vernon Hills, IL, USA). For this purpose, a drop of distilled water (10 μL) was applied to the surface of the films, and then the contact angle of wetting was measured. The values were averaged over the corners obtained from 10 drops per film (Figure S6). The 6 films were analyzed for each polymer sample.

2.6. In Vitro Degradation Experiment

The study of the degradation of PHA films was carried out as follows. The plates were incubated in 15 mL of phosphate-buffered saline (PBS) and in 15 mL of a solution of porcine pancreatic lipase in PBS (Sigma L3126) having a pH of 7.4 at 37 °C in a thermostat for 183 days. The lipase concentration in the phosphate buffer solution was 0.25 mg/mL. The concentration of pancreatic lipase was selected based on earlier obtained experimental data (Figure S7). The pH was monitored with an Orion 420 + pH meter (Thermo Fisher Scientific, Waltham, MA, USA). To assess the changes in the mass of the polymer plates, the test plates were withdrawn from the lipase solution after 1 week, 1 month, 3 months, and 6 months, dried, and weighed on a scale. The average mass of the plates was 15–25 mg. The change in the mass of the plates during the degradation was determined gravimetrically on the AL-64 scales (Max = 60 g, d = 0.1 mg, ACCULAB, Bohemia, NY USA). To prevent bacterial contribution to the degradation of polymers, sodium azide (2 g/L) was added to the buffer solution, and the buffer solution was replaced twice per week [37,44].

2.7. Statistical Analysis

For statistical analysis a one-way ANOVA was applied. In the tables and figures, the data were presented as mean values and standard deviation (M ± SD) at a significance level of $P < 0.05$.

3. Results

3.1. The Decrease in Mass of PHA Films

Biodegradation of PHA occurs as a result of a combination of hydrolytic and enzymatic degradation. This leads to a change in the mass of the samples and their physical and chemical properties [1,2,12,44,45]. The analysis of the degradation kinetic curves (Figure 1) showed that during the first week, the mass of all the samples under study was decreased.

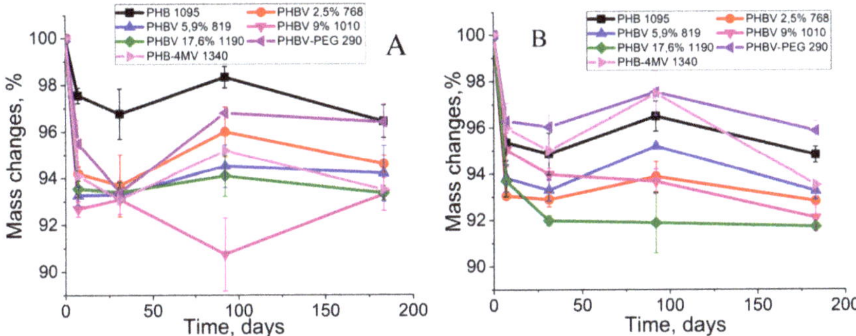

Figure 1. The change in the mass of poly(3-hydroxybutyrate) (PHB), poly(3-hydroxybutyrate-co-3-hydroxyvalerate) (PHBV) (with a different molar content of hydroxyvalerate (HV)), PHB-4 methylvalerate (MV), PHBV-polyethylene glycol (PEG) films during degradation for 6 months in phosphate buffer solution (**A**) and in the same phosphate buffer solution supplemented with the pancreatic lipase (0.25 mg/mL) (**B**).

All the investigated PHA films did not show the visual erosion during the entire degradation process. Subsequently, the dry mass of the PHA films placed in the lipase solution did not demonstrate significant changes even after 180 days. The mass of all the samples did not decrease more than to 92% of the initial mass, indicating a rather slow degradation in the lipase solution. The greatest loss (~ 8–9%) of the mass was observed in the PHBV 17.6% 1190 and PHBV 9% 1010.

3.2. Changes in Molecular Mass (MM)

MM is one of the most significant parameters of polymer degradation. The value of the molecular mass of natural PHAs can be specified by the controlled bacterial biosynthesis and polymer chemical processing [46]. When studying the biopolymers degradation, considerable attention is paid to the MM, because it has a great effect on the other physicochemical parameters of biopolymers [47]. In addition, the MM values are the most indicative degradation parameter. It is extremely sensitive to the polymer backbone destruction. It was found that with a slight change in the mass of the samples, MM underwent more significant changes. The special features of MM changes during biodegradation make it possible to understand the type of its mechanism: not autocatalytic or autocatalytic [42,48].

Within 6 months of incubation in PBS and in lipase, the greatest loss of MM is observed in sample PHB 1095 that was up to 80% of the initial MM (Figure 2). For the other polymer samples, a decrease in MM was also observed, but no dependence on the molar content of HV was detected. Only two polymers from this group in the first week increased their MM–PHBV 17.6% 1190 and PHBV 5.9% 819.

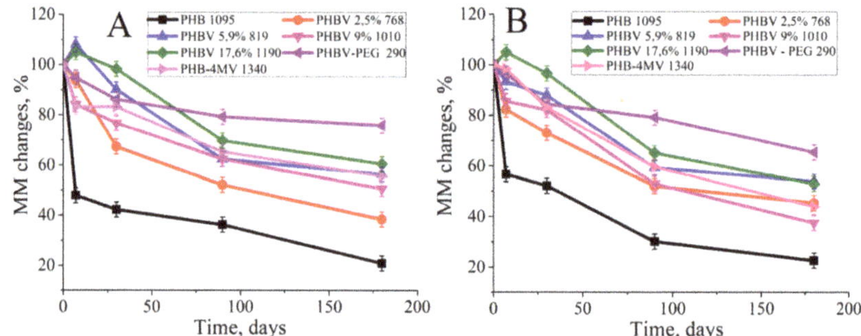

Figure 2. The change in the molecular mass of PHB, PHBV (with a different molar content of HV), PHB-4MV, PHBV-PEG films during hydrolytic (**A**) and enzymatic degradation (**B**) for 6 months.

To analyze the curves of the MM decrease during the degradation, the decomposition models of partially crystalline polymers were applied [42,49]. To evaluate the applicability of the specific model, the curves of degradation were calculated via the statistical correlation coefficients. For this object, the plots of 1/MM and ln(MM) versus degradation time, reflecting not autocatalytic (not autocatalytic for enzymatic degradation) and autocatalytic mechanism, respectively (Figure 3A,B).

Figure 3A shows that the analyzed curves are aligned with the increase in the molar content of HV in the PHBV chain (from homopolymer (line #1) to copolymer with 17.6 mol% of HV (line #5).

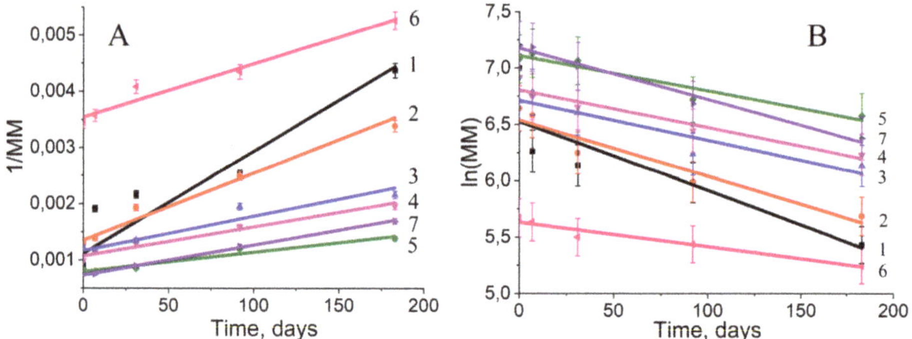

Figure 3. Model of not autocatalytic degradation (**A**) and autocatalytic degradation (**B**), constructed on the basis of the results of changes in the molecular mass of polyhydroxyalkanoates (PHA) in the process of hydrolytic degradation. 1 – PHB 1095, 2 – PHBV 2.5% 768, 3 – PHBV 5.9% 819, 4 – PHBV 9% 1010, 5 – PHBV 17.6% 1190.

By presenting the tangent of the slope of the curves as a function of the HV content, the graph of the slope angle dependence for the not autocatalytic model was obtained (Figure 4).

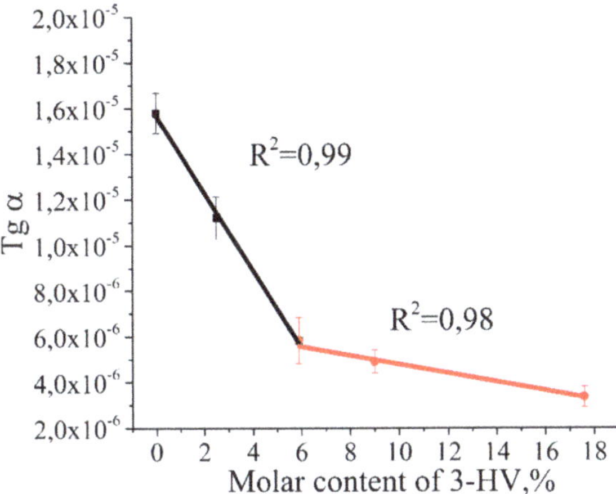

Figure 4. The graph of the slope angle on HV content for the not autocatalytic model of PHBV hydrolytic degradation.

The intersection point for two lines corresponds to 5.9 mol. % HV in PHBV. These data confirm that, at this molar content of PHBV, the polymer structure and corresponding mechanical properties could be modified.

In accordance with the described models, correlation coefficients were determined for each curve (Table 2). In Table 2, it is shown the correlation coefficients for hydrolytic and enzymatically catalytic models.

Table 2. Correlation coefficients of a not-autocatalytic and autocatalytic degradation models.

Sample	Hydrolytic Degradation		Enzymatic Degradation	
	R^2 (Not-Autocatalytic Model)	R^2 (Autocatalytic Model)	R^2 (Not-Autocatalytic Model)	R^2 (Autocatalytic Model)
PHB 1095	0.88	0.68	0.95	0.83
PHBV 2.5% 768	0.99	0.93	0.93	0.88
PHBV 5.9% 819	0.96	0.89	0.92	0.90
PHBV 9% 1010	0.96	0.90	0.99	0.97
PHBV 17.6% 1190	0.97	0.93	0.96	0.95
PHBV-PEG 290	0.92	0.83	0.99	0.98
PHB-4MV 1340	0.97	0.91	0.99	0.98

The correlation coefficients of the not-autocatalic model were quite high (>90%) for the both hydrolytic and enzymatic degradation, while the correlation coefficients of the autocatalic model were higher for the enzymatic degradation with the exception of polymers PHB 1095 and PHBV 2.5% 768.

3.3. Degree of Crystallinity

The crystallinity degree of PHBV was calculated on the basis of the melting heat for the completely crystalline PHB (146.6 J/g) [43]. Comparing the graphs, Figure 5A,B, you can see that in the initial period of time the changes in crystallinity in different solutions are similar. In both cases, in the period up to 1-month, strong changes in the degree of crystallinity were observed.

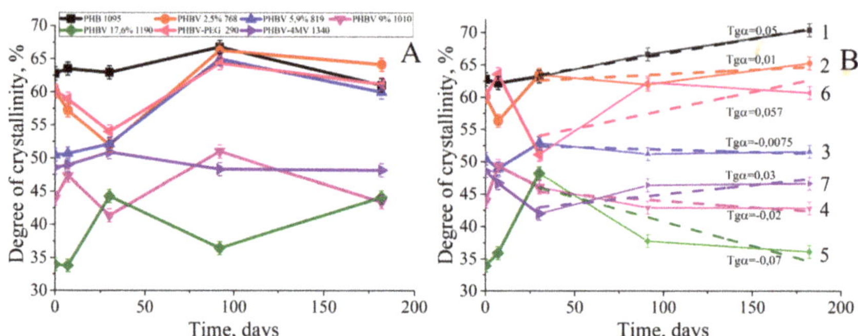

Figure 5. The change in the crystallinity of PHB, PHBV (with a different molar content of HV), PHB-4MV, PHBV-PEG films during hydrolytic (**A**) and enzymatic degradation (**B**) for 6 months: 1 – PHB 1095, 2 – PHBV 2.5% 768, 3 – PHBV 5.9% 819, 4 – PHBV 9% 1010, 5 – PHBV 17.6% 1190, 6 – PHBV-PEG 290, 7 – PHB-4MV 1340.

It was shown that the degree of crystallinity for the copolymers is less than for the sample of PHB 1095. Thus, if the degree of crystallinity for intact PHB 1095 before incubation was 63%, the degree of crystallinity of the PHBV copolymer with the highest molar HV content (17.6 mol. %) at the initial moment was 34%. It is almost half that of PHB 1095. It can also be noted that the degree of crystallinity decreases with increasing molar content of 3-hydroxyvalerate chains.

With further degradation of the polymers for 6 months, incubation of PHA films in PBS resulted in a wave-shape change in the degree of crystallinity (Figure 5A). However, the degree of crystallinity changed differently when incubated in a solution of phosphate buffer with pancreatic lipase. The wave-shape change in crystallinity did not occur (Figure 5B), and its values were observed only in PHB 1095 and PHBV 2.5% 768. It is also important to note that, unlike hydrolytic degradation, there was a clear tendency (Figure 5B) to decrease the degree of crystallinity as the function of the molar content of 3-HV in the homopolymer chain. In addition, the slope of the curves (the values of tgα in Figure 5B) also changed. During degradation, the crystallinity of PHB 1095 increased over 6 months, however, the crystallinity of the PHBV 17.6% 1090 decreased through 6 months. So, it is possible to calculate the molar content of 3-HV in the PHBV chain, which begins to affect the structural and mechanical properties of the entire polymer, which leads to a different character of polymer decomposition (Figure 6).

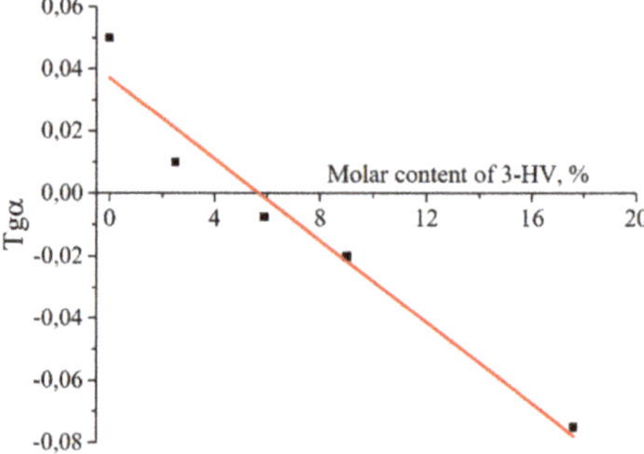

Figure 6. The effect of the 3-HV molar content in the PHBV chain on the crystallinity degree of the copolymers in the degradation process in the lipase solution for 6 months, in the coordinates of the slope of the incubation time.

Figure 6 shows that after incubation in a lipase solution by the 6th month of degradation, the degree of crystallinity was decreased if the percentage of 3-HV in the PHBV chain exceeded 5.7%. This is very close to the value determined earlier in at the analysis of MM changes for the copolymer with HV content equals to 5.9%.

3.4. The Change in Mechanical Properties of PHA Films

Changes in the mechanical properties of the copolymers as a result of degradation were measured by the nanoindentation method (Figure 7). It should be noted that Young's modulus of the homopolymer (2.2 ± 0.06 GPa) before degradation was higher than that of the copolymers (~1 GPa). This is due to the fact that the copolymers have a lower degree of crystallinity and, consequently, stiffness.

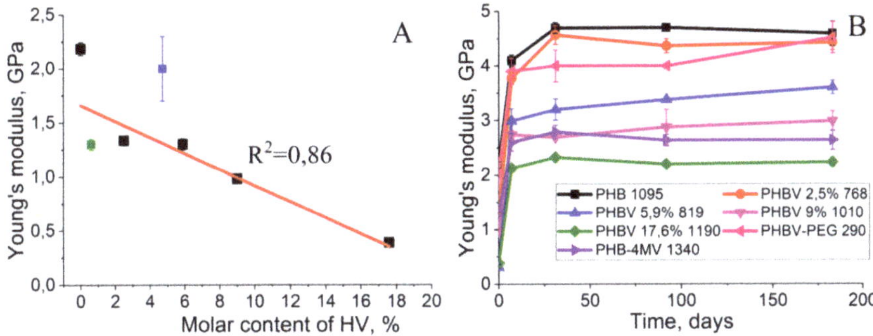

Figure 7. (**A**) Graphs of the dependence of the young's modulus on the content of 3-HV. The green square corresponds to the PHB-4MV 1340 copolymer, and the blue square corresponds to the PHBV-PEG 290 copolymer. (**B**) Changing Young's modulus of the polymers in the process of enzymatic degradation.

During the first week, Young's modulus of all polymers was sharply increased (Figure 7). The largest value of Young's modulus was observed for PHB 1095 kDa. Over a month, Young's modulus of the homopolymer increased to 4.7 ± 0.1 GPa and remained at this level throughout the entire experiment (6 months). Young's modulus of the other copolymers also increased during the first week and did not change significantly in the future. However, the stiffness of the copolymers is still less than that of the homopolymer, for comparison, Young's modulus of the PHBV 17.6% 1190 for the 6-month degradation was 2.2 ± 0.07 GPa, which was half that of Young's modulus of homopolymers.

3.5. The Change in Hydrophobicity of PHA Films

The balance between hydrophobicity and hydrophilicity of the surface is one of the main characteristics indicating biocompatibility of the surface. Biocompatibility is one of the most important properties of polymers that can be used in medicine, so the degree of hydrophilicity of the surface of the polymer affects the growth of cells [3].

During the biodegradation, the contact angle between the standard drop of water and the surface of the polymer film decreased, which indicates that the hydrophobicity degree for the copolymers decreases (Figure 8).

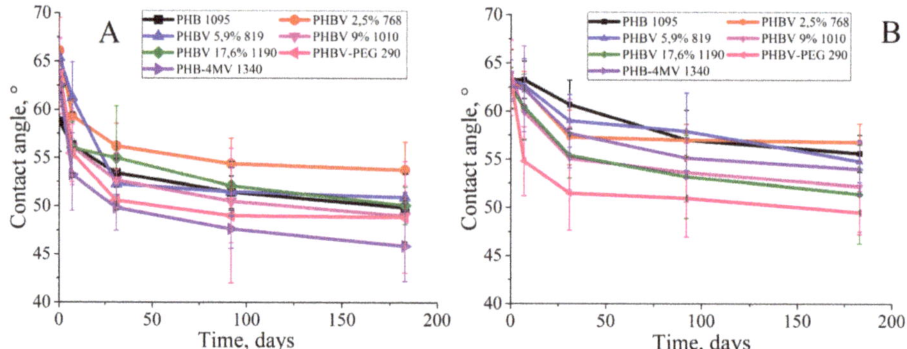

Figure 8. The changes of the contact angle at the water/film boundary of a PHA with a different HV molar content in the PHB chain during enzymatic degradation (**A**) and hydrolytic degradation (**B**) for 6 months.

In general, during biodegradation, the contact angle for all copolymers contacted with lipase decreased by an average value of 23% compared to 17% for hydrolytic degradation (the differences are statistically significant, p ≤ 0.01).

4. Discussion

Thanks to their unique properties, PHAs are the most promising polymers for use in various fields, such as ecology, biomedicine, and packaging. PHAs are biopolymer family that is obtained using microorganisms. The *Azotobacter chroococcum 7B* strain was used for the biosynthesis of PHAs. The polymers synthesized by this producing bacterial strain have low polydispersity, a widespread spectrum of the 3-HB/3-HV monomer ratio in the PHB chain, and MM variability. By varying these parameters during the synthesis of copolymers, their properties, such as Young's modulus, biodegradation rate, and biocompatibility, can also be varied.

The physicochemical parameters of PHAs are changed in different ways during degradation in solutions of PBS and PBS with lipase. The initial mass of polymer films of various copolymers remained virtually unchanged during the entire 183 days of the experiment (Figure 1). The decrease in film mass during the first week can be explained by the dissolution of water-soluble oligomers and monomers during their desorption from the film into aqueous media.

The change in MM is an important characteristic describing the degradation of polymers. As mentioned earlier, using different methods of growing polymer producers, it is possible to vary the molecular mass of the biosynthesized product. This procedure allows development of devices from PHAs with a programmable rate of degradation. The articles presented recently [1,29,44,45] described the hydrolytic degradation of polymers in the presence of various agents. But, in these works, explanations of reaction mechanisms were not presented. This work is one of the first which to attempt to compare the different behavior of biopolyesters in PBS and in the medium containing pancreatic lipase as one of the most important components involved in the decomposition of PHA in the human body. The study of molecular mass changes during degradation allows us to elucidate this mechanism of the process.

Surprisingly, an increase in molecular mass was observed for PHBV 17.6% 1190 and PHBV 5.9% 819 (Figure 2A,B). The initial increase in molecular mass is associated with leaching of the low molecular mass polymer fraction. This means that, in the manufacture of a polymer film, in its volume, and on the surface, there are short-chain polymer residues that dissolve relatively well in water. When studying the molecular mass of polymers prior to incubation in experimental solutions, short-chain residues will affect the average molecular mass. Upon further placement of the polymer films in aqueous solutions, these residues dissolve and are washed out of the polymer film. This leads to a slight increase in molar mass during subsequent measurement. The hydrolysis of the amorphous polymer component also occurs, which is located on the surface of the film, because, according to the literature, the amorphous component decomposes 20 times faster than crystalline [47]. The subsequent strong decrease in MM testified to the fact that the degradation of these polymers was carried out not only on the surface of the polymer film but partially in volume. The absence of mass reduction of polymer films and Young's modulus change (due to the process of secondary crystallization of polymers, and also, probably, to the elution of the amorphous component) (Figures 1 and 7) prompt the conclusion that our polymers will retain their structural and mechanical properties over time and perform their function. Thus, the synthesized polymers can find application in the field of bone tissue engineering for the manufacture of implants. The use of calculation models (Table 2) suggested that, in the case of lipase addition, the process of autocatalysis began to effect the rate of degradation, in contrast to hydrolysis in phosphate buffer. From the Table 2 data, it could be concluded that all copolymers and homopolymer, PHB, are better described by a notautocatalytic degradation model, since the values of R^2 correspond to it. The similar behavior of partially crystalline polymers was presented in the study of Han et al. [45]. In addition, when constructing models and calculating the slope of the curves, a transition point of 5.9% of 3-HV in the copolymer was obtained. Probably, the presence of a more branched radical in the copolymer leads to the fact that it is more difficult for water molecules to hydrolyze the ester bonds. And these steric hindrances begin to affect the entire polymer precisely when the content of 3-HB is greater than 5.9%. In the future, taking this factor into account can help to more correctly calculate the rate of degradation of medical materials based on PHB and its homologues, PHA.

According to published data, the degree of crystallinity of PHA is quite high and varies between 40–70% [42,50]. These data are confirmed by our studies. The degree of crystallinity of the homopolymer was 63%, and the PHBV copolymer 17.6% 1340 was almost two times less −34%. Differences in the degree of crystallinity of polymer films during degradation in PBS and lipase solutions were also demonstrated for the first time. A wave-like evolution in the degree of crystallinity during degradation in PBS was proposed recently [44,49]. In these articles, it was assumed that the changes would not go linearly. We showed that the degree of crystallinity during hydrolysis will have a wave-shape change. All of this is due to crystallization and recrystallization processes. This means the following. During degradation, the amorphous component decomposes faster than the crystalline component 20 times [47]. This leads to an increase in the degree of crystallinity in the polymer. However, hydrolysis occurs non-directionally, that is, the crystalline part also decomposes. Upon decomposition of the crystalline part, an amorphous component is formed—weaving—that leads to a decrease in the degree of crystallinity. Further, the decomposition of the amorphous component or its secondary crystallization occurs, which again leads to an increase in the degree of crystallinity. All of this will have a wave-like appearance, which we have shown in our work. It should be noted that the PHB-4MV copolymer with a 4-MV content of 0.6%, in its physical properties, in particular, crystallinity, is between the PHBV with a content of 5.9% and 9%. It seems that 4-MV makes significant conformational changes in the three-dimensional structure of the copolymer. At the same time, the addition of PEG to the composition of the copolymer does not explicitly affect the polymer structure—the degree of crystallinity of PHBV-PEG is between the copolymer with a 3-HV content of 2.5% to 5.9%. The 3-HV molar content in PHB-PEG is 4.86%. Preliminary, we observed three types of morphological changes in ultrathin PHB films under enzymatic degradation: the emergence of new lamellar structures, fragmentation of lamellar structures, and the disappearance of lamellar structures (Figures S8–S10). However, the degree of crystallinity of the polymers changed differently (from PBS) in the presence of a lipase. But, more important is the result that describes the changes in the behavior of the degree of crystallinity during degradation, depending on the content of HV included in the composition (Figure 9).

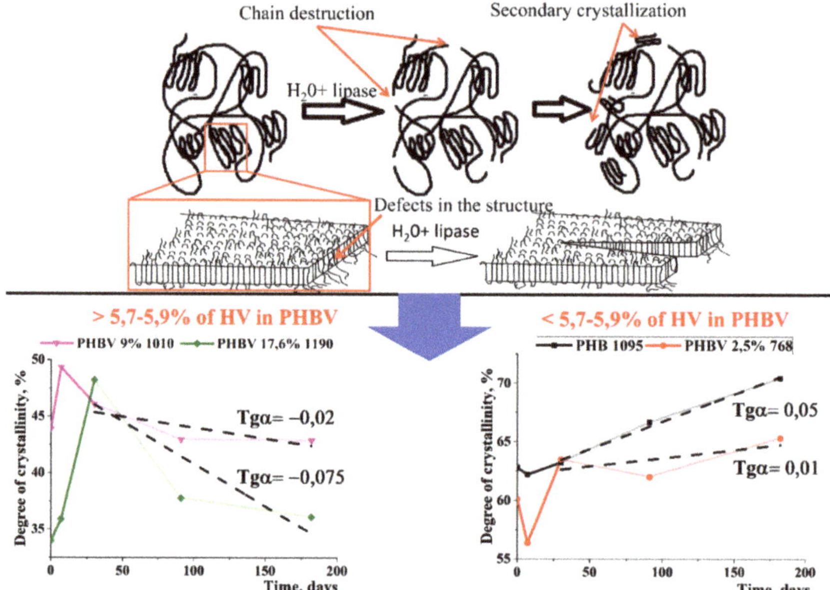

Figure 9. The model of in vitro degradation of PHB and PHBV. It was found that in the process of degradation, the degree of crystallinity increases if the content of 3-HV in the copolymer is less than 5.9% and decreases if it is more than 5.9%.

Based on the obtained data, it can be concluded that when HV comonomer is included in the molecular chain less than 5.9% of PHBV, there is the increase in degree of crystallinity for the films during degradation. As the percentage of HV increases, the reverse process occurs—the decrease in the degree of crystallinity. This is probably due to the influence of HV groups while the small HV content does not prevent the crystallization of newly formed chains of PHB. Probably, HV comonomer creates steric hindrances for folding into the perfect crystals, which leads to a decrease in crystallinity. Owing to its importance, the above phenomenon requires further study since it can affect the change in mechanical and transport (barrier) properties during degradation. In addition, this phenomenon must be taken into account in the future when predicting the properties of products based on PHA that are contacted with tissues during applications as implants or food packaging.

5. Conclusions

Thus, a comprehensive study of the changes in the physicochemical properties of PHB and its copolymers with 3-hydroxyvalerate with different monomer content during the long-term enzymatic and hydrolytic biodegradation under in vitro model conditions was performed. In vitro biodegradation of polymers was studied in PBS and in PBS in the presence of pancreatic lipase at 37 °C for 183 days. An insignificant drop in the mass of the polymers was revealed. However, the change in molecular mass was more significant: a molecular mass decreases up to 80% was found in PHB 1095 kDa in both solutions. In addition, it was shown that, during the enzymatic degradation, the effect of autocatalysis was observed, which was not observed during the hydrolytic degradation. It was also found that Young's modulus of the copolymers was lower than that of the homopolymer. The incubation of polymer films in both solutions led to an increase in Young's modulus by more 2-fold. The changes in the degree of crystallinity of the polymers were wave-shape. The films based on biopolymers became more hydrophilic during biodegradation. It was found that the value of ~5.7–5.9% of HV content in the PHBV copolymer was a transition point for changes in the structural and mechanical properties of the PHBV during its degradation.

Supplementary Materials: The following are available online at http://www.mdpi.com/2073-4360/12/3/728/s1, Figure S1: ^{1}H-NMR spectra of a PHBV copolymer with a content of HV 17% (A); HV 2.5% (B) and homopolymer PHB (C), I – CH(b), II – CH$_2$(b), III – CH$_2$(s)HV, IV – CH$_3$(s)HB, V – CH$_3$(s) HV, s – side chain; b – polymer backbone. Figure S2: ^{1}H-NMR spectra of a PHBV copolymer with a content of PHB4MV (a) PHB chain: 1 – CH$_3$(s), 2 – CH(b), 3 – CH$_2$(b), PHB-4MV chain (b): 4 – CH$_2$(s), 5 – CH$_3$(s), 6 –CH(b), 7 – CH$_2$(b), s – side chain; b – polymer backbone. Figure S3: ^{1}H-NMR spectra of a PHBV-PEG: (a) – PHB chain: 1 – CH$_3$(s), 2 – CH(b), 3 – CH$_2$(b), PHBV chain: 4 – CH$_2$(s), 5 – CH$_3$(s), 6 – CH(b), 7 – CH$_2$(b), s – side chain; b – polymer backbone.; enlarged plot of the graph is shown in the inset (b); (b) PEG chain: «a» -O–CH$_2$ (4.24 ppm), «b» - CH$_2$ (3.73), «c» - common signal from the middle groups [–O–CH$_2$–CH$_2$-] (3.66 ppm), «e» and «d» end groups –CH$_2$- (3.70 ppm) and –CH$_2$-OH (3.61 ppm). Figure S4: Thermogram of the PHBV copolymer 17.6% 1190 obtained by the DSC method. The thermogram shows two heating curves (red and purple lines) and two cooling curves of the sample (white blue and deep blue lines). Figure S5: Image of a characteristic curve in the coordinates of the dependence of the applied force on the indenter displacement obtained using the nanoindentation method. Figure S6: The image of a drop of water on the surface of a polymer film. Figure S7: Diagrams of changes in the weight of plates in (A) a buffer simulating blood plasma (SBF) and in (B) phosphate buffer (PSB) depending on the concentration of pancreatic lipase for a week and a month. Figure S8: The phase images of PHB film before (A) and after (B) hydrolysis in lipase solution. Arrows indicate the new lamellas. Figure S9: The phase images of PHB film before hydrolysis in lipase solution (A) and after (B). Arrows indicate the fragmentation of lamellae. Figure S10: The phase images of PHB film before (A) and after (B) hydrolysis in lipase solution. Arrows indicate the disappearance of lamellar structures. Table S1: Biosynthesis of copolymers of sodium phosphate buffer (PBS) with the A. chroococcum 7B producer strain in the culture medium with sucrose as the main carbon source and salts of carboxylic acids as additional sources of carbon and precursors for biosynthesis of the copolymers of poly(3-hydroxybutyrate) (PHB). HV = hydroxyvalerate. PEG = polyethylene glycol. Table S2: Molecular mass of newly synthesized poly (3-hydroxybutyrate) obtained by various purification methods.

Author Contributions: Conceptualization, V.A.Z., Y.V.Z.; methodology, V.A.Z., A.P.B., T.K.M., V.V.V., V.L.M., G.A.B., A.R., A.U.; validation, A.A.B., A.L.I. and A.P.B.; formal analysis, Y.V.Z.; investigation, V.A.Z.; resources, T.K.M., V.L.M., G.A.B.; data curation, A.P.B.; writing—original draft preparation, V.A.Z.; writing—review and editing, A.P.B, Y.V.Z., A.L.I., A.P.B.; supervision, A.P.B.; project administration, G.A.B. A.P.B. and A.L.I. contributed equally to this work. All authors have read and agreed to the published version of the manuscript.

Funding: This work was supported by the Russian Science Foundation, project No 17-74-20104 (in part of PHBV enzymatic degradation); by Russian Foundation of Basic Research, project No 18-29-09099 (in part of PHB hydrolytic and enzymatic degradation) in all other parts in the framework of government assignment of the Ministry of Science and Higher Education of the Russian Federation. The equipment used in this work was from the User Facilities Center of M.V. Lomonosov Moscow State University and the User Facilities Center of Research Center of Biotechnology of Russian Academy of Sciences.

Conflicts of Interest: The authors declare no conflict of interest.

References

1. Artsis, M.I.; Bonartsev, A.P.; Iordanskii, A.L.; Bonartseva, G.A.; Zaikov, G.E. Biodegradation and Medical Application of Microbial Poly(3-Hydroxybutyrate). *Mol. Cryst. Liq. Cryst.* **2012**, *555*, 232–262. [CrossRef]

2. Chen, G.-Q.; Wu, Q. The application of polyhydroxyalkanoates as tissue engineering materials. *Biomaterials* **2005**, *26*, 6565–6578. [CrossRef] [PubMed]

3. Domínguez-Díaz, M.; Meneses-Acosta, A.; Romo-Uribe, A.; Peña, C.; Segura, D.; Espin, G. Thermo-mechanical properties, microstructure and biocompatibility in poly-β-hydroxybutyrates (PHB) produced by OP and OPN strains of Azotobacter vinelandii. *Eur. Polym. J.* **2015**, *63*, 101–112. [CrossRef]

4. Douglas, T.E.L.; Krawczyk, G.; Pamula, E.; Declercq, H.A.; Schaubroeck, D.; Bucko, M.M.; Balcaen, L.; Van Der Voort, P.; Bliznuk, V.; van den Vreken, N.M.F.; et al. Generation of composites for bone tissue-engineering applications consisting of gellan gum hydrogels mineralized with calcium and magnesium phosphate phases by enzymatic means. *J. Tissue Eng. Regen. Med.* **2016**, *10*, 938–954. [CrossRef] [PubMed]

5. Kašpárková, V.; Humpolíček, P.; Capáková, Z.; Bober, P.; Stejskal, J.; Trchová, M.; Rejmontová, P.; Junkar, I.; Lehocký, M.; Mozetič, M. Cell-compatible conducting polyaniline films prepared in colloidal dispersion mode. *Colloids Surf. B Biointerfaces* **2017**, *157*, 309–316. [CrossRef]

6. Moisenovich, M.M.; Malyuchenko, N.V.; Arkhipova, A.Y.; Kotlyarova, M.S.; Davydova, L.I.; Goncharenko, A.V.; Agapova, O.I.; Drutskaya, M.S.; Bogush, V.G.; Agapov, I.I.; et al. Novel 3D-microcarriers from recombinant spidroin for regenerative medicine. *Dokl. Biochem. Biophys.* **2015**, *463*, 232–235. [CrossRef]

7. Pramanik, N.; Das, R.; Rath, T.; Kundu, P.P. Microbial Degradation of Linseed Oil-Based Elastomer and Subsequent Accumulation of Poly(3-Hydroxybutyrate-co-3-Hydroxyvalerate) Copolymer. *Appl. Biochem. Biotechnol.* **2014**, *174*, 1613–1630. [CrossRef]

8. Ramiro-Gutiérrez, M.L.; Will, J.; Boccaccini, A.R.; Díaz-Cuenca, A. Reticulated bioactive scaffolds with improved textural properties for bone tissue engineering: Nanostructured surfaces and porosity. *J. Biomed. Mater. Res. Part A* **2014**, *102*, 2982–2992. [CrossRef]

9. Raucci, M.G.; Alvarez-Perez, M.A.; Demitri, C.; Sannino, A.; Ambrosio, L. Proliferation and osteoblastic differentiation of hMSCs on cellulose-based hydrogels. *J. Appl. Biomater. Funct. Mater.* **2012**, *10*, 302–307. [CrossRef]

10. Stevanović, M.; Pavlović, V.; Petković, J.; Filipič, M.; Uskoković, D. ROS-inducing potential, influence of different porogens and in vitro degradation of poly (D,L-lactide-co-glycolide)-based material. *Express Polym. Lett.* **2011**, *5*, 996–1008. [CrossRef]

11. Zhang, Y.; Xu, J.; Ruan, Y.C.; Yu, M.K.; O'Laughlin, M.; Wise, H.; Chen, D.; Tian, L.; Shi, D.; Wang, J.; et al. Implant-derived magnesium induces local neuronal production of CGRP to improve bone-fracture healing in rats. *Nat. Med.* **2016**, *22*, 1160–1169. [CrossRef] [PubMed]

12. Bonartsev, A.P.; Bonartseva, G.A.; Shaitan, K.V.; Kirpichnikov, M.P. Poly(3-hydroxybutyrate) and poly(3-hydroxybutyrate)-based biopolymer systems. *Biochem. Suppl. Ser. B Biomed. Chem.* **2011**, *5*, 10–21. [CrossRef]

13. Andreeva, N.V.; Bonartsev, A.P.; Zharkova, I.I.; Makhina, T.K.; Myshkina, V.L.; Kharitonova, E.P.; Voinova, V.V.; Bonartseva, G.A.; Shaitan, K.V.; Belyavskii, A.V. Culturing of Mouse Mesenchymal Stem Cells on Poly (3-hydroxybutyrate) Scaffolds. *Bull. Exp. Biol. Med.* **2015**, *159*, 567–571. [CrossRef] [PubMed]

14. Bonartsev, A.P.; Bonartseva, G.A.; Makhina, T.K.; Myshkina, V.L.; Luchinina, E.S.; Livshits, V.A.; Boskhomdzhiev, A.P.; Markin, V.S.; Iordanskii, A.L. New poly(3-hydroxybutyrate)-based systems for controlled release of dipyridamole and indomethacin. *Appl. Biochem. Microbiol.* **2006**, *42*, 625–630. [CrossRef]

15. Bonartsev, A.P.; Zharkova, I.I.; Yakovlev, S.G.; Myshkina, V.L.; Makhina, T.K.; Zernov, A.L.; Kudryashova, K.S.; Feofanov, A.V.; Akulina, E.A.; Ivanova, E.V.; et al. 3D-scaffolds from poly(3-hydroxybutyrate)-poly(ethylene glycol) copolymer for tissue engineering. *J. Biomater. Tissue Eng.* **2016**, *6*, 42–52. [CrossRef]

16. Gredes, T.; Gedrange, T.; Hinüber, C.; Gelinsky, M.; Kunert-Keil, C. Histological and molecular-biological analyses of poly(3-hydroxybutyrate) (PHB) patches for enhancement of bone regeneration. *Ann. Anat.-Anat. Anz.* **2015**, *199*, 36–42. [CrossRef]

17. Misra, S.K.; Ansari, T.I.; Valappil, S.P.; Mohn, D.; Philip, S.E.; Stark, W.J.; Roy, I.; Knowles, J.C.; Salih, V.; Boccaccini, A.R. Poly(3-hydroxybutyrate) multifunctional composite scaffolds for tissue engineering applications. *Biomaterials* **2010**, *31*, 2806–2815. [CrossRef]

18. Olkhov, A.A.; Staroverova, O.V.; Bonartsev, A.P.; Zharkova, I.I.; Sklyanchuk, E.D.; Iordanskii, A.L.; Rogovina, S.Z.; Berlin, A.A.; Ishchenko, A.A. Structure and properties of ultrathin poly-(3-hydroxybutirate) fibers modified by silicon and titanium dioxide particles. *Polym. Sci. Ser. D* **2015**, *8*, 100–109. [CrossRef]

19. Reyes, A.P.; Torres, A.M.; Carreón, P.; Rogelio, J.; Talavera, R.; Muñoz, S.V.; Manuel, V.; Aguilar, V. Novel Poly (3-hydroxybutyrate-g- vinyl alcohol) Polyurethane Scaffold for Tissue Engineering. *Nat. Publ. Group* **2016**, *6*, 1–8. [CrossRef]

20. Ribeiro-Samy, S.; Silva, N.A.; Correlo, V.M.; Fraga, J.S.; Pinto, L.; Teixeira-Castro, A.; Leite-Almeida, H.; Almeida, A.; Gimble, J.M.; Sousa, N.; et al. Development and characterization of a PHB-HV-based 3D scaffold for a tissue engineering and cell-therapy combinatorial approach for spinal cord injury regeneration. *Macromol. Biosci.* **2013**, *13*, 1576–1592. [CrossRef]

21. Shishatskaya, E.I.; Kamendov, I.V.; Starosvetsky, S.I.; Vinnik, Y.S.; Markelova, N.N.; Shageev, A.A.; Khorzhevsky, V.A.; Peryanova, O.V.; Shumilova, A.A. An in vivo study of osteoplastic properties of resorbable poly (3-hydroxybutyrate) in models of segmental osteotomy and chronic osteomyelitis. *Artif. Cells Nanomed. Biotechnol.* **2014**, *42*, 344–355. [CrossRef] [PubMed]

22. Silva, M.A.C.; Oliveira, R.N.; Mendonça, R.H.; Lourenço, T.G.B.; Colombo, A.P.V.; Tanaka, M.N.; Tude, E.M.O.; da Costa, M.F.; Thiré, R.M.S.M. Evaluation of metronidazole-loaded poly(3-hydroxybutyrate) membranes to potential application in periodontitis treatment. *J. Biomed. Mater. Res. Part B Appl. Biomater.* **2016**, *104*, 106–115. [CrossRef] [PubMed]

23. Bonartsev, A.P.; Yakovlev, S.G.; Filatova, E.V.; Soboleva, G.M.; Makhina, T.K.; Bonartseva, G.A.; Shaitan, K.V.; Popov, V.O.; Kirpichnikov, M.P. Sustained release of the antitumor drug paclitaxel from poly(3-hydroxybutyrate)-based microsphere. *Biochem. Suppl. Ser. B Biomed. Chem.* **2012**, *6*, 42–47. [CrossRef]

24. Bonartsev, A.P.; Zernov, A.L.; Yakovlev, S.G.; Zharkova, I.I.; Myshkina, V.L.; Mahina, T.K.; Bonartseva, G.A.; Andronova, N.V.; Smirnova, G.B.; Borisova, J.A.; et al. New poly(3-hydroxybutyrate) microparticles with paclitaxel sustained release for intraperitoneal administration. *Anticancer. Agents Med. Chem.* **2016**, *17*, 434–441. [CrossRef] [PubMed]

25. Filatova, E.V.; Yakovlev, S.G.; Bonartsev, A.P.; Makhina, T.K.; Myshkina, V.L.; Bonartseva, G.A. Prolonged release of chlorambucil and etoposide from poly-3-oxybutyrate-based microspheres. *Appl. Biochem. Microbiol.* **2012**, *48*, 598–602. [CrossRef]

26. Livshits, V.A.; Bonartsev, A.P.; Iordanskii, A.L.; Ivanov, E.A.; Makhina, T.A.; Myshkina, V.L.; Bonartseva, G.A. Microspheres based on poly(3-hydroxy)butyrate for prolonged drug release. *Polym. Sci. Ser. B* **2009**, *51*, 256–263. [CrossRef]

27. Biazar, E.; Heidari Keshel, S. Electrospun poly (3-hydroxybutyrate-co-3-hydroxyvalerate)/hydroxyapatite scaffold with unrestricted somatic stem cells for bone regeneration. *ASAIO J.* **2015**, *61*, 357–365. [CrossRef]

28. Levine, A.C.; Heberlig, G.W.; Nomura, C.T. Use of thiolene click chemistry to modify mechanical and thermal properties of polyhydroxyalkanoates (PHAs). *Int. J. Biol. Macromol.* **2016**, *83*, 358–365. [CrossRef]

29. Miroiu, F.M.; Stefan, N.; Visan, A.I.; Nita, C.; Luculescu, C.R.; Rasoga, O.; Socol, M.; Zgura, I.; Cristescu, R.; Craciun, D.; et al. Composite biodegradable biopolymer coatings of silk fibroin—Poly(3-hydroxybutyric-acid-co-3-hydroxyvaleric-acid) for biomedical application. *Appl. Surf. Sci.* **2015**, *355*, 123–1131. [CrossRef]

30. Lenz, R.W.; Marchessault, R.H. Bacterial polyesters: Biosynthesis, biodegradable plastics and biotechnology. *Biomacromolecules* **2005**, *6*, 1–8. [CrossRef]

31. Bonartsev, A.P.; Yakovlev, S.G.; Zharkova, I.I.; Boskhomdzhiev, A.P.; Bagrov, D.V.; Myshkina, V.L.; Makhina, T.K.; Kharitonova, E.P.; Samsonova, O.V.; Feofanov, A.V.; et al. Cell attachment on poly(3-hydroxybutyrate)-poly(ethylene glycol) copolymer produced by Azotobacter chroococcum 7B. *BMC Biochem.* **2013**, *14*, 12. [CrossRef] [PubMed]

32. Myshkina, V.L.; Ivanov, E.A.; Nikolaeva, D.A.; Makhina, T.K.; Bonartsev, A.P.; Filatova, E.V.; Ruzhitsky, A.O.; Bonartseva, G.A. Biosynthesis of Poly-3-Hydroxybutyrate–3-Hydroxyvalerate Copolymer by Azotobacter chroococcum Strain 7B. *Appl. Biochem. Microbiol.* **2010**, *46*, 289–296. [CrossRef]

33. Shirazi, R.N.; Aldabbagh, F.; Ronan, W.; Erxleben, A.; Rochev, Y.; McHugh, P. Effects of material thickness and processing method on poly(lactic-co-glycolic acid) degradation and mechanical performance. *J. Mater. Sci. Mater. Med.* **2016**, *27*, 154. [CrossRef] [PubMed]

34. Iwata, T.; Doi, Y.; Tanaka, T.; Akehata, T.; Shiromo, M.; Teramachi, S. Enzymatic Degradation and Adsorption on Poly[(R)-3-hydroxybutyrate] Single Crystals with Two Types of Extracellular PHB Depolymerases from Comamonas acidovorans YM1609 and Alcaligenes faecalis T1. *Macromolecules* **1997**, *30*, 5290–5296. [CrossRef]

35. Song, X.; Liu, F.; Yu, S. Kinetics of poly(3-hydroxybutyrate) hydrolysis using acidic functionalized ionic liquid as catalyst. *Catal. Today* **2016**, *276*, 145–149. [CrossRef]

36. Marois, Y.; Zhang, Z.; Vert, M.; Deng, X.; Lenz, R.; Guidoin, R. Mechanism and rate of degradation of polyhydroxyoctanoate films in aqueous media: A long-term *in vitro* study. *J. Biomed. Mater. Res.* **2000**, *49*, 216–224. [CrossRef]

37. Mitomo, H.; Barham, P.J.; Morimoto, H. Crystallization and morphology of poly(?-hydroxybutyrate) and its copolymer. *Polym. J.* **1987**, *11*, 1241–1253. [CrossRef]

38. Toldrá, F.; Torrero, Y.; Flores, J. Simple test for differentiation between fresh pork and frozen/thawed pork. *Meat Sci.* **1991**, *29*, 177–181. [CrossRef]

39. Chung-Wang, Y.J.; Bailey, M.E.; Marshall, R.T. Reduced oxidation of fresh pork in the presence of exogenous hydrolases and bacteria at 2 degrees C. *J. Appl. Microbiol.* **1997**, *82*, 317–324. [CrossRef]

40. Kaneniwa, M.; Yokoyama, M.; Murata, Y.; Kuwahara, R. Enzymatic hydrolysis of lipids in muscle of fish and shellfish during cold storage. *Adv. Exp. Med. Biol.* **2004**, *542*, 113–119.

41. Qu, C.; Wang, H.; Liu, S.; Wang, F.; Liu, C. Effects of microwave heating of wheat on its functional properties and accelerated storage. *J. Food Sci. Technol.* **2017**, *54*, 3699–3706. [CrossRef] [PubMed]

42. Weir, N.A.; Buchanan, F.J.; Orr, J.F.; Dickson, G.R. Degradation of poly-L-lactide. Part 1: In vitro and in vivo physiological temperature degradation. *Proc. Inst. Mech. Eng. Part H* **2004**, *218*, 307–319. [CrossRef] [PubMed]
43. Barham, P.J.; Keller, A.; Otun, E.L.; Holmes, P.A. Crystallization and morphology of a bacterial thermoplastic: Poly (3-hydroxybutyrate). *J. Mater. Sci.* **1984**, *19*, 2781–2794. [CrossRef]
44. Boskhomdzhiev, A.P.; Bonartsev, A.P.; Makhina, T.K.; Myshkina, V.L.; Ivanov, E.A.; Bagrov, D.V.; Filatova, E.V.; Iordanskii, A.L.; Bonartsev, G.A. Biodegradation kinetics of poly(3-hydroxybutyrate)-based biopolymer systems. *Biochem. Suppl. Ser. B Biomed. Chem.* **2010**, *4*, 177–183. [CrossRef]
45. Han, J.; Wu, L.P.; Bin Liu, X.; Hou, J.; Zhao, L.L.; Chen, J.Y.; Zhao, D.H.; Xiang, H. Biodegradation and biocompatibility of haloarchaea-produced poly(3-hydroxybutyrate-co-3-hydroxyvalerate) copolymers. *Biomaterials* **2017**, *139*, 172–186. [CrossRef] [PubMed]
46. Khandal, D.; Pollet, E.; Averous, L. Polyhydroxyalkanoate-based Multiphase Materials. *RSC Green Chem.* **2015**, *30*, 119–140.
47. Sudesh, K.; Abe, H.; Doi, Y. Synthesis, structure and properties of polyhydroxyalkanoates: Biological polyesters. *Prog. Polym. Sci.* **2000**, *25*, 1503–1555. [CrossRef]
48. Lyu, S.; Untereker, D. Degradability of Polymers for Implantable Biomedical Devices. *Int. J. Mol. Sci.* **2009**, *10*, 4033–4065. [CrossRef]
49. Pan, J. Modelling degradation of semi-crystalline biodegradable polyesters. In *Modelling Degradation of Bioresorbable Polymeric Medical Devices*; Elsevier: Amsterdam, The Netherlands, 2015; pp. 53–69.
50. Bugnicourt, E.; Cinelli, P.; Lazzeri, A.; Alvarez, V. Polyhydroxyalkanoate (PHA): Review of synthesis, characteristics, processing and potential applications in packaging. *Express Polym. Lett.* **2014**, *8*, 791–808. [CrossRef]

Article

Gas Transport Phenomena and Polymer Dynamics in PHB/PLA Blend Films as Potential Packaging Materials

Valentina Siracusa [1,*], Svetlana Karpova [2], Anatoliy Olkhov [2,3], Anna Zhulkina [3], Regina Kosenko [3] and Alexey Iordanskii [3]

[1] Department of Chemical Science (DSC), University of Catania, Viale A. Doria 6, 95125 Catania, Italy
[2] Plekhanov Russian University of Economics, Stremyanny per. 36, 117997 Moscow, Russian Federation; karpova@sky.chph.ras.ru (S.K.); aolkhov72@yandex.ru (A.O.)
[3] Semenov Institute of Chemical Physics, Kosygin str. 4, 119991 Moscow, Russian Federation; annazhulkina@gmail.com (A.Z.); vadim-parfenov5@rambler.ru (R.K.); aljordan08@gmail.com (A.I.)
* Correspondence: vsiracus@dmfci.unict.it; Tel.: +39-3387275526

Received: 14 February 2020; Accepted: 10 March 2020; Published: 12 March 2020

Abstract: Actually, in order to replace traditional fossil-based polymers, many efforts are devoted to the design and development of new and high-performance bioplastics materials. Poly(hydroxy alkanoates) (PHA$_S$) as well as polylactides are the main candidates as naturally derived polymers. The intention of the present study is to manufacture fully bio-based blends based on two polyesters: poly (3-hydroxybutyrate) (PHB) and polylactic acid (PLA) as real competitors that could be used to replace petrol polymers in packaging industry. Blends in the shape of films have been prepared by chloroform solvent cast solution methodology, at different PHB/PLA ratios: 1/0, 1/9, 3/7, 5/5, 0/1. A series of dynamic explorations have been performed in order to characterize them from a different point of view. Gas permeability to N_2, O_2, and CO_2 gases and probe (TEMPO) electron spin resonance (ESR) analyses were performed. Blend surface morphology has been evaluated by Scanning Electron Microscopy (SEM) while their thermal behavior was analyzed by Differential Scanning Calorimetry (DSC) technique. Special attention was devoted to color and transparency estimation. Both probe rotation mobility and N_2, O_2, and CO_2 permeation have monotonically decreased during the transition from PLA to PHB, for all contents of bio-blends, namely because of transferring from PLA with lower crystallinity to PHB with a higher one. Consequently, the role of the crystallinity was elucidated. The temperature dependences for CO_2 permeability and diffusivity as well as for probe correlation time allowed the authors to evaluate the activation energy of both processes. The values of gas transport energy activation and TEMPO rotation mobility are substantially close to each other, which should testify that polymer segmental mobility determines the gas permeability modality.

Keywords: poly(3-hydroxybutyrate) (PHB); polylactic acid (PLA); biomaterials; gas permeability; gas diffusion; segmental dynamics; electron spin resonance (ESR); scanning electron microscopy (SEM); differential scanning calorimetry (DSC)

1. Introduction

The biodegradable biopolyester family comprising polylactides (PLA)s and polyhydroxyalcanoates (PHA)s has to be considered as an attractive opportunity to the family of conventional petrol-based polymers. Owing to their specific transport behavior [1–3], appropriate mechanical properties [4,5], enhanced functionality [6,7], as well as in virtue of controlled biodegradability [8], both biopolymer categories are designated to replace the synthetic plastics in innovative areas such as biomedicine, environmental improving, and especially the packaging industry [9,10]. The high upsurge in human population and hence the corresponding food product expenditure enlarge exogenous pollutions, as well as a waste cluster area. The extremely important impact on enhancing the anthropogenic pollution belongs to the short-term plastic packaging such as a custom care and one-off household utensils. Therefore, PLAs and PHAs elaboration as novel packaging materials could minimize the persistent harmful wastes in aquatic and terrestrial environments [11,12].

Poly(3-hydroxybutyrate) (PHB) is the principal and a most widely used member of the PHAs family that demonstrates the higher potentiality for replace fossil-based synthetic packaging [13]. In homopolymeric status, PHB shows a high stereo-regularity, leading to an extremely large crystallinity up to 70–75 wt.%. The optimal combination of biopolymer crystalline entities (crystallites, lamellae, spherulites) and its intercrystalline transitional chains promote favorable mechanical properties such as high elasticity modulus of around 2.5–3 GPa and tensile strength at break of 35–40 MPa [14]. Thanks to its structure, this biopolymer displays thermophysical and mechanical characteristics similar to polystyrene and isotactic polypropylene [15]. The extensional lamellae and spherulyte morphology, furthermore, constitute a good barrier structure with appropriate low permeability values for atmospheric gaseous components [13,16,17] and water vapor [13,14,18,19].

As well reported in literature [1,20–23], having molecular permeability and diffusivity dependent on biopolymer morphology, their knowledge is identified as crucial factors for determining the inherent packaging behavior, including shelf life terms, food saving, and antibacterial resistance.

The PLA family, with polymers of various proportion in D-lactic and L-lactic stereoisomers, dominates on the biodegradable polymers' market due to their appropriate characteristics such as processability, high transparency, rigidity, and relatively low cost (today it is about 2.5 $/kg). Nonetheless, along with tailoring characteristics like traditional petro-based thermoplastics such as iso-propylene (i-PP), polystyrene (PS), and poly(ethyleneterphtalate) (PET), its high glass transition temperature (T_g) provides a more redundant brittleness, impairing barrier behavior for packaging. Beyond the above positive attributes, PLA is characterized by a number of noticeable flaws that confines its implementation as constructional and packaging materials chiefly. According to the barrier characteristics against gas, water, and low-molecular organic substances, PLA is evaluated as packaging with intermediate permeability coefficient values between PS and PET [24]. Actually, the inherent confinements include the low-speed crystallization, high values of T_g leading to extra brittleness, and low toughness. Furthermore, via electrospray ionization mass spectrometry, researchers such as Bor, Alin, and Hakkarainen [25] have discovered lactide acid oligomers release from PLA films into the fat food mimetic media (96% ethanol solution), which may spoil the food quality to a certain extent.

To avoid the characteristic disadvantages of PHB and PLA, there are several updating approaches that include copolymerization [26], physical action on crystallization via nucleation agents loading or/and annealing [27,28], plasticization [29,30], surface modification [31], filler bulk embedding [32], and so on. The most common technique to tailor relevant packaging features of the biopolyesters includes their blending under appropriate conditions through melt or co-solution procedures [30,33–35].

Using the similarity of molecular structure for two biopolyesters, PHB/PLA blends and composites have been prepared and carefully investigated [23,29,34,36–40]. One of the principal motivations for creating PHB/PLA compositions is to obtain new packaging materials that are free from the above homopolymers disadvantages and especially in such areas as controlled transport, tailored mechanical behavior, and operated biodegradability in post-exploitation landfill conditions.

At the molecular level, both biopolyesters with analogous terminal (-OH, -COOH) and polymer backbone (-C-O(C=O)-C-) groups are conducive to the interaction that reflects trans-esterification reactions [41]. Enabling the conditions for the reactions of that type contributes to solid-state polymerization [42] and reactive compatibility between PHB and PLA using spacers [34,43]. However, the new polymeric adduct, where the macromolecules are chemically connected, is transformed into novel matrix with diffusional, dynamic, and mechanical properties, which are essentially different from the initial homopolymers. On the other hand, the additives like plasticizers and compatibilizers e.g., from low-molecular acetyl tri-n-butyl citrate [44] to oligomers [45], lead to compatibility improvement as well. In the elaboration of polymer packaging, the plasticizer migration, desorption, and leaching [45,46] should be carefully assessed. In the framework of this publication, the developing work of Yeo et al. [27] is of great interest. Here, by structural modification, the mechanical features of PHB/PLA composition have been dramatically improved. Authors elaborated the PHB-based filler particles with effective nucleation activity and extremely high toughening of PLA matrix.

While numerous publications were devoted to the mechanical behavior of the PHB/PLA blend films [47–49], their gas transport characteristics have been represented relatively poorly. The literature mainly displays the data on homopolymer gas permeability, predominantly with oxygen and carbon dioxide, and water vapor permeability [1,13,17,19,38,50–52], as also reported by authors of this paper on PHB [53] and PLA [31] homopolymers, also describing the corresponding data on segmental mobility (obtained by electron spin resonance (ESR) technique), under the assumption that the chain mobility controls the gas diffusion mechanism in bio-polymers.

In this paper, expanded feasibility studies were performed for permeability as well as gas diffusivity and solubility, namely the assessment of features that determine O_2, N_2, and CO_2 transfer. These findings, in combination with the structure-morphology pattern, could be interpreted as ground barrier packaging characteristics, which are necessary for a cohesive interpretation of more elaborated packaging with active functions such as antimicrobial performance and food-modifier controlled release.

2. Materials and Methods

PHB and PLA are the most common bio-based polymers. Their biodegradable behavior is well utilized to solve waste and plastic pollution environmental concerns. As a consequence, in order to develop and extend their applications as blends, it is important to understand their properties related to their chemistry.

2.1. Materials

PHB was kindly supplied by Biomer Co (Krailing, Germany), personally by Dr. Urs Haenggi. Initially, the biopolymer was presented as white powder with particle size of 3–7 μm, MW = 2.06 × 10^5 Da. PLA as micropellets (MW = 120,000) was purchased from Nature Works, LLC (Minnetonka, MN, USA), product 3801X, with > 1% of d-isomer. All materials were used as received, without any further purification.

2.2. Biopolymers Film Preparation

PHB and PLA films were prepared by solvent casting procedure. PHB powder and PLA pellets were weighed, dissolved in chloroform (7% wt/v), and heated on a hot plate with magnetic stirrer (Heidolph Instruments GmbH & CO. KG, Schwabach, Germany), under vigorous stirring at 70 °C (±0.1 °C), for 12 h, until complete dissolution. The obtained solutions were cooled at room temperature and, subsequently, 11 ± 0.5 ml were plated in glass petri-dishes (7 cm diameter) and allowed to slow evaporation (for 48 h) at room temperature (25 ± 1°C) and 60 ± 2% of relative humidity. To ensure a complete solvent removal, the obtained films were dried under vacuum for 48 h (instrument used), in order to reach a constant weight. The dried films were detached from the petri-plates carefully by peeling and then stored at least 2 days at 40 °C temperature, to remove the effect of residual structural relaxation, before characterization.

For PHB/PLA blends, a total concentration of 7% wt/v was dissolved in chloroform, under vigorous stirring at 7 °C, for 12 h. Blends of PHB/PLA with the following weight ratio were obtained: 1/0 (PHB), 9/1, 7/3, 5/5, and 0/1 (PLA). All films were allowed to dry before characterization. Films were stored in a desiccator (with silica gel) prior to testing other parameters.

2.3. Scanning Electron Microscopy

The surface morphology of the PHB/PLA films was studied by SEM technique with the JSM-6510LV JEOL LLC scanning electron microscope (Tokyo, Japan) on samples coated with vapor-deposited gold (Au). The samples were mounted onto Al stud and coated with Au using a sputter (Polaron E5200, Denton Vacuum, Moorestown, NJ, USA), set at 25 mA, for 10 s.

2.4. Differential Scanning Calorimetry (DSC) Measurements

Differential scanning calorimetry (DSC) tests, to evaluate the characteristic temperatures and the corresponding phase change enthalpies, were carried out on a DSC Q-20 instrument (TA Instruments, New Castle, De, USA) in a nitrogen atmosphere at a heating rate of 10 °C/min, equipped with a liquid sub-ambient accessory and calibrated with high purity standards. Polymer films were cut into small pieces of about 2 mm^2 and 10–12 mg in weight and placed in a 50 μL sealed aluminum pan. After an isotherm of 5 min at 20 °C, samples were heated with a scanning rate of 10 °C/min, from 20 to 180 °C (first scan) and then, after a further isotherm of 2 min at 180 °C, were cooled to 20 °C at a rate of 10 °C/min. Finally, after an isotherm of 3 min, samples were reheated from −20 °C/min to 180 °C/min at 10 °C/min (second scan). All the experiments were performed under nitrogen flow (20 cm^3/min). The melting temperature (T_m) was determined as the peak value of the endothermic phenomena in the DSC curve. The melting enthalpy (ΔHm) of the crystal phase was calculated from the area of the DSC endothermic peak. T_m and ΔH_m values were collected from the first scan. The average error in thermal effect measurements was approximately ±3%.

2.5. ESR Characterization

The X-band ESR spectra were recorded on an EPR automated spectrometer (EPR-B Instrument producted in ICPRAS, Moscow, RF). The microwave power in the resonator did not exceed 7 mW, in order to avoid signal saturation effects. When recording the spectra, the modulation amplitude was always substantially smaller than the width of the resonance line and did not exceed 0.5 G. The probe was a stable nitroxyl radical TEMPO synthesized in Emanuel Biochemical Physics Institute, RAS. The radical was introduced into the films from the vapor phase at a temperature of 60 °C. The concentration of the radical in the polymer did not exceed 10^{-3} mol/L. The experimental spectra of the spin probe in the region of slow motion (the correlation time of probe rotation $\tau > 10^{-9}$ s) was analyzed within the framework of the isotropic Brownian rotation model using the program described in [54]. For modeling the spectra, we used the following main values of the g tensor and the hyperfine radical interaction tensor: g_{xx} =2.0096, g_{yy} = 2.0066, g_{zz} = 2.0025, A_{xx} = 7.0 G, A_{yy} =5.0 G, A_{zz} = 35.0 G. Note that the value of Azz was determined experimentally from the EPR spectra of the nitroxyl radical in the polymer at 77 K; it did not differ much from the value given in [55]. Correlation times of probe rotation τ in the region of fast rotations ($5 \times 10^{-11} < \tau < 10^{-9}$ s) were determined from the EPR spectra by the Equation (1) [56]:

$$\tau = \Delta H_+ \times [(I_+/I_-)^{0.5} - 1]\, 6.65 \times 10^{-10}, [s] \tag{1}$$

where ΔH_+ is the width of the spectrum component located in a weak field and I_+/I_- is the ratio of the intensities of the components in weak and strong fields, respectively. The experimental error of τ measured is within ±5% for pristine polymers and ±7% for their blends.

2.6. Barrier Properties Evaluation

Permeability tests were performed by a manometric method with a Permeance Testing Device, type GDP-C (Brugger Feinmechanik GmbH, Munich, Germany), according to ASTM 1434-82 (Standard test Method for Determining Gas Permeability Characteristics of Plastic Film and Sheeting), DIN 53 536 in accordance with ISO/DIS 15105-1, and according to Gas Permeability Testing Manual (Registergericht München HRB 77020, Brugger Feinmechanik GmbH, Munich, Germany). Method A was employed in the analysis, as described in the Brugger manual, with evacuation of both top and bottom chambers.

The film sample with a surface area of 0.785 cm^2 was placed between the two analyses chambers, as previously described by Siracusa and Ingrao [57]. A film mask of aluminum was used to cover the rest of the permeation chamber. The amount of gas flowing through the membrane was determined from the pressure variation due to the gas accumulation in the closed downstream chamber. Gas Transmission Rate (GTR, in *cm^3/cm^2 d bar*) was determined considering the pressure increase in relation to the time and the volume of the device. Time lag (t_L, *sec*), diffusion coefficient (D, in *cm^2/sec*), and solubility (S, in *cm^3/cm^3 bar*) of the test gases were also measured. The mathematical relations used for the calculations were those already reported in the literature [58–61]. Measurements were carried out at 8, 15, and 23 °C with a gas stream of 100 cm^3 min^{-1}, 0% of gas RH. Chamber and sample temperature were controlled by an external thermostat, KAAKE-Circulator DC10-K15 type (Thermoscientific, Selangor Darul Ehsan, Malaysia). O$_2$ and CO$_2$ 100% pure food grade gases were used, and all experiments were run in triplicate; results were provided as the average ± standard deviation.

2.7. Thickness Determination

The film thickness was determined by using the Sample Thickness Tester DM-G with a digital dial indicator (model MarCartor 1086 type, Mahr GmbH, Esslingen, Germany) with associated DM-G software. The reading was made twice per second, with a resolution of 0.001 mm. The minimum, maximum, and average of each reading was recorded in triplicates, in 10 different positions of each film, at room temperature and reported as mean thickness value, expressed in microns.

2.8. Colour Characterization

The color and transparency of samples were measured using a HunterLab ColorFlex EZ 45/0° TM (mod. A60-1010-615) color spectrophotometer (Hunterlab, Reston, VA, USA) with D65 illuminant, 10° observer, according to ASTM E308 (*Practice For Computing The Colors Of Objects By Using The CIE System*). Measurements were made using CIE Lab scale. The instrument was calibrated with a black and white tile before the measurements. Results were expressed as L* (lightness), a* (red/green) and b* (yellow/blue) parameters. The total color difference (ΔE) was calculated using the following Equation (2):

$$\Delta E = [(\Delta L)^2 + (\Delta a)^2 + (\Delta b)^2]^{0.5} \tag{2}$$

where ΔL, Δa, and Δb are the differentials between a sample color parameter (L*, a*, b*) and the color parameter of a standard white plate used as the film background (L′= 66.52, a′= -0.71, b′= 1.16). Chromaticity (C*) and hue angle (h_{ab}) were calculated in accordance to the following Equation (3), as previously reported [62–64]:

$$C^* = [(a^*)^2 + (b^*)^2]^{0.5} \, , \, h_{ab} = \tan^{-1}(b^*/a^*) \tag{3}$$

Measurements were carried out in triplicate at random positions over the film surface. Average values are reported.

3. Results and Discussion

3.1. Scanning Electron Microscopy

SEM micrographs of the surfaces for the homopolymer films PHB (A) and PLA (D), as well as for PHB/PLA blend films (B, C), recorded in order to evaluate the surface homogeneity and structure, are illustrated in Figure 1 (micrograph of PHB/PLA 1/9 sample was omitted due to poor quality image).

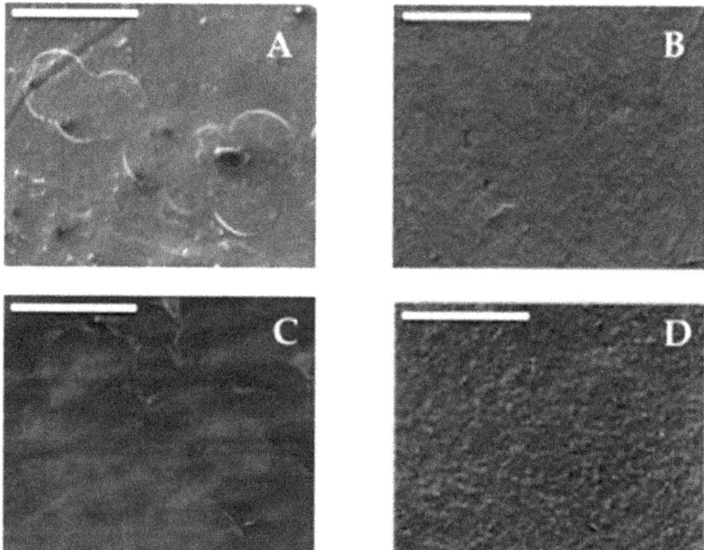

Figure 1. SEM images of Poly(3-hydroxybutyrate) (PHB)/polylactides (PLA) blend surfaces for the films with the biopolymer ratios: (**A**) (1/0), (**B**) (3/7), (**C**) (5/5), and (**D**) (0/1). Scale bar for all micrographs is of 100 μm.

Under given magnification, it could be seen that the surface of homopolymers PHB and PLA displayed, respectively, the textured patterns with coalesced globular elements and finely dispersed entities, forming the rough plane surface. At micrometric scale, micrographs of the blend specimens look like relatively smooth. However, for them the poor relief with the PHB traces of the slightly globular (B) and fibrillar (C) constituents appeared after blending. It is worth noting that the radius of the globules in blend microphotographs (C) was approximately close to 40 μm, which corresponds to the associates of native cell globules of PHB presented in Figure 1A.

Both globular and fibrillar structures could testify the poor blending of two polyesters and partly heterogeneous character of the films' surfaces. At the same time, as a consequence of PLA content growth in the films, the transition of globular structures (B) to fibrillar ones (C) occurred, which may be connected with partial interaction between the PHB macromolecules and the PLA ones. The morphology structure was supported by further characterization, such as thermal and gas barrier properties, pointing out that a different structural arrangement of the components in the blend could affect the final behavior.

3.2. Thermal Characterization

The micro-inhomogeneity of PHB/PLA blends brought the structural evolution, with consequent variation in crystalline structure and thermal characteristics. To examine the thermal features, such as melting temperature (T_m, °C) and enthalpy of fusion (ΔH_m, J/g), in response to crystalline structure changing, DSC technique was applied. Figure 2 reports the DSC heating curves for homopolymers and for PHB/PLA blend films in the area of melting, while Table 1 reports the data in a wider temperature range (20–185 °C).

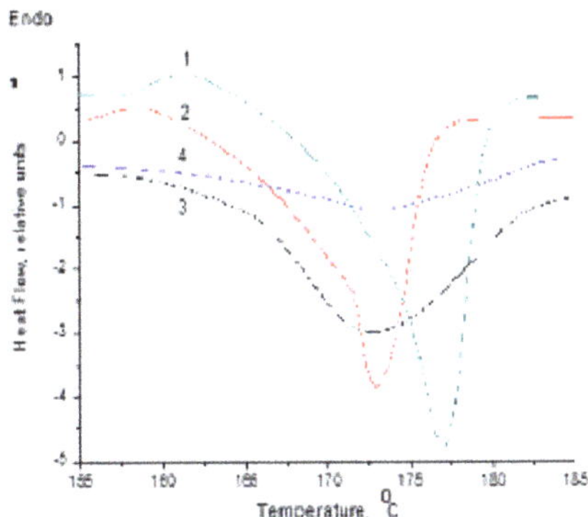

Figure 2. The DSC endotherms in the area of melting for the homopolymers PHB (1) and PLA (2) as well as their blends with the ratio PHB/PLA 5/5 (3) and 3/7 (4) (no data were recorded for sample 1/9 due to poor reproducibility).

Table 1. Differential scanning calorimetry data for homopolymers PHB and PLA and for PHB/PLA blends.

Sample PHB/PLA (w/w)	T_g (°C)	T_{cc} (°C)	T_m (°C)	ΔH_{cc} (J/g)	ΔH_m (J/g)	X_c* (%)
1/0	-	-	177		93.0	64
3/7	40.0	69.9	173	2.3	30.7	50
5/5	39.5	67.8	172	2.6	38.2	49
0/1	43.5	-	172.5	-	54.8	51

* Calculated from referenced specific enthalpy of melting: 107 J/g for PLA [65] and 146 J/g for PHB [66].

Figure 2 and data reported in Table 1 reflect the basic thermophysical transition of the constitutive polymer components in the PHB/PLA films such as crystalline phase melting temperature (T_m) with corresponding specific melting enthalpy (ΔH_m), glassy state transition temperature (T_g), and cold crystallization (T_{cc}) with specific cold crystallization enthalpy (ΔH_{cc}). As can be observed, biodegradable polyesters PHB and PLA differed in thermal behavior expressed by DSC nonisothermal thermograms. Both biopolyesters had relatively close T_m values such as 177 and 172.5 °C for PHB and PLA, respectively. For their blends, the single broadened peak of melting was observed as well, but as the PLA content grew, it shifted from 177 to 173°C, practically to the melting temperature of the PLA homopolymer. The respective values of ΔH_m for the 3/7 and 5/5 blends presented the integral thermal effect of both polyesters with an initial crystallinity degree of 64% (PHB) and 51% (PLA), respectively. Therefore, at the lack of melting peaks resolution on the thermograms of the blends, we were not able to calculate the direct contribution of each component to the total endothermic process. In this case, we calculated the crystallinity degree of the PHB/PLA blends (X_c) according to the Equation (4):

$$X_c = \frac{\Delta H'_m - \Delta H_{cc}}{\Delta H_m(0) W_{(PHB)}} \tag{4}$$

where $\Delta H'_m$ is apparent heat of fusion, ΔH_{cc} is the enthalpy of second crystallinity, $\Delta H_m(0)$ is crystallization enthalpy of pure PHB crystal (100% of crystallinity), and $W_{(PHB)}$ expresses the PHB content (the PHB/PLA weight ratio).

The decrease in ΔH_m values below the values of the homopolymers could attest to their deviation from the additivity principle, indicating the intermolecular interaction for the polyesters.

Because the T_m for PHB slightly exceeded the same characteristic for PLA, the melting of the latter occurred in the solid medium formed by PHB molecules when the segmental interactions between both biopolyesters molecules impeded dynamically. Along with that, for the PLA melted at the lower temperature (173 °C), its crystalline phase fusion could be essentially hindered by the rigid molecular carcass of PHB that is reflected in a sharp drop of total specific enthalpy (see the column in Table 1). Additionally, it is worth noting that authors [65] have recently reported that at temperatures below 150 °C, the PLA could exist in the metastable crystalline state designated as the α'-form with specific enthalpy equals to 107 J/g. This value was used for PLA crystallinity degree calculation.

The cold crystallization temperatures observed for the PHB/PLA blends only had practically the same values and belonged to the PHB predominantly as to the polymer with the slower rate of crystallization [66].

Based on the DSC technique, the immiscibility degree of the polymers in the binary blend can be estimated on the basis of differences in their glass transition temperatures (T_g) [67]. The veracity of such approach is that the higher it is, the greater the T_g difference of the homopolymers. By contrast, in the event of complete miscibility, only T_g should be observed, and for immiscible blends two specific T_g belonging to each of the biopolyesters must exist [68,69]. For PHB and PLA, the constant glass transition was observed, which shifted by −4 °C relative to T_g of PLA (see Table 1). The decrease of T_g and the constancy of its values in the content interval 0.7–0.5 leave open the question of interaction between PHB and PLA molecules. The ability of such interaction will be considered further in the section devoted to segmental dynamics of the biopolyesters investigated by the probe ESR method.

3.3. ESR Spectral Characterization of PHB/PLA Films

Over two recent decades, temporal (dynamic) and spatial (structural) heterogeneity of biopolymers has been coherently investigated by modern spectroscopic methods and particularly by the probe-radical ESR technique [70–72]. The insight into the correlation between blend biopolymer structure and its dynamic molecular behavior is the key step to the design of functional materials for active packaging, biomedicine applications including drug delivery, and environmental problem solution. The previous sections have introduced morphological and thermophysical patterns revealing some specific characteristics of PHB/PLA blend films derived from binary polymer solutions in chloroform particularly. The two succeeding sections will report segmental dynamics of two polyesters that were evaluated by radical-probe ESR technique and gas permeability-diffusion features as principal characteristics of barrier materials for packaging. Atmosphere gas transport in the combination with PHB/PLA segmental mobility indicated by the TEMPO probe rotational diffusion opens the way of creation of environmentally friendly biodegradable material packaging.

The ESR spectra of TEMPO-radicals loaded in the PHB/PLA films are represented in Figure 3 for PLA, PHB homopolymers, and their 5/5 blend.

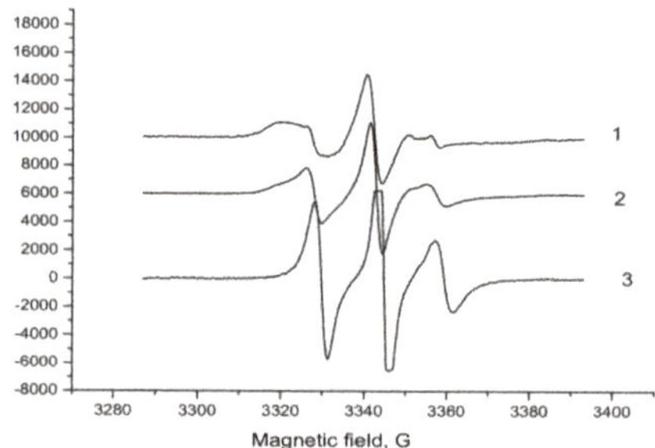

Figure 3. Electron spin resonance (ESR) spectra of PHB, PLA, and PHB/PLA blend films at different polymer ratio: 1 - 1/0, 2 -1/1, and 3 - 0/1.

As in case of the PHB/PLA nanofibers [73], the analogous films are inherently described by the superposition of two single spectra reflecting two different modes of radical rotation with the characteristic correlation times, τ_1 and τ_2. Here, the intrinsic correlation time, τ_1, designates the status of the radical in the denser amorphous fields with slow rotation mobility whilst τ_2 does the fast, radical rotation in the less dense amorphous fields of the PHB/PLA films. The existence of two TEMPO populations in the amorphous phase of PHB and PLA with distinctive rotation frequencies indicates heterogeneous structure of the intercrystalline area in the biopolymer films. The state of the intercrystalline quasi-amorphous area could be approximated by bimodal model that was earlier proposed for several semicrystalline polymers such as PHB, PLA, and (PET) [74–76] and was confirmed for the PHB/PLA electrospun fibers [73].

Figure 4 reports the dependence of correlation time (τ_C) versus the concentration of active component dipyridamole (DPD) used as modulator in segmental mobility (A) and in respect to PHB content, without DPD content (B).

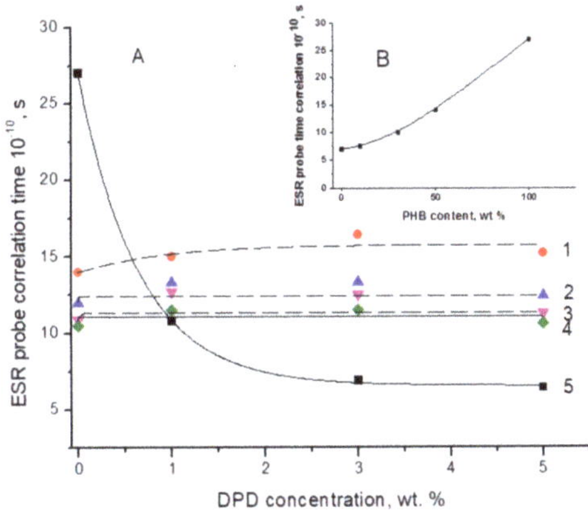

Figure 4. (**A**) dependence of correlation time (τ_C) on the concentration of active component at the different ratio of biopolyesters PHB/PLA: 0/1 (1), 1/9 (2), 3/7 (3), 5/5 (4), and 1/0 (5). (**B**) dependence of τ_C on the PHB content in the pristine blend films without active component.

In contrast to the PHB/PLA electrospun fibers, the dependence of correlation time on the PHB/PLA content ratio in the corresponding films has no explicit minimum (see Figure 5 in reference [77]) and monotonically increased (the probe rate of oscillation decreases) with PHB content rise. The minimal and maximal τ_C values were observed for homopolymers PHB and PLA, but the intermediate values of τ_C belong to their blends in the sequence $\tau_C(1/9) > \tau_C(3/7) > \tau_C(5/5)$ where, in parentheses, the PHB/PLA fractions are given. In Figure 4A, the largest span in the probe mobility complies with the PHB films containing the low-molecular additive, dipyridamole (DPD), at concentrations in the range from 0 to 5 wt.% as modulator segmental mobility. Actually, when the TEMPO-probe could not penetrate the crystalline entities due to steric hindrances, the increase in the concentration of the high-crystalline PHB should decrease an averaged distance among dispersed crystallites and, hence, enhance the restrictive impact on segmental dynamics of intercrystalline macromolecules. The availability of the third component (DPD) in the polymer films affected the mobility of the probe in a complex way, and dramatically decreased the range of the τ_C values for PHB as maximally crystallized biopolymer and weakly influenced the τ_C-concentration dependence. The maximal crystallinity degree of PHB led to the minimal fraction of amorphous (intercrystalline) volume. The PLA molecules, being embedded there, disturbed the conformation of PHB molecules and because of interaction with polyester functional groups, confined segmental dynamics as well. For the blends with relatively low values of ΔH_m and therefore with low crystalline degree (see ΔH_m values in Table 2), the amorphous space among the crystallites was enlarged. Therefore, the probability of intermolecular interaction between the pair of ester groups for PHB and PLA became essentially lower, and in the absence of such interactions, the segmental mobility was higher.

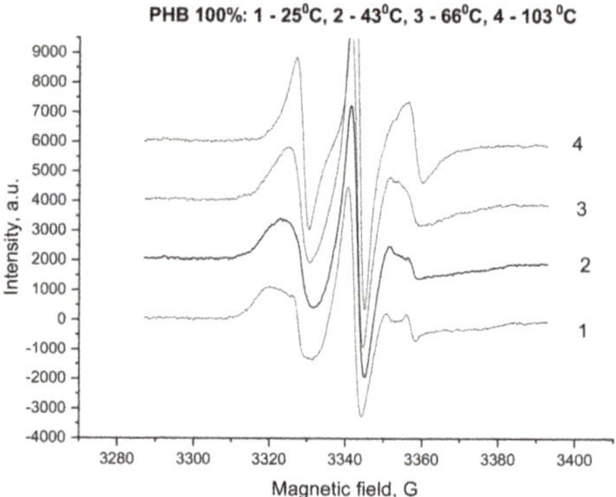

Figure 5. Temperature dependence of the ESR spectra for probe TEMPO in PHB films.

Table 2. Energy activations of CO_2 permeability (E_P) and diffusion (E_D) combined with CO_2 molar enthalpy of solubility (ΔH_S) for pristine PHB and PLA and their blends.

PHB/PLA sample (w/w)	E_P (kJ/mol)	E_D (kJ/mol)	ΔH_S (KJ/mol)
1/0	22.8 ± 4.4	50.6 ± 4.5	−27.8 ± 4.5
3/7	41.4 ± 2.4	54.4 ± 4.9	−13.1 ± 3.5
5/5	32.5 ± 1.5	40.6 ± 4.2	−8.1 ± 3.8
0/1	24.9 ± 4.5	53.6 ± 3.0	−28.7 ± 3.7

In Figure 5, the temperature variation of probe ESR spectra for PHB is presented in the range 25–105 °C. The temperature span was chosen to exclude the glassy-state transition that typically occurs between 5 and 20 °C and thereby avoid the change in the mechanism of probe rotation from the "slow" in the glassy state to the "fast" characterizing PHB elastic state.

With the temperature increase, the broad spectra changed into sharper ones with the narrower peak separation. This effect indicates that the spin probes begin to rotate with higher rate due to the growth of segmental mobility. For the obtained ESR spectra, the correlation time τ_c was estimated using the conventional formula, which is reasonable for the specimens above glassy state temperature, (see Equation (1) reported in the experimental section).

Figure 6 summarizes the effective activation energies for the rotational correlation times (E_τ) that were obtained via the fitting in an Arrhenius semilogarithmic coordinates, analogously to the previous procedure performed for pristine PLA [31] and PHB [53].

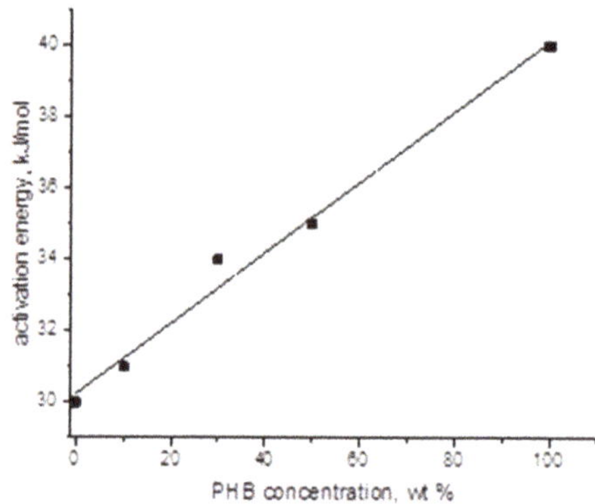

Figure 6. Energy activation of TEMPO probe rotation in PHB/PLA blends.

In these works, it has been shown that there are two different segmental dynamic regimes: (a) At the high-temperature range $T > T_g$, where the values of $\ln(\tau_c)$ [s/rad] decreased quickly with reciprocal temperature (in K^{-1}); and (b) at the low-temperature range $T < T_g$. For the PHB/PLA blends given, however, no dramatic change in mobility and activation energy was observed in the range of T_g values between 39 and 43.5 °C. The monotonically increasing values of $E_{\tau c}$ with the PHB content demonstrate the enhancing hindrance to diffusion of radical moving because of the crystallinity growth in the blends.

Earlier, a number of comprehensive works [70,78–80] were devoted to polymer micro heterogeneity exploration by probe ESR technique. In particular, this phenomenon was carefully investigated for PLA as a potential packaging material [24]. The pioneer papers [68,70] stated that the dimension of structural inhomogeneity may differ depending on which method is used for its evaluation. If the custom DSC data for T_g measuring are based on micro-spatial averaging in a polymer volume, the averaged dimension of heterogeneity in the same object are comparative to the size of the probe. For the TEMPO, it should take about 1nm [70]. To avoid this discrepancy between the quasi-thermodynamic and dynamic approach, Kusumoto and collaborators [81] and Bullock, Cameron, and Miles [82] introduced the special correlation characteristic T_{5mT} that is physically determined in the framework of free volume theory of Bueche [83]. In our subsequent planned works, we will try to apply this fruitful approach for the consideration of dynamic behavior at essentially lower temperatures ($T < T_g$) to connect thermal and dynamical data of the blends under investigation.

3.4. Gas permeability Characteristics

3.4.1. Isothermal Permeability

Gas transport in biodegradable plastic packaging is subject to the penetrant chemical structure, external conditions such as pressure, temperature, humidity, etc., as well as the polymer features including crystallinity, morphology, heterogeneity, the pores' architecture, and others.

The measurement results for transmission of basic atmospheric gases such as N_2, O_2, and CO_2 through PHB, PLA, and their blends with various contents are presented in Figure 7.

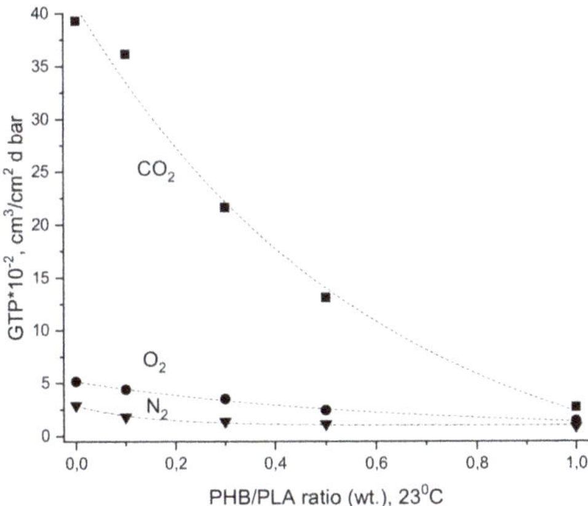

Figure 7. Gas transport characteristic dependences on ratio of PHB/PLA film sample, at 23°C.

As seen for all compositions, the gas transmission rates (GTR) for these gases correspond to the following sequence

$$GTR(CO_2) > GTR(O_2) > GTR(N_2)$$

The transmission of all gases through pristine PHB was lower than that for pristine PLA.

It is well recognized that at the molecular level, gas transmission is driven by gas solubility and diffusion, and the gaseous permeability coefficient represents the product of diffusivity and solubility coefficient [84–86]. Gas diffusional transport and utmost sorption capacity (thermodynamic solubility) depend on polymer segmental mobility and thermodynamic affinity of functional groups for gas molecules, respectively. Further, the fraction of crystallinity and the morphology of crystals have a big impact on gas diffusion. Particularly, during gas transport, the gas molecules have to increase a diffusion pass trajectory, tortuosity, to move around impervious objects in the form of polymer crystalline entities. At high degree of crystallinity, which is representative for PHB, the distance between any pair of crystals (crystallites or lamella) is critically reduced, leading to a segmental mobility decrease of PHB transition chains situated in the intercrystalline area accessible for gas transmission.

As glass transition temperatures (T_g) for PHB and PLA vary by several tens of degrees, namely $\approx$ 20 °C and $\approx$ 60 °C, respectively, at room temperature, each of these biopolymers is in a different physical state. Namely, PHB with T_g that just below ambient temperature displays viscoelastic behavior limited by high crystallinity while PLA is rather characterized by glassy-state features. Restricted molecular mobility for both polymers should reflect failure of both mechanical and transport characteristics.

For the high content of PLA, Figure 7 shows the selectivity of the gas pairs CO_2/N_2 and CO_2/O_2, which were much higher than those for PHB. As it was noted in our previous works [31,53], CO_2 has the highest polarity and its affinity to the polar ester groups has to be the biggest one as compared with O_2 and N_2. For all PHB/PLA compositions, the CO_2 gas transmission essentially exceeds the analogous characteristics of O_2 and N_2. The significant difference in the CO_2 transport characteristics for pristine polymers PHB and PLA is due to the inequality in ester group concentration accessed to the penetrant molecules' interaction. For PLA with relatively low crystallinity, this parameter was essentially larger than for PHB owing to the higher ester concentration, more specifically due to the higher ratio of ester/hydrocarbon groups. In pristine PLA, the O_2/N_2 perm-selective ratio was larger when compared to all other samples. Since O_2 is a molecule with smaller critical volume in comparison to N_2, it can more easily diffuse by activation mechanism. The main idea of the influence of PHB/PLA ratio on CO_2 diffusivity and solubility is reflected in Figure 8 where the decrease of two fundamental characteristics (D and S) is clearly revealing.

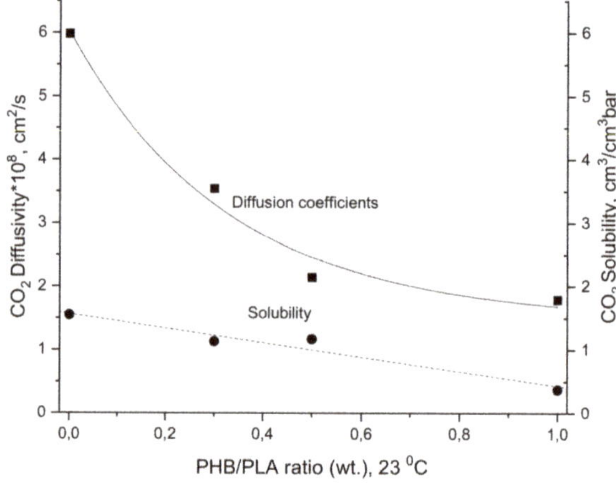

Figure 8. Diffusivity and solubility dependences versus PHB/PLA ratio.

In contrast to P and D, the gas-penetrant solubility coefficient (S) is a thermodynamic characteristic that is determined by penetrant condensability as well as by intensity of biopolymer–penetrant interaction. Its values are in the same order of $CO_2 > O_2 > N_2$. The analogous trend in CO_2 sorption in PHB/PLA blend films is in good conformity with earlier published papers [87–89]. The good accordance is found to be related to the critical temperature of the gas penetrants, whereby CO_2 with a critical temperature of 304.15 K is highly condensable within the polymeric matrix in comparison to oxygen (154.55 K) and, subsequently, nitrogen (126.2 K) [90,91]. Solubility always favors those of higher critical temperature since it indicates tendency in gas penetrants to condense within the polymer.

3.4.2. Temperature Dependency of Gas Permeability

On the way to implementing food packaging, the permeability dependency on temperature is a major subject that should be taken into consideration for the food safety control. In concordance with our above data in Sections 2 and 3, the temperature interval given in Figure 9 does not include thermal transition points for both biopolyesters, PHB and PLA, and as it follows from this figure, the CO_2 transport through blend polymer films satisfies the Arrhenius Equation (5) in conventional semilogarithmic coordinates:

$$P = P_0 \times exp\left(-\frac{E_p}{RT}\right) \tag{5}$$

where P_0, E_p, and R are the pre-exponential factor of permeation related to transport entropy, the transport activation energy, and the gas constant, respectively.

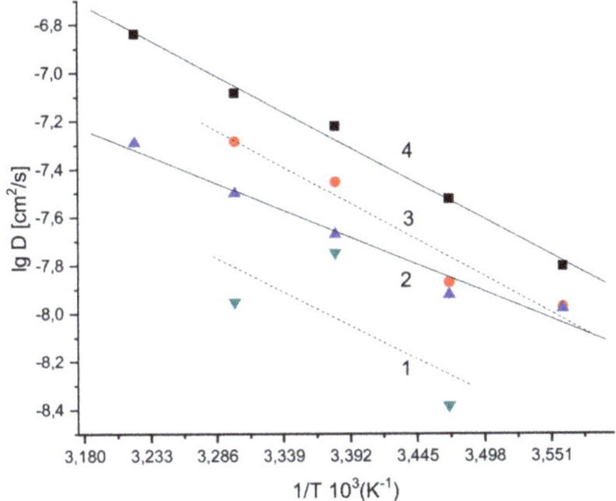

Figure 9. CO_2 diffusivity in PHB/PLA blend films as the function of reciprocal absolute temperature. PHB/PLA ratios are: (1) PHB; (2) PHB/PLA 5/5; (3) PHB/PLA 3/7; (4) PLA.

The analogous expression could be written for the energy activation of diffusion (Equation (6)):

$$D = D_0 \times exp(-E_d/RT) \tag{6}$$

where D_o and E_d are the pre-exponential factor of diffusivity and the diffusion activation energy, respectively.

The temperature dependence of thermodynamic solubility, ΔH_s, of carbon dioxide in the PHB/PLA blend matrices is described by the Van't Hoff expression that was presented in Equation (7):

$$S = S_0 \times exp\left(-\frac{\Delta H_s}{RT}\right) \tag{7}$$

where S_0 is a pre-exponential factor and ΔH_s is the apparent molar heat of equilibrium gas sorption.

The comparison of rotation diffusivity energy activation for the radical probe TEMPO (32–42 kJ/mol) (see Figure 6), with CO_2 permeation energy activation for molecules (40–53 kJ/mol) in PHB/PLA reported in Table 2 shows that both characteristics are close enough. In the given temperature range, consequently, the primary mechanism of CO_2 activated transport is segmental mobility of the blended biomacromolecules. With decreasing the permeability temperature and approaching the glass transition point, the enhancement in the discrepancy increment for E_τ and E_D could be related to the exchange of segmental motion mechanism, namely the predominance of the pendant group mobility in comparison with segmental translation.

3.5. Color Evaluation

When polymer materials are used for food packaging application, the film transparency and color are important requisites. In order to evaluate the food quality, color is one of the key factors evaluated, especially from consumers, and is considered as one of the primary criteria affecting food appearance, taste, and freshness [92]. In general, in order to attract the consumers' attention, packaging has to interfere as little as possible with the food color. Our film surface color results are reported in Table 3, for PHB-PLA samples, compared to white standard.

Table 3. Lightness coefficient (L*), a*, b*, total color difference (ΔE), C*, and h_{ab} of PHB-PLA films.

PHB/PLA Sample	L*	a*	b*	ΔE	C*	h_{ab}
White standard	67.16 ± 0.02	−0.69 ± 0.01	1.25 ± 0.01	-	0.93	138
1/0 (PHB)	34.49 ± 0.67	−0.24 ± 0.03	−1.69 ± 0.13	32.70	1.71	98
1/9	35.60 ± 0.01	−0.43 ± 0.01	−1.10 ± 0.01	31.61	1.18	111
3/7	35.02 ± 0.04	−0.36 ± 0.01	−1.96 ± 0.01	32.24	1.99	100
5/5	34.31 ± 0.02	−0.34 ± 0.02	−1.77 ± 0.01	32.94	1.80	101
0/1 (PLA)	33.66 ± 0.31	−0.67 ± 0.01	−3.14 ± 0.01	33.71	3.21	102

$h_{ab} = 0°$, red–purple; $h_{ab} = 90°$, yellow; $h_{ab} = 180°$, green; $h_{ab} = 270°$, blu.

All films appeared clear, showing a slightly different transparency, despite their different crystallinity. In order to explain the results, as previously reported by Siracusa et al. and by R.G. McGuire [53,92] the CIE Lab Color scale was considered. In the scale, the lightness coefficient (L*) ranges from black (0) to white (100) and, for any L* value, a* and b* values allocate the color on a rectangular coordinate grid, perpendicular to the L* axis. At the axis origin (a* = 0 and b* = 0), the color is achromatic (grey). Considering the horizontal axis, a positive a* value indicates a hue of red-purple and a negative a* value indicates a green hue. Considering the vertical axis, a positive b* value indicates a yellow hue and a negative b* value indicates a blue hue. All films showed similar L* value, showing the same translucency, whereas a* and b* indicated a faint tendency toward a yellowish color (h_{ab} over 90°). The low C* values recorded means low color saturation and, consequently, a good transparence of the films, despite the different copolymer compositions.

4. Conclusions

The transformation from custom packaging materials based on petrochemical non-biodegradable polymers to bio-based biodegradable ones as the innovative gas barriers with selective permeability encourages intensive scientific and industrial explorations. Here, the multifaceted approach combining dynamic (gas permeation and probe-radical ESR technique) and structural (SEM, DSC, and color estimation) methods is given to describe the complex behavior of fully bio-based blends based on two polyesters: poly (3-hydroxybutyrate) (PHB) and polylactic acid (PLA). Both polymers were considered as the real competitors that have to replace petrol polymers in the packaging industry.

Both probe rotation mobility and atmospheric gas permeation as well as their diffusivity monotonically decreased during the transition from PLA to PHB for all component ratios, namely because of transferring from PLA with lower crystallinity to PHB with higher one. The impermeable crystalline entities impede gas transport due to diffusion pass extension. The temperature dependences for CO_2 permeability and diffusivity as well as for probe correlation time allow the authors to evaluate the activation energy of both processes. The values of gas transport energy of activation and the same characteristic for TEMPO rotation mobility are substantially close to each other, which evidently support that polymer molecular mobility determines the gas permeability.

Author Contributions: Conceptualization: V.S. and A.I.; methodology: S.K., R.K., and A.Z.; software: A.O.; data validation: A.O., S.K., and A.Z.; formal analysis: R.K., S.K., and V.S.; investigation: S.K., R.K., A.Z., and V.S.; resources: A.O.; data curation: A.I. and A.O.; writing and original draft preparation: A.I.; writing, review and editing: A.I. and V.S.; visualization: A.Z.; supervision: V.S. and A.I.; funding acquisition: A.I. All authors have read and agreed to the published version of the manuscript.

Funding: This research was funded by the Russian Foundation for Basic Research, grant number 18-29-05017-mk.

Acknowledgments: The authors are grateful to the Biomer company (Krailing, Germany) and personally to U. Heanggi for providing certified PHB.

Conflicts of Interest: The authors declare no conflict of interest. The funders had no role in the design of the study; in the collection, analyses, or interpretation of data; in the writing of the manuscript, or in the decision to publish the results.

References

1. Sonchaeng, U.; Iñiguez-Franco, F.; Auras, R.; Selke, S.; Rubino, M.; Lim, L.-T. Poly(lactic acid) mass transfer properties. *Prog. Polym. Sci.* **2018**, *86*, 85–121. [CrossRef]
2. Pankova, Y.; Shchegolikhin, A.; Iordanskii, A.; Zhulkina, A.; Ol'Khov, A.; Zaikov, G. The characterization of novel biodegradable blends based on polyhydroxybutyrate: The role of water transport. *J. Mol. Liq.* **2010**, *156*, 65–69. [CrossRef]
3. Gårdebjer, S.; Larsson, M.; Gebäck, T.; Skepö, M.; Larsson, A. An overview of the transport of liquid molecules through structured polymer films, barriers and composites—Experiments correlated to structure-based simulations. *Adv. Colloid Interface Sci.* **2018**, *256*, 48–64. [CrossRef] [PubMed]
4. Farah, S.; Anderson, D.G.; Langer, R. Physical and mechanical properties of PLA, and their functions in widespread applications—A comprehensive review. *Adv. Drug Deliv. Rev.* **2016**, *107*, 367–392. [CrossRef]
5. Larrañaga, A.; Pompanon, F.; Gruffat, N.; Palomares, T.; Alonso-Varona, A.; Varga, A.L.; Yagüe, M.F.; Biggs, M.; Sarasua, J.-R. Effects of isothermal crystallization on the mechanical properties of a elastomeric medium chain length polyhydroxyalkanoate. *Eur. Polym. J.* **2016**, *85*, 401–410. [CrossRef]
6. Dinjaski, N.; Prieto, M.A. Smart polyhydroxyalkanoate nanobeads by protein based functionalization. *Nanomed. Nanotechn. Biol. Med.* **2015**, *11*, 885–899. [CrossRef]
7. Han, J.; Xiong, L.; Jiang, X.; Yuan, X.; Zhao, Y.; Yang, D. Bio-functional electrospun nanomaterials: From topology design to biological applications. *Prog. Polym. Sci.* **2019**, *91*, 1–28. [CrossRef]
8. Laycock, B.; Nikolic, M.; Colwell, J.M.; Gauthier, E.; Halley, P.; Bottle, S.E.; George, G. Lifetime prediction of biodegradable polymers. *Prog. Polym. Sci.* **2017**, *71*, 144–189. [CrossRef]
9. Siracusa, V.; Rocculi, P.; Romani, S.; Rosa, M.D. Biodegradable polymers for food packaging: A review. *Trends Food Sci. Technol.* **2008**, *19*, 634–643. [CrossRef]
10. Rodríguez-Rojas, A.; Ospina, A.A.; Rodríguez-Vélez, P.; Arana-Florez, R. What is the new about food packaging material? A bibliometric review during 1996–2016. *Trends Food Sci. Technol.* **2019**, *85*, 252–261. [CrossRef]
11. Bradney, L.; Wijesekara, H.; Palansooriya, K.N.; Obadamudalige, N.; Bolan, N.S.; Ok, Y.S.; Rinklebe, J.; Kim, K.-H.; Kirkham, M. Particulate plastics as a vector for toxic trace-element uptake by aquatic and terrestrial organisms and human health risk. *Environ. Int.* **2019**, *131*, 104937. [CrossRef] [PubMed]
12. Raddadi, N.; Fava, F. Biodegradation of oil-based plastics in the environment: Existing knowledge and needs of research and innovation. *Sci. Total Environ.* **2019**, *679*, 148–158. [CrossRef] [PubMed]
13. Fabra, M.J.; Sánchez, G.; López-Rubio, A.; Lagarón, J.M. Microbiological and ageing performance of polyhydroxyalkanoate-based multilayer structures of interest in food packaging. *LWT* **2014**, *59*, 760–767. [CrossRef]
14. Yeo, J.C.C.; Muiruri, J.; Thitsartarn, W.; Li, Z.; He, C. Recent advances in the development of biodegradable PHB-based toughening materials: Approaches, advantages and applications. *Mater. Sci. Eng. C* **2018**, *92*, 1092–1116. [CrossRef] [PubMed]

15. Melendez-Rodriguez, B.; Castro-Mayorga, J.L.; Reis, M.A.; Sammon, C.; Cabedo, L.; Torres-Giner, S.; Lagaron, J.M. Preparation and Characterization of Electrospun Food Biopackaging Films of Poly(3-hydroxybutyrate-co-3-hydroxyvalerate) Derived From Fruit Pulp Biowaste. *Front. Sustain. Food Syst.* **2018**, *2*, 38. [CrossRef]

16. Ragaert, P.; Buntinx, M.; Maes, C.; Vanheusden, C.; Peeters, R.; Wang, S.; D'Hooge, D.R.; Cardon, L. Polyhydroxyalkanoates for Food Packaging Applications. In *Reference Module in Food Science*; Elsevier BV: Amsterdam, The Netherlands, 2019.

17. Sanchez-Garcia, M.D.; Gimenez, E.; Lagaron, J.M. Morphology and barrier properties of nanobiocomposites of PHB and layered silicates. *J. Appl. Polym. Sci.* **2008**, *108*, 2787–2801. [CrossRef]

18. Miguel, O.; Fernández-Berridi, M.J.; Iruin, J. Survey on transport properties of liquids, vapors, and gases in biodegradable poly(3-hydroxybutyrate) (PHB). *J. Appl. Polym. Sci.* **1997**, *64*, 1849–1859. [CrossRef]

19. Miguel, O.; Iruin, J.J. Water transport properties in poly(3-hydroxybutyrate) and poly(3-hydroxybutyrate-co-3-hydroxyvalerate) biopolymers. *J. Appl. Polym. Sci.* **1999**, *73*, 455–468. [CrossRef]

20. Wolf, C.; Angellier-Coussy, H.; Gontard, N.; Doghieri, F.; Guillard, V. How the shape of fillers affects the barrier properties of polymer/non-porous particles nanocomposites: A review. *J. Membr. Sci.* **2018**, *556*, 393–418. [CrossRef]

21. Figoli, F.G.A.A.; Briceno, K.; Marino, T.; De Gisi, S.; Christensen, K.; Figoli, A. Advances in biopolymer-based membrane preparation and applications. *J. Membr. Sci.* **2018**, *564*, 562–586.

22. Sanchez-Garcia, M.D.; López-Rubio, A.; Lagaron, J.M. Natural micro and nanobiocomposites with enhanced barrier properties and novel functionalities for food biopackaging applications. *Trends Food Sci. Technol.* **2010**, *21*, 528–536. [CrossRef]

23. Nofar, M.; Sacligil, D.; Carreau, P.J.; Kamal, M.R.; Heuzey, M.-C. Poly (lactic acid) blends: Processing, properties and applications. *Int. J. Biol. Macromol.* **2019**, *125*, 307–360. [CrossRef] [PubMed]

24. Courgneau, C.; Vitrac, O.; Ducruet, V.; Riquet, A.-M. Local demixion in plasticized polylactide probed by electron spin resonance. *J. Magn. Reson.* **2013**, *233*, 37–48. [CrossRef]

25. Bor, Y.; Alin, J.; Hakkarainen, M. Electrospray Ionization-Mass Spectrometry Analysis Reveals Migration of Cyclic Lactide Oligomers from Polylactide Packaging in Contact with Ethanolic Food Simulant. *Packag. Technol. Sci.* **2012**, *25*, 427–433. [CrossRef]

26. Yu, I.; Ebrahimi, T.; Hatzikiriakos, S.G.; Mehrkhodavandi, P. Star-shaped PHB–PLA block copolymers: Immortal polymerization with dinuclear indium catalysts. *Dalton Trans.* **2015**, *44*, 14248–14254. [CrossRef] [PubMed]

27. Yeo, J.C.C.; Muiruri, J.; Tan, B.H.; Thitsartarn, W.; Kong, J.; Zhang, X.; Li, Z.; He, C. Biodegradable PHB-Rubber Copolymer Toughened PLA Green Composites with Ultrahigh Extensibility. *ACS Sustain. Chem. Eng.* **2018**, *6*, 15517–15527. [CrossRef]

28. Simmons, H.; Tiwary, P.; Colwell, J.E.; Kontopoulou, M. Improvements in the crystallinity and mechanical properties of PLA by nucleation and annealing. *Polym. Degrad. Stab.* **2019**, *166*, 248–257. [CrossRef]

29. Arrieta, M.P.; Samper, M.D.; Aldas, M.; López-Martínez, J. On the Use of PLA-PHB Blends for Sustainable Food Packaging Applications. *Materials* **2017**, *10*, 1008. [CrossRef]

30. Ma, Y.; Li, L.; Wang, Y. Development of PLA-PHB-based biodegradable active packaging and its application to salmon. *Packag. Technol. Sci.* **2018**, *31*, 739–746. [CrossRef]

31. Siracusa, V.; Rosa, M.D.; Iordanskii, A. Performance of Poly(lactic acid) Surface Modified Films for Food Packaging Application. *Materials* **2017**, *10*, 850. [CrossRef]

32. Sorrentino, A.; Gorrasi, G.; Vittoria, V. Permeability in Clay/Polyesters Nano-Biocomposites. In *Green Energy and Technology*; Springer Science and Business Media LLC: Amsterdam, The Netherlands, 2012; pp. 237–264.

33. Vieira, M.G.A.; Da Silva, M.A.; Santos, L.O.; Beppu, M.M. Natural-based plasticizers and biopolymer films: A review. *Eur. Polym. J.* **2011**, *47*, 254–263. [CrossRef]

34. Imre, B.; Pukánszky, B. Compatibilization in bio-based and biodegradable polymer blends. *Eur. Polym. J.* **2013**, *49*, 1215–1233. [CrossRef]

35. Arrieta, M.P.; Lopez, J.; López, D.; Kenny, J.M.; Peponi, L. Effect of chitosan and catechin addition on the structural, thermal, mechanical and disintegration properties of plasticized electrospun PLA-PHB biocomposites. *Polym. Degrad. Stab.* **2016**, *132*, 145–156. [CrossRef]

36. Mlalila, N.; Hilonga, A.; Swai, H.; Devlieghere, F.; Ragaert, P. Antimicrobial packaging based on starch, poly(3-hydroxybutyrate) and poly(lactic-co-glycolide) materials and application challenges. *Trends Food Sci. Technol.* **2018**, *74*, 1–11. [CrossRef]

37. Zembouai, I.; Bruzaud, S.; Kaci, M.; Benhamida, A.; Corre, Y.M.; Grohens, Y.; Taguet, A.; Lopez-Cuesta, J.-M. Poly(3-hydroxybutyrate-co-3-hydroxyvalerate)/polylactide blends: Thermal stability, flammability and thermo-mechanical behavior. *J. Polym. Environ.* **2014**, *22*, 131–139. [CrossRef]

38. D'Amico, D.; Montes, M.I.; Manfredi, L.B.; Cyras, V.P. Fully bio-based and biodegradable polylactic acid/poly(3-hydroxybutirate) blends: Use of a common plasticizer as performance improvement strategy. *Polym. Test.* **2016**, *49*, 22–28. [CrossRef]

39. Bartczak, Z.; Galeski, A.; Kowalczuk, M.; Sobota, M.; Malinowski, R. Tough blends of poly(lactide) and amorphous poly([R,S]-3-hydroxy butyrate)—Morphology and properties. *Eur. Polym. J.* **2013**, *49*, 3630–3641. [CrossRef]

40. Li, L.; Huang, W.; Wang, B.; Wei, W.; Gu, Q.; Chen, P. Properties and structure of polylactide/poly (3-hydroxybutyrate-co-3-hydroxyvalerate) (PLA/PHBV) blend fibers. *Polymer* **2015**, *68*, 183–194. [CrossRef]

41. Dawidziuk, K.A. Peroxide-initiated Modification of Polylactic acid (PLA) and Poly(3-hydroxyalkanoates) (PHAs) in the Presence of Allylic and Acrylic Coagents. Master's Thesis, Queen's University, Kingston, ON, Canada, 2018.

42. Dawin, T.P.; Ahmadi, Z.; Taromi, F.A. Biocompatible PLA/PHB coatings obtained from controlled solid state polymerization. *Prog. Org. Coat.* **2019**, *132*, 41–49. [CrossRef]

43. Melendez-Rodriguez, B.; Torres-Giner, S.; Aldureid, A.; Cabedo, L.; Lagaron, J.M. Reactive Melt Mixing of Poly(3-Hydroxybutyrate)/Rice Husk Flour Composites with Purified Biosustainably Produced Poly(3-Hydroxybutyrate-co-3-Hydroxyvalerate). *Materials* **2019**, *12*, 2152–2173. [CrossRef]

44. Arrieta, M.P.; López-Martínez, J.; López, D.; Kenny, J.M.; Peponi, L. Development of flexible materials based on plasticized electrospun PLA–PHB blends: Structural, thermal, mechanical and disintegration properties. *Eur. Polym. J.* **2015**, *73*, 433–446. [CrossRef]

45. Hoppe, M.; de Voogt, P.; Franz, R. Identification and quantification of oligomers as potential migrants in plastics food contact materials with a focus in polycondensates—A review. *Trends Food Sci. Technol.* **2016**, *50*, 118–130. [CrossRef]

46. Rahman, M.; Brazel, C.S. The plasticizer market: An assessment of traditional plasticizers and research trends to meet new challenges. *Prog. Polym. Sci.* **2004**, *29*, 1223–1248. [CrossRef]

47. Zhang, L.; Xiong, C.; Deng, X. Miscibility, crystallization and morphology of poly(β-hydroxybutyrate) / poly(d,l-lactide) blends. *Polymer* **1996**, *37*, 235–241. [CrossRef]

48. Zhang, M.; Thomas, N. Blending polylactic acid with polyhydroxybutyrate: The effect on thermal, mechanical, and biodegradation properties. *Adv. Polym. Technol.* **2011**, *30*, 67–79. [CrossRef]

49. Arrieta, M.P.; Fortunati, E.; Dominici, F.; Rayón, E.; López-Martínez, J.; Kenny, J.M. PLA-PHB/cellulose based films: Mechanical, barrier and disintegration properties. *Polym. Degrad. Stab.* **2014**, *107*, 139–149. [CrossRef]

50. Shogren, R. Water vapor permeability of biodegradable polymers. *J. Polym. Environ.* **1997**, *5*, 91–95. [CrossRef]

51. Razavi, S.M.; Dadbin, S.; Frounchi, M. Oxygen barrier properties of PLA-PVAcAl blends as biodegradable films. Special Issue: Biopolymers and Renewably Sourced Polymers. *J. Appl. Polym. Sci.* **2012**, *125*, E20–E26. [CrossRef]

52. Zhang, T.; Yu, Q.-Y.; Yang, H.-T.; Wang, J.-J.; Li, X.-B. Understanding the oxygen barrier property of highly transparent poly(lactic acid)/benzoxazine composite film by analyzing the UV-shielding performance. *J. Appl. Polym. Sci.* **2019**, *136*, 47510. [CrossRef]

53. Siracusa, V.; Ingrao, C.; Karpova, S.G.; Olkhov, A.A.; Iordanskii, A. Gas transport and characterization of poly(3 hydroxybutyrate) films. *Eur. Polym. J.* **2017**, *91*, 149–161. [CrossRef]

54. Budil, D.; Lee, S.; Saxena, S.; Freed, J.H. Nonlinear-Least-Squares Analysis of Slow-Motion EPR Spectra in One and Two Dimensions Using a Modified Levenberg–Marquardt Algorithm. *J. Magn. Reson. Ser. A* **1996**, *120*, 155–189. [CrossRef]

55. Timofeev, V.P.; Misharin, A.Y.; Tkachev, Y.V. Simulation of EPR spectra of the radical TEMPO in water-lipid systems in different microwave ranges. *Biophysics* **2011**, *56*, 407–417. [CrossRef]

56. Buchachenko, A.L.; Vasserman, A.M. *Stable Radicals*; Khimiya: Moscow, Russia, 1973.

57. Siracusa, V.; Ingrao, C. Correlation amongst gas barrier behaviour, temperature and thickness in BOPP films for food packaging usage: A lab-scale testing experience. *Polym. Test.* **2017**, *59*, 277–289. [CrossRef]

58. Mangaraj, S.; Goswami, T.K.; Mahajan, P.V. Applications of Plastic Films for Modified Atmosphere Packaging of Fruits and Vegetables: A Review. *Food Eng. Rev.* **2009**, *1*, 133–158. [CrossRef]

59. Siracusa, V.; Blanco, I.; Romani, S.; Tylewicz, U.; Rocculi, P.; Rosa, M.D. Poly(lactic acid)-modified films for food packaging application: Physical, mechanical, and barrier behavior. *J. Appl. Polym. Sci.* **2012**, *125*, e390–e401. [CrossRef]

60. Mrkić, S.; Galić, K.; Ivankovic, M.; Hamin, S.; Ciković, N. Gas transport and thermal characterization of mono- and di-polyethylene films used for food packaging. *J. Appl. Polym. Sci.* **2005**, *99*, 1590–1599. [CrossRef]

61. Siracusa, V. Food Packaging Permeability Behaviour: A Report. *Int. J. Polym. Sci.* **2012**, *2012*, 1–11. [CrossRef]

62. Galus, S.; Lenart, A. Development and characterization of composite edible films based on sodium alginate and pectin. *J. Food Eng.* **2013**, *115*, 459–465. [CrossRef]

63. Syahidad, K.; Rosnah, S.; Noranizan, M.A.; Zaulia, O.; Anvarjon, A. Quality changes of fresh cut cantaloupe (Cucumis melo L. var Reticulatus cv. Glamour) in different types of polypropylene packaging. *Int. Food Res. J.* **2015**, *22*, 753–760.

64. Genovese, L.; Soccio, M.; Gigli, M.; Lotti, N.; Gazzano, M.; Siracusa, V.; Munari, A. Gas permeability, mechanical behaviour and compostability of fully-aliphatic bio-based multiblock poly(ester urethane)s. *RSC Adv.* **2016**, *6*, 55331–55342. [CrossRef]

65. Righetti, M.C.; Gazzano, M.; Di Lorenzo, M.L.; Androsch, R. Enthalpy of melting of α- and β-crystals of poly(L-lactic acid. *Eur. Polym. J.* **2015**, *70*, 215–220. [CrossRef]

66. Lin, K.-W.; Lan, C.-H.; Sun, Y.-M. Poly[(R)3-hydroxybutyrate] (PHB)/poly(L-lactic acid) (PLLA) blends with poly(PHB/PLLA urethane) as a compatibilizer. *Polym. Degr. Stab.* **2016**, *134*, 30–40. [CrossRef]

67. Cavalcante, M.P.; Toledo, A.L.M.M.; Rodrigues, E.; Neto, R.C.; Tavares, M.I. Correlation between traditional techniques and TD-NMR to determine the morphology of PHB/PCL blends. *Polym. Test.* **2017**, *58*, 159–165. [CrossRef]

68. Qiu, Z.; Yang, W.; Ikehara, T.; Nishi, T. Miscibility and crystallization behavior of biodegradable blends of two aliphatic polyesters. Poly(3-hydroxybutyrate-co-hydroxyvalerate) and poly(ε-caprolactone. *Polymer* **2005**, *46*, 11814–11819. [CrossRef]

69. Lovera, D.; Marquez, L.; Balsamo, V.; Taddei, A.; Castelli, C.; Müller, A.J. Crystallization, Morphology, and Enzymatic Degradation of Polyhydroxybutyrate/Polycaprolactone (PHB/PCL) Blends. *Macromol. Chem. Phys.* **2007**, *208*, 924–937. [CrossRef]

70. Veksli, Z.; Andreis, M.; Rakvin, B. ESR spectroscopy for the study of polymer heterogeneity. *Prog. Polym. Sci.* **2000**, *25*, 949–986. [CrossRef]

71. Kanemoto, K.; Mizutani, N.; Muramatsu, K.; Hashimoto, H.; Baba, M.; Yamauchi, J. ESR investigations on doped conjugated polymers diluted in a solid matrix. *Chem. Phys. Lett.* **2010**, *494*, 41–44. [CrossRef]

72. Bartoš, J.; Švajdlenková, H. On the mutual relationships between spin probe mobility, free volume and relaxation dynamics in organic glass-formers: Glycerol. *Chem. Phys. Lett.* **2017**, *670*, 58–63. [CrossRef]

73. Cao, K.; Liu, Y.; Olkhov, A.; Siracusa, V.; Iordanskii, A.L. PLLA-PHB fiber membranes obtained by solvent-free electrospinning for short-time drug delivery. *Drug Deliv. Transl. Res.* **2017**, *8*, 291–302. [CrossRef]

74. Baudet, C.; Grandidier, J.-C.; Cangémi, L. A two-phase model for the diffuson-mechanical behaviour of semicrystalline polymers in gaseous environment. *Int. J. Sol. Struc.* **2009**, *46*, 1389–1401. [CrossRef]

75. Shimada, S. ESR studies on molecular motion and chemical reactions in solid polymers in relation to structure. *Prog. Polym. Sci.* **1992**, *17*, 1045–1106. [CrossRef]

76. Ma, Q.; Pyda, M.; Mao, B.; Cebe, P. Relationship between the rigid amorphous phase and mesophase in electrospun fibers. *Polymer* **2013**, *54*, 2544–2554. [CrossRef]

77. Iordanskii, A.; Karpova, S.; Olkhov, A.; Borovikov, P.; Kildeeva, N.; Liu, Y. Structure-morphology impact upon segmental dynamics and diffusion in the biodegradable ultrafine fibers of polyhydroxybutyrate-polylactide blends. *Eur. Polym. J.* **2019**, *117*, 208–216. [CrossRef]

78. Švajdlenková, H.; Arrese-Igor, S.; Nógellová, Z.; Alegria, A.; Bartos, J. Molecular dynamic heterogeneity in relation to free volume and relaxation dynamics in organic glass-formers: Oligomeric cis-1,4-poly(isoprene. *Phys. Chem. Chem. Phys.* **2017**, *19*, 15215–15226. [CrossRef] [PubMed]

79. Barashkova, I.I.; Bermeshev, M.V.; Wasserman, A.M.; Yampolskii, Y.P. Mobility of the spin probe TEMPO in glassy metathesis polymers with substituents containing flexible Si-O-Si groups. *Polym. Sci. Ser. A* **2015**, *57*, 279–284. [CrossRef]

80. Kurzbach, D.; Junk, M.J.N.; Hinderberger, D. Nanoscale Inhomogeneities in Thermoresponsive Polymers. *Macromol. Rap. Com.* **2013**, *34*, 119–134. [CrossRef]

81. Kusumoto, N.; Sano, S.; Zaitsu, N.; Motozato, Y. Molecular motions and segmental size of vulcanized natural and acrylonitrile-butadiene rubbers by the spin-probe method. *Polymer* **1976**, *17*, 448–454. [CrossRef]

82. Bullock, A.T.; Cameron, G.G.; Miles, I.S. Segment size in synthetic polymers by the spin-probe method. *Polymer* **1982**, *23*, 1536–1539. [CrossRef]

83. Bueche, F. Viscosity, Self-diffusion, and Allied Effects in Solid Polymers. *J. Chem. Phys.* **1952**, *20*, 1959. [CrossRef]

84. Siracusa, V.; Genovese, L.; Ingrao, C.; Munari, A.; Lotti, N. Barrier Properties of Poly(Propylene Cyclohexanedicarboxylate) Random Eco-Friendly Copolyesters. *Polymer* **2018**, *10*, 502. [CrossRef]

85. Hannesschläger, C.; Horner, A.; Pohl, P. Intrinsic Membrane Permeability to Small Molecules. *Chem. Rev.* **2019**, *119*, 5922–5953. [CrossRef] [PubMed]

86. Craster, B.; Jones, T.G. Permeation of a Range of Species through Polymer Layers under Varying Conditions of Temperature and Pressure: In Situ Measurement Methods. *Polymer* **2019**, *11*, 1056. [CrossRef] [PubMed]

87. Aitken, C.L.; Koros, W.J.; Paul, D.R. Effect of structural symmetry on gas transport properties of polysulfones. *Macromolecules* **1992**, *25*, 3424–3434. [CrossRef]

88. Costello, L.M.; Koros, W.J. Temperature dependence of gas sorption and transport properties in polymers: Measurement and applications. *Ind. Eng. Chem. Res.* **1992**, *31*, 2708–2714. [CrossRef]

89. Matteucci, S.; Yampolskii, Y.; Freeman, B.D.; Pinnau, I. *Transport of Gases and Vapors in Glassy and Rubbery Polymers*; Wiley: Hoboken, NJ, USA, 2006; pp. 1–47.

90. De Oliveira, M.J. *Equilibrium Thermodynamics*; Springer Science and Business Media LLC: Berlin/Heidelberg, Germany, 2013.

91. Lock, S.; Keong, L.K.; Mash'Al, A.-A.B.; Shariff, A.M.; Yeong, Y.F.; Mei, I.L.S.; Ahmad, F. Elucidation on the Effect of Operating Temperature to the Transport Properties of Polymeric Membrane Using Molecular Simulation Tool. In *E-Business and Telecommunication Networks*; Springer Science and Business Media LLC: Berlin/Heidelberg, Germany, 2017; Volume 752, pp. 456–471.

92. McGuire, R.G. Reporting of Objective Color Measurements. *HortScience* **1992**, *27*, 1254–1255. [CrossRef]

 *polymers*

Article

Effect of Natural Rubber in Polyethylene Composites on Morphology, Mechanical Properties and Biodegradability

Elena Mastalygina [1,2,*], Ivetta Varyan [1,2], Natalya Kolesnikova [2], Maria Isabel Cabrera Gonzalez [3] and Anatoly Popov [1,2]

1 Scientific Laboratory "Advanced Composite Materials and Technologies", Plekhanov Russian University of Economics, 36 Stremyanny per., 117997 Moscow, Russia; ivetta.varyan@yandex.ru (I.V.); anatoly.popov@mail.ru (A.P.)

2 Department of Biological and Chemical Physics of Polymers, Emanuel Institute of Biochemical Physics, Russian Academy of Sciences, 4 Kosygina str., 119334 Moscow, Russia; kolesnikova@sky.chph.ras.ru

3 Biología Molecular y Bioquímica, Universidad de Malaga, 2 Cervantes str., 29071 Malaga, Spain; mariacabreragonzalez@uma.es

* Correspondence: elena.mastalygina@gmail.com; Tel.: +7-910-444-2364

Received: 28 December 2019; Accepted: 7 February 2020; Published: 13 February 2020

Abstract: Compounding natural additives with synthetic polymers allows developing more eco-friendly materials with enhanced biodegradability. The composite films based on low-density polyethylene (PE) with different content of natural rubber (NR) (10–30 wt%) were investigated. The influence of NR content on structural features, water absorption and mechanical properties of the composites were studied. The 70PE/30NR composite is characterized by the uniform distribution and the smallest size of NR domains (45 ± 5 μm). A tensile test was satisfied by the mechanical properties of the biocomposites, caused by elasticity of NR domains. The tensile strength of 70PE/30NR composite film is 5 ± 0.25 MPa. Higher water absorption of PE/NR composites (1.5–3.7 wt%) compared to neat PE facilitates penetrating vital activity products of microorganisms. Mycological test with mold fungi and full-scale soil test detected the composite with 30 wt% of NR as the most biodegradable (mass loss was 7.2 wt% for 90 days). According to infrared spectroscopy and differential scanning calorimetry analysis, NR consumption and PE structural changes in the biocomposites after exposure to soil occurred. The PE/NR composites with enhanced biodegradability as well as satisfied mechanical and technological properties have potential applications in packaging and agricultural films.

Keywords: packaging material; bio-based polymer composite; polyethylene; natural rubber; water absorption; mycological test; biodegradability; mechanical properties

1. Introduction

Synthetic polymers widely used for packaging and agricultural applications are characterized by high resistance to degradation under environmental conditions. As a result, the problem of polymer wastes accumulation has become relevant. One of the most promising approaches for solving this environmental problem is developing biodegradable polymeric materials, including naturally occurring biodegradable polymers, biodegradable polymers derived from renewable resources, biodegradable composites based on petroleum and biodegradable polymers [1].

Among the listed biodegradable polymer materials, the direction of developing biodegradable polymer composites are deemed cost-effective by passing an expensive step of the synthesis [2]. Modification of the synthetic polymer matrix by introducing additives initiating degradation allows obtaining new materials with enhanced biodegradability after the life cycle. In this case, the synthetic polymer being a part of the composite determines the required performance and technological properties, as well as the possibility of recycling. The use of poly(olefins), particularly poly(ethylene), as the polymer matrix for such composites is caused by importance of their utilization and a great deal of wastes based on them [3].

Currently, oxo-additives based on transition metal salts [4,5] are commonly used as degrading additives for poly(olefins). Such additives could initiate the material photo-oxidative degradation by oxygen and sunlight, which is rarely realized in terms of waste disposal.

It is known a series of polyolefin compositions with the addition of naturally occurring biodegradable polymers, such as poly(hydroxyalkanoates) [6,7] or poly(lactic acid) [8,9] required the specific composting conditions. For example, the biodegradability of poly(lactic acid) is fully realized only at an elevated temperature (50–60 °C) [10]. In turn, the materials based on poly(hydroxyalkanoates) have high biodegradability and biocompatibility, but mechanical and physical properties of such materials are too poor [11] that confine their scope of use. Furthermore, the technological difficulties of obtaining polyesters by biosynthesis determine a high cost of the materials based on them. To improve mechanical properties, biopolymer blends transformed into micro- or nanofibrillar biocomposites in which higher melting point biopolymer acts as reinforcement and lower melting point biopolymer acts as a matrix have been developed. For instance, there are studies devoted to microinjection molded poly(lactic acid)-based composites with 10–40 wt% of polybutylene succinate nanofibrils [12], 3–10 wt% of poly(butylene adipate-co-terephthalate), and 20 wt % poly(ε-caprolactone) [13].

The addition of natural fillers to the synthetic polymer matrix allows getting materials with the capability of being decomposed more rapidly under the environmental conditions [14]. The use of natural fillers derived from renewable resources allows, on the one hand, to replace non-renewable petrochemical raw materials with renewable those; on the other hand, to reduce the production cost by using manufacturing and agricultural waste products. The numerous research achievements have been devoted to the composites based on low-density or high-density poly(ethylene) filled with cellulose [15] and cellulosic materials such as wood flour, flax shive [16], rice husk [17], corn [18], sisal and hemp [19] fibers, as well as banana flour [20] and soy protein [21]. The high ability of natural polymers to be assimilated by microorganisms determines the intensification of biofouling on the hybrid composite polymeric materials. Unfortunately, the main weakness of such materials is related to lower mechanical properties compared with traditional polyolefins.

The additive of natural rubber appears to have a great potential for using as a component initiating the biodegradation of poly(ethylene) composites. Natural rubber had been reported to be susceptible to biodegradation by a wide range of bacteria [22] and different cultures of mold fungi [23]. The main component of natural rubber is poly(isoprene). According to the previous studies, poly(isoprene) component forms flexible nano-scale or micro-scale drops in the matrix of poly(ethylene) [24]. The size of elastomeric drops in the poly(ethylene) matrix depends on natural rubber content and blending technology. Due to its low modulus of elasticity and low tensile strength, elastomeric natural rubber cannot act as reinforcing filler for poly(ethylene). In this study, the investigation of structure, physical-mechanical properties and biodegradability of the composites based on low-density poly(ethylene) with natural rubber was carried out. According to the previous results [25], the significant modification of the poly(ethylene) crystalline phase in the poly(ethylene)/natural rubber composites affected by moisture and environmental factors have been determined. This paper is focused mostly on investigating of thermophysical and mechanical properties of the composite films based on poly(ethylene) and natural rubber. The biodegradability of the materials under investigation was analyzed by a microbiological test with mold fungi and a full-scale test in the soil medium.

2. Materials and Methods

2.1. Materials

The objects of the study were low-density poly(ethylene) (PE, grade 15803-020 from Neftekhimsevilen, OJSC, Kazan, Russia), natural rubber (NR, grade SVR 3L, Dong Xoai, Vietnam), synthetic isoprene rubber (SRI, grade SRI-3, Sintez Kauchuk, OJSC, Sterlitamak, Russia) and double composites of poly(ethylene)/natural rubber (PE/NR) containing 10, 20, 30 wt% of NR. The characteristics of low-density poly(ethylene) used in this study are given in Table 1 [15].

Table 1. Main characteristics of low-density poly(ethylene) (PE) [2,15].

Parameter	Method	Characterization
Molecular mass characteristics	Gel filtration chromatography, at 140 °C, 1,2,4-trichlorobenzene as a solvent	$M_w = 1.0 \times 10^5$, $M_n = 1.5 \times 10^4$, $M_w/M_n = 7.03$
MFI (melt flow index)	Capillary viscometry, at 190 °C and load of 2.16 kg	1.6 ± 0.1 g·10 min^{-1}
Density	Hydrostatic weighing, in ethanol 95 vol%	0.923 ± 0.001 g·sm^{-3}

The natural rubber under investigation consists of poly(cis-1,4-isoprene), with minor impurities of other compounds and water (Table 2) [26].

Table 2. Chemical composition of natural rubber (NR), grade SVR 3L [27].

Substance	Content, wt%
Poly(cis-1,4-isoprene)	91–96
Protein and amino acids	2–3
Resins	2–3
Soot indicator	0.5
Volatiles	0.8
Transition metal compounds	<1
Other	<1

All the blends for this study were prepared by a Brabender mixing system (ICHP, Moscow, Russia) in an argon atmosphere (State Standard GOST 10157-79) at a temperature of (140 ± 2) °C and a rotor rotational speed of 30 rev·min^{-1}. Before processing the composites, natural rubber was granulated. The initial size of NR pellets was 3–5 mm. A weighted amount of PE was placed in the mixing chamber, and 2 minutes later, NR was added. Mixing of polymers was performed for 5 min. An inert argon atmosphere was used to reduce the oxidation of polymers. After cooling, the samples were milled using a knife mill RM-120 (Vibrotechnik, LLC, St. Petersburg, Russia) and then compressed by a manual hydraulic press PRG-10 (VNIR, LLC, Moscow, Russia) at a temperature of (190 ± 2) °C and under a load of 0.78 kN on a cellophane substrate followed by quenching in water at (20 ± 2) °C. As a result, the film round samples with a diameter of 8 cm and a thickness of (120 ± 10) μm were obtained.

2.2. Methods

2.2.1. Optical Microscopy

The NR domain size and the uniformity of NR particle distribution in the PE matrix were investigated under an optical microscope Axio Imager Z2m (Carl Zeiss, Oberkochen, Germany) in the transmitted and reflected light (at a magnification of 100× and 200×). The average value of domains' diameter (D) was obtained through measurements of 10 domains in the 10 microscopic fields. Based on the conducted analysis, the size distribution of NR domains in the composites with different content of

NR was obtained. The intensity of microorganisms' growth was investigated in the transmitted light (at a magnification of 100×).

2.2.2. Mechanical Properties (Tensile Test)

The mechanical properties were studied using a tensile machine GP UG 5 DLC-0,5 (DVT Devotrans, Istanbul, Turkey) at a rate of 100 mm/min in accordance with BS EN ISO 527-1 and BS EN ISO 527-3 (7 measurements for each point). Test samples corresponded to 1B type (EN ISO 527-3, the working length was 40 mm) were prepared from the films by a punching press. The number of samples of each composition was 7. The extreme values were excluded from the results. The modulus of elasticity (Young's modulus) was determined by the stress-strain curves in the field of elastic deformation (the interval between two strains was from 0.1% to 0.3%).

2.2.3. Differential Scanning Calorimetry (DSC)

The behavior of PE in the composites during melting and crystallization was studied using a differential scanning calorimeter DSM-10M (IBI, Pushchino, Russia) in a temperature range from 40 to 150 °C at a scanning rate of 8 °C·min^{-1}. The sample weight was (10 ± 0.1) mg. The temperature scale was calibrated by an indium standard (T_m = 156.6 °C, ΔH = 28.44 J·g^{-1}). To obtain cooling curves, the samples of PE/NR blends were heated up to 150 °C, stored at this temperature for 5 min, then cooled at a rate of 8 °C·min^{-1} to 40 °C. The temperatures of melting and crystallization (T_m and T_c) were determined by an endothermic maximum of the melting peak and an exothermic maximum of the crystallization peak on DSC thermograms, respectively. The enthalpy of fusion (ΔH_i) was calculated from the area of a melting peak limited by the baseline. Each value of the parameters ΔH_i, T_m, T_c was obtained by averaging of 5 measurements. To calculate the degree of PE crystallinity (χ), Equation (1) was used [27], where ΔH_i is the specific heat of melting calculated per the content of PE in the composite, ΔH_0 = 293 J·g^{-1}—the specific heat of melting of a completely crystalline PE [27].

$$x = \frac{\Delta H_i}{\Delta H_0} \times 100. \tag{1}$$

2.2.4. Fourier-Transform Infrared Spectroscopy (FTIR)

The analysis of chemical composition was performed by Fourier-transform infrared spectroscopy using an FTIR spectrometer Spectrum 100 (Perkin Elmer, Beaconsfield, United Kindom) at a temperature of (22 ± 2) °C in the wavenumber range of 4600 ≤ n ≤ 650 cm^{-1} in the transition light and by an attenuation total reflection method (ATR). The changes in the chemical composition of the materials after exposure in the aqueous and soil media, as well as after fungi biodeterioration were studied.

2.2.5. Kinetics of Water Absorption

The water absorption degree of the films was determined into distilled water at (30 ± 2) °C for 45 days according to DIN EN ISO 62:2008-05. The method consists of determining the change in sample mass and appearance after exposure in the model medium (distilled water) at the stated intervals. The measurements were carried out for 45 days until an equilibrium water absorption. Additionally, changes in the PE crystalline structure and the degree of extractable substances were determined.

2.2.6. Mycological Test with Mold Fungi

The ability of materials to support the fungal growth was evaluated using different cultures of fungi according to ISO EN 846:1997. Mold fungi strains from the collection of the Department of Mycology and Algology (Lomonosov Moscow State University) were used: I—*Trichoderma harzianum Rifai*; II—*Penicillium chrysogenum Thorn*; III—*Fiisarhim moniliforme Sheld*; IV—*Chaetommm glohosimi Kunze*; V—*Trichoderma asperellum Samuels Lieckf and Nireberg*.

Two series of experiments were carried out: with potato nutrient culture medium as a substrate for fungi cultivation and with a saturated solution of potassium chloride (samples were used as a sole source of nutrition). Incubation of the samples (size of 1 × 1 cm) inoculated by the fungi spore suspension or by dry process (5×10^4 spores/mL) was carried out at a relative humidity greater than 90% at (28 ± 2) °C using a moist chamber for 25 days with an intermediate inspection after 4 and 14 days.

The morphology and intensity of fungal growth were evaluated by an optical microscope according to ISO EN 846:1997. The criteria of biofouling assessment used in this work are shown in Table 3.

Table 3. Assessment of microorganisms' growth (ISO EN 846:1997).

Intensity of Growth (Points)	Evaluation
0	No growth was seen under the microscope. The material does not contain any nutritive component.
1	No growth was seen by the naked eye but was visible under the microscope.
2	Growth was visible to the naked eye. 25% of the test sample surface was covered with microorganisms growth. The material contains nutritive components providing a slight growth of microorganisms.
3	Growth was visible to the naked eye. 50% of the test sample surface was covered with microorganisms growth.
4	Growth was visible to the naked eye. More than 50% of the test sample surface was covered with microorganisms growth. The material contains enough nutritive components enabling the growth of microorganisms.
5	Heavy growth covered the entire surface of the test sample.

After the exposure for 25 days, the biomass gain and the fungal intensity of growth were assessed by optical microscopy [28]. The values of hyphae length and spores' quantity were measured.

The average length value of hyphae (cm) was calculated by an Equation (2), where a—value of hyphae length in one microscopic field (units of ocular scale); $b = 9.091$ μm—unit value of ocular scale (μm); $S_1 = 108$ μm^2—specimen area; $S_2 = 2.34 \times 10^7$ μm^2—area of the microscopic field.

$$L = a \times b \times \frac{S_1}{S_2} \times 10^{-4} \tag{2}$$

The biomass value (fungal mycelium weight) was calculated using Equation (3), where L—average length value of hyphae calculated according to Equation (2); $r = 2.5 \times 10^{-4}$ cm—value of hyphae radius; $\rho = 1.05$ g·cm^{-3}—weight density of mycelium.

$$P = L \times \rho \times \pi \times r^2. \tag{3}$$

2.2.7. Biodegradability Test in Soil Medium

The biodegradation test was carried out by exposure of the film samples into the soil for 45 and 90 days. The biodegradation test on the recovered soil was carried out in accordance with ASTM D 5899 to investigate the biodegradability of the material. The soil was made of sand, garden soil, and horse manure taken in equal amounts. The resulting soil was held for two months at (20 ± 5) °C with daily stirring and humidity maintained at (60 ± 10)%. The film samples were immersed vertically in the prepared soil and exposure at (22 ± 3) °C and relative humidity of (60 ± 10)% for 45 and 90 days. The change in appearance, weight and chemical composition (by FTIR) of the materials were investigated after exposure in the soil medium. Three specimens of each composite were tested.

The work was carried out using scientific equipment of the Center of Shared Usage «New Materials and Technologies of Emanuel Institute of Biochemical Physics» and Joint Research Center of Plekhanov Russian University of Economics.

3. Results

3.1. Structure and Properties

The addition of filler to the polymeric matrix leads to significant changes in the morphology of the composite material and the macromolecular mobility of the boundary layers. By means of optical microscopy, the microstructure of composite materials was analyzed. Figure 1 shows the selected microphotographs of the PE/NR composites. When compounding PE with NR, the formed system represented a PE matrix with domains of NR dispersed inside. Having an elastomeric nature, NR behaved as a flexible dispersed filler in the PE matrix. In this case, PE/NR systems cannot be considered as a blend of two thermoplastic polymers. The rubber domains in the composites were characterized by a size of 10–100 μm. It was found that with an increase in the content of NR in the PE matrix, more uniform dispersion of domains occurred with a decrease in their sizes (Figure 1).

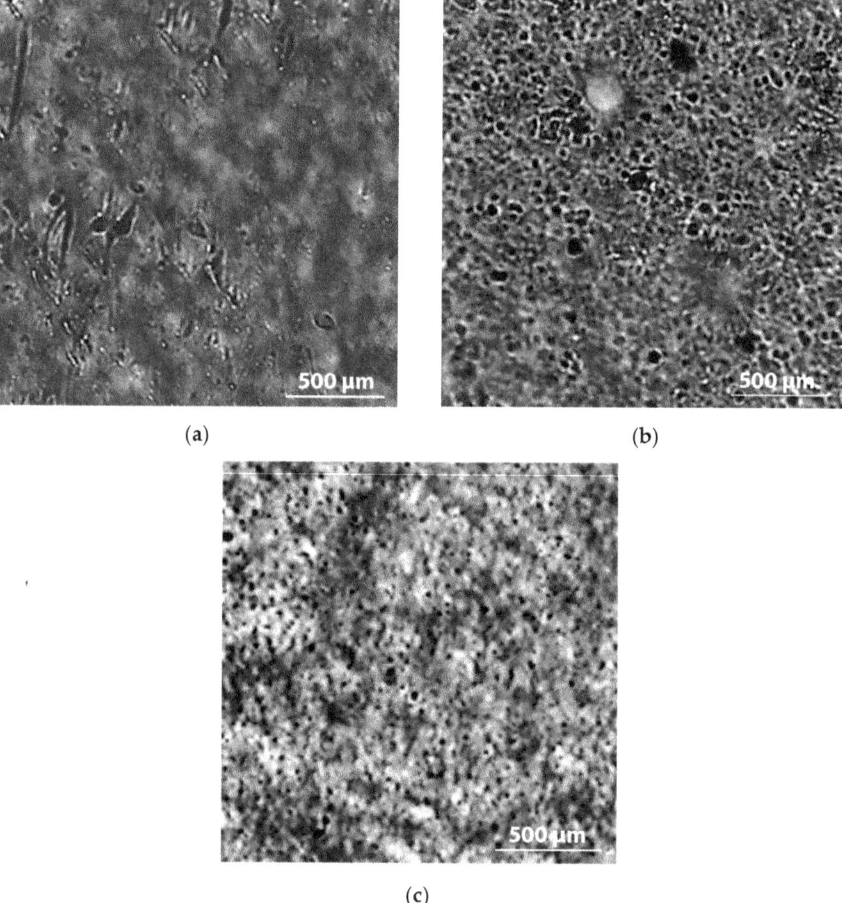

Figure 1. The microphotographs of the PE/NR composite samples containing 10 (**a**), 20 (**b**) and 30 (**c**) wt% of NR (transmitted light, at a magnification of 200×).

The size distribution of NR domains in the PE matrix is presented in Figure 2. According to the obtained results, the composites PE/NR = 70/30 had a more uniform distribution of rubber particles with an average size of (45 ± 5) µm. The composites with 20 wt% of NR in the PE matrix had a wider distribution throughout domains sizes. In the case of small NR additives, an inhomogeneous structure with sufficiently large inclusions of the NR phase was observed in PE/NR composites.

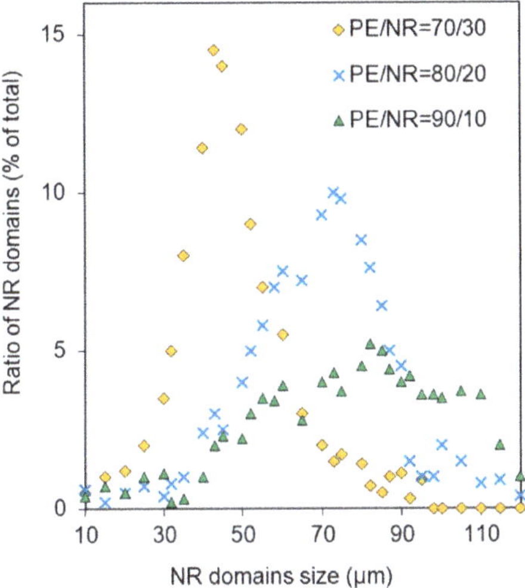

Figure 2. The size distribution of NR domains in the PE/NR composites containing 10, 20 and 30 wt% of NR.

The microstructure of the PE/NR composites determines their behavior under tension. Figure 3 demonstrates the basic parameters of the mechanical properties of the investigated composite films. Adding NR to PE contributes to reducing the relative elongation at break. When 10 wt% of NR was added to PE, the value of relative elongation at break reduced to four-fold the relative elongation of neat PE. As the content of NR increased, no further changes were observed. It was defined that the tensile strength and Young's modulus for the PE/NR composites were about two times lower than these values for neat PE.

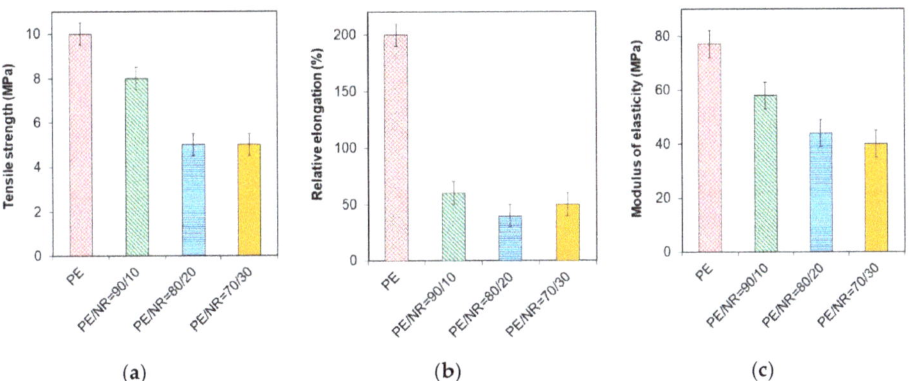

Figure 3. The parameters of tensile strength at break (**a**), relative elongation at break (**b**), modulus of elasticity (**c**) under tension for neat PE and the PE/NR composites with NR content of 10, 20, 30 wt%.

Using DSC analysis, the melting and crystallization behavior of PE as a part of the composites was studied. The analysis was conducted for the initial samples, as well as the samples after aging in the aqueous and soil media. The resulting parameters are presented in Table 4. According to DSC results, as the content of NR increased, the melting peak of PE shifted in the region of lower temperatures. The exposure to water had an insignificant influence on the PE component of the composites. The degree of PE crystallinity increased from 29% to 34% after exposure to the soil medium for 90 days.

Table 4. The thermophysical parameters of PE (melting point (T_m), degree of crystallinity (χ), peak width at half height ($\Delta T\frac{1}{2}$)) for initial samples and samples after aging in the aqueous and soil media.

NR, wt%	Initial Samples			Samples after Exposure in Aqueous Medium for 14 Days			Samples after Exposure in Soil Mediumfor 90 Days	
	T_m, °C ($\Delta \pm 0.2$ °C)	χ, % ($\Delta \pm 1\%$)	$\Delta T\frac{1}{2}$, °C ($\Delta \pm 0.2$ °C)	T_m, °C($\Delta \pm 0.2$ °C)	χ, %($\Delta \pm 1\%$)	$\Delta T\frac{1}{2}$, °C ($\Delta \pm 0.2$ °C)	T_m, °C ($\Delta \pm 0.2$ °C)	χ, % ($\Delta \pm 1\%$)
0	110.0	29	10.6	110.0	29	10.5	110.0	29
10	109.5	29	10.4	106.5	30	11.5	108.0	29
20	105.5	29	10.0	105.5	32	13.0	108.0	32
30	106.0	29	9.8	106.0	31	12.0	107.5	34

3.2. Degradation Study

Figure 4 shows the data of equilibrium water absorption and the water absorption curves for the composites under investigation. NR was characterized by a high degree of water absorption $(36 \pm 0.2)\%$, which made it susceptible to soil microorganisms. For synthetic poly(isoprene) rubber SKI-3, the degree of equilibrium water absorption was lower $(18 \pm 0.2)\%$ than for NR. This fact could be related to the difference in the structure and chemical composition of the materials. Natural rubber consists almost entirely of the cis-1,4-polyisoprene with different accessory substances. SKI-3 contains only trans-1,4-polyisoprene. According to the data, the values of equilibrium water absorption for the PE/NR composites increased with decreasing PE content. The water absorption for the composite PE/NR = 70/30 was 3.7 wt% after 45 days in the aqueous medium.

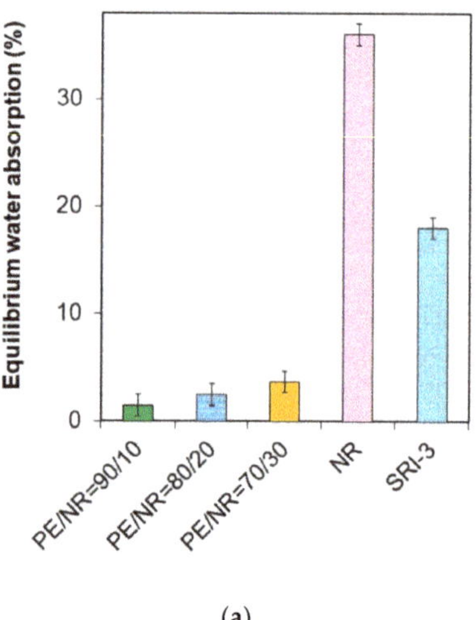

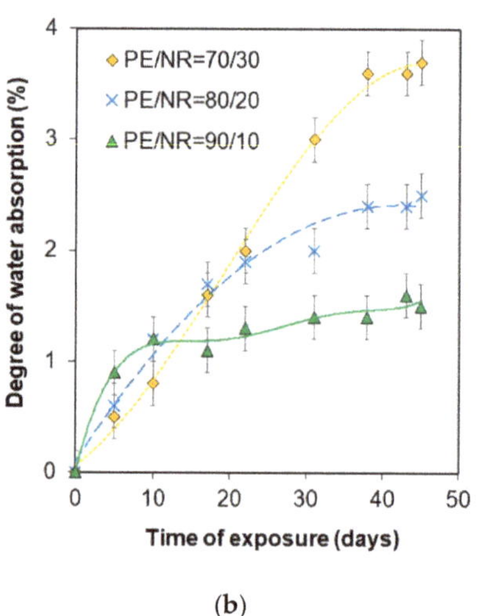

Figure 4. (**a**) The equilibrium degree of water absorption of PE/NR composites, NR and SRI-3 after 45 days in the aqueous medium; (**b**) The water absorption curves for the PE/NR composites with NR content of 30 (1), 20 (2), 10 (3) wt%.

Using the fungal test, two parameters were estimated: the intensity of samples biofouling in dynamics and the biomass value after 25 days of incubation. By way of an example, the photographs of a composite sample based on PE/NR = 70/30 after 25 days from inoculation by fungal spores (potato nutrient culture medium) are presented (Figure 5).

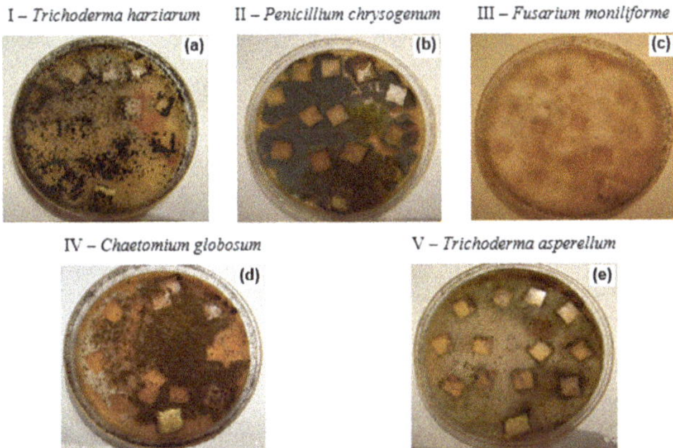

Figure 5. The photographs of composite samples of PE/NR = 70/30 after 25 days from the inoculation by mold fungi spores of *Trichoderma harzianum Rifai* (**a**), *Penicillium chrysogenum Thorn* (**b**), *Fiisarhim moniliforme Sheld* (**c**), *Chaetommm glohosimi Kunze* (**d**), *Trichoderma asperellum Samuels Lieckf* and *Nireberg* (**e**) (potato nutrient culture medium).

It was determined that most fungal stains had the common tendencies of growth depending on NR content in the composites. As an example, the samples biofouling assessment according to five-scoring scale (ISO EN 846:1997) for a test culture of *T. harzianum* is shown in Figure 6a. The composite with 30 wt% of NR was characterized by a higher rate of growth compared to the composites with a lower content of NR. The cumulative biomass gain after 25 days from inoculation with fungal spores allowed to determine the hybrid composite of PE/NR = 70/30 as the most biodegradable among the investigated composites (Figure 6b). Moreover, neat NR and PE composite with 30 wt% of NR has a similar cumulative biomass gain values.

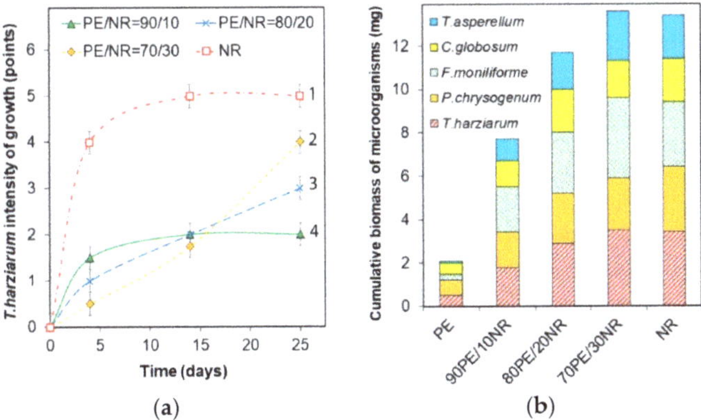

Figure 6. (**a**) The intensity of biofouling of NR and PE/NR composites by a test culture *Trichoderma harzianum* (potassium chloride medium); (**b**) The cumulative biomass gain value ($\Delta \pm 0.1$ mg) after 25 days from the inoculation of NR and PE/NR composites by fungal spores (potassium chloride medium).

The biodegradability of the PE/NR composites under environmental conditions was investigated by a full-scale soil test. Preliminarily neat natural rubber was tested under the soil medium. The reduction in weight of neat NR samples after exposure for 90 days in soil was 38.3% that indicated a high biodegradability of natural rubber (Table 5). The mass loss of PE/NR = 70/30 composite after 90 days in soil was about 7%.

Table 5. The changes in composite materials after exposure in the soil medium.

NR, wt%	Mass Loss, % (after Exposure in Soil for 45 Days)	Mass Loss, % (after Exposure in Soil for 90 Days)	D_{836}/D_{1463}* Initial Samples)	D_{836}/D_{1463} (after Exposure in Soil for 90 Days)	D_{1376}/D_{1463}** (Initial Samples)	D_{1376}/D_{1463} (after Exposure in Soil for 90 Days)
0	0	0	0	0	0	0
10	1.3	1.3	–	–	–	–
20	1.5	1.5	0.0920	0.0850	0.2310	0.2180
30	2.7	7.2	0.1668	0.1029	0.3409	0.2448
100	16.2	38.3	–	–	–	–

* D_{836}/D_{1463}—The absorbance ratio of the bands at 836 and 1463 cm^{-1}; ** D_{1376}/D_{1463}—The absorbance ratio of the bands at 1376 and 1463 cm^{-1}.

After exposure of the composites to both aqueous and soil media, the appearance, structure and chemical composition changes of the materials were analyzed. The degree of crystallinity and the range of crystalline particles of PE increased after both kinds of treatment (Table 5). The higher the content of NR in the composites was, the more intense structural changes by aqueous and soil media occurred.

Table 5 shows the results of the evaluation of different absorbance ratios from FTIR for analyzing the changes in the chemical composition of the materials after a soil test. The peak at 1463 cm^{-1} corresponds to the bending deformation vibrations of the CH$_2$–group in polyethylene. The peaks at 836 cm^{-1} and 1368 cm^{-1} are characteristic for C–H bending vibrations in =CH and –CH$_3$ groups of cis-1,4-polyisoprene, correspondently [29]. Based on the absorbance ratios of the bands at 836 and 1463 cm^{-1} (D_{836}/D_{1463}) and the bands at 1376 and 1463 cm^{-1} (D_{1376}/D_{1463}), the quantitate analysis of changing NR content in the materials compared to PE was conducted. According to the results, the changes in chemical composition were revealed for the PE/NR composites containing 20 and 30 wt% of NR exposed to the soil medium. The intensity of bands at 838 cm^{-1} and 1376 cm^{-1}, which are specific for NR, decreased by 38% and 29%, correspondently, for the composite with 30 wt% of natural rubber after 90 days in the soil medium. The observed loss in NR content by FTIR (ATR-method) was higher than the mass loss of the PE/NR composites due to the higher rate of degradation for the surface layers in comparison with the internal layer.

After 45 and 90-day soil tests, visual inspection and analysis by optical microscopy were conducted. There were no changes in the appearance of the samples of neat PE after exposure in the soil medium. Figure 7a shows photographs of the initials sample of neat PE and the sample of PE after 90 days in the soil medium. Figure 7b demonstrates the microphotograph of neat PE exposed for 90 days in soil. No defects, staining and biological fouling were observed on the polyethylene samples.

The obtained photographs and microphotographs of NR and the PE/NR composites are presented in Figures 8 and 9, respectivley. Figure 8 demonstrates appearance changes in the films based on neat NR after 45 and 90 days in soil medium. Numerous defects, color darkening and deterioration by soil microorganisms of NR samples are visible by the naked eye.

For the PE/NR composites, a loss of transparency and appearance of colored spots after the soil test were detected by the naked eye. By virtue of optical microscopy, biofouling of the PE/NR composites by mold fungi with mycelium growth was revealed. The microphotographs of the composite based on PE/NR = 70/30 are shown in Figure 9. The intensity of biofouling of the PE/NR = 70/30 sample after a 90-day soil test was 4 points (according to ISO EN 846:1997). More than 50% of this sample surface was covered with microorganisms' growth that indicates the material contained enough nutritive components enabling for the growth of soil microorganisms.

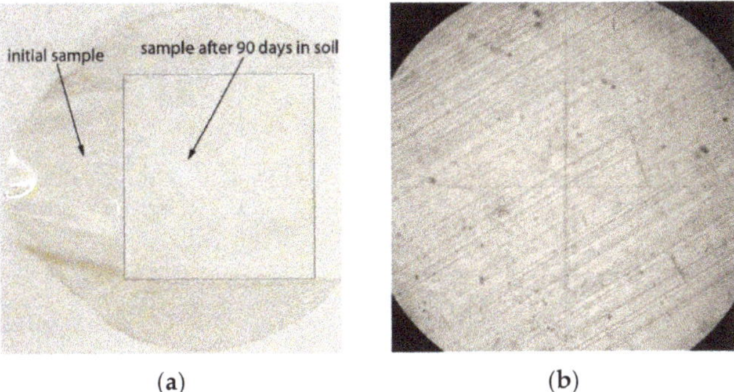

(a)   (b)

Figure 7. (**a**) The appearance changes of the films based on neat PE (initial sample and sample after 90 days in the soil medium); (**b**) The microphotograph of the neat PE after exposure in the soil medium for 90 days (transmitted light, at a magnification of 100×).

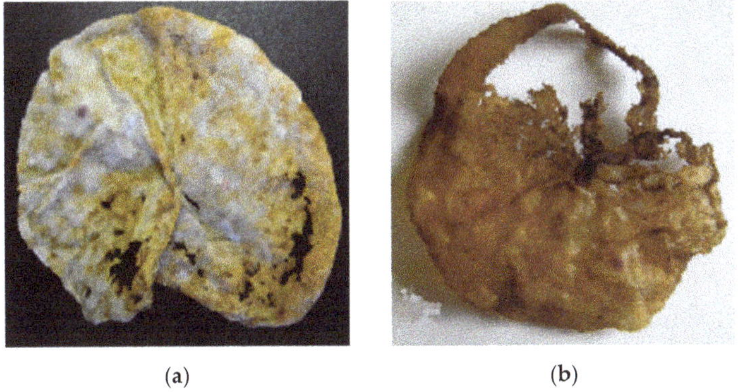

(a)   (b)

Figure 8. The appearance changes of the films based on neat NR after 45 (**a**) and 90 (**b**) days in the soil medium.

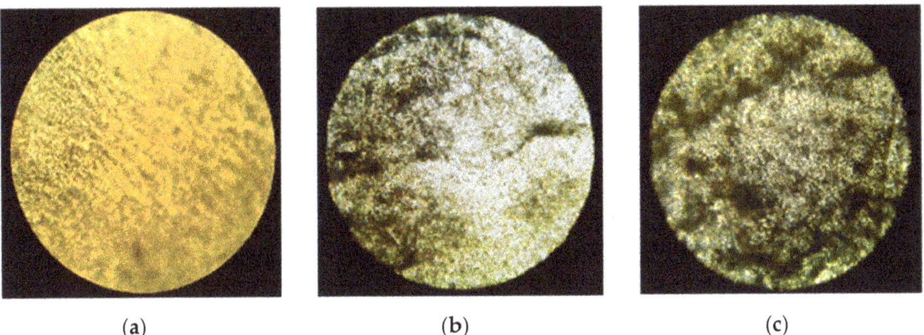

(a)   (b)   (c)

Figure 9. The microphotographs of the composite of PE/NR = 70/30: initial sample (**a**), sample after exposure in the soil for 45 days (**b**), sample after exposure in the soil for 90 days (**c**) (transmitted light, at a magnification of 100×).

Figure 10 shows the DSC curves of the melting of the initial samples, as well as the samples exposed to the aqueous and soil media. There were no changes in the crystalline structure of neat

PE. In contrast, the crystalline structure of PE in the composites with NR was changed under the influence of the aqueous and soil media, which was apparent in the bimodality of the PE melting peak. An additional low temperature melting peak of PE was observed for the PE/NR = 70/30 composite samples exposed in the moist soil for 90 days (Figure 10b, curve 2). At the second melting of this PE/NR composite, the PE endothermic peak on the DSC curve becomes monomodal (Figure 10b, curve 3).

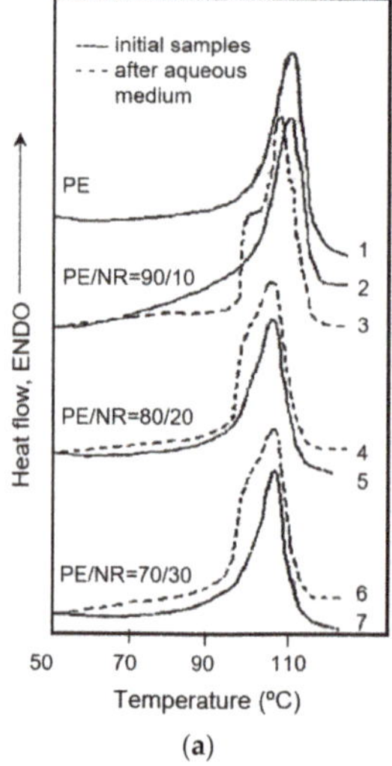

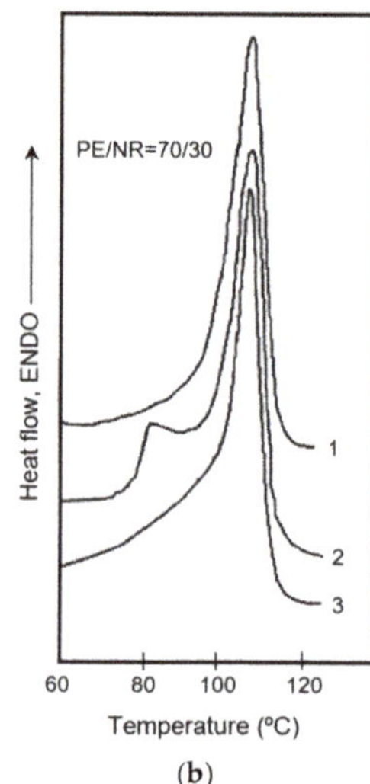

Figure 10. The DSC melting curves: (**a**) PE (1) and PE/NR composites with NR content of 10 (2,3), 20 (4,5), 30 (6,7) wt%: solid lines 1, 2, 5, 7—initial samples; dashed lines 3, 5, 7—samples after exposure in the aqueous medium for 45 days; (**b**) PE/NR = 70/30 composite: 1—initial sample; 2—sample after exposure in soil for 90 days (first heating); 3—sample after exposure in soil for 90 days (second heating).

4. Discussion

The phase structure of the PE/NR composites including the character of distribution of NR domains has a significant effect on performance properties and mechanism of degradation of the composite materials. It was determined that the composites with NR content of 30 wt% were characterized by a smaller and narrower size distribution than those with a lower NR content. Decreasing in NR domain size could be evidence of forming an abnormal overstrained system with excess free energy. According to the literature, such polymeric systems are highly susceptible to destructive factors [30]. In addition, a more uniform distribution of NR domains in the PE matrix for the composite of PE/NR = 70/30 indicates better interaction of components in the melt.

The entirely new structure of composite material is responsible for new mechanical properties. The addition of natural filler to the polymer matrix provides defective zones at the interface region between the polymer and filler particles [31]. As a result, the plastic yield under the tension of the filled composites decreased significantly. Nevertheless, PE/NR composites under investigation are characterized by satisfying elastic properties compared to another composite based on polyethylene with dispersed fillers. The good mechanical properties of the PE/NR composites could be caused by high elasticity and uniform distribution of rubber particles. In contrast, in case of adding 10–30 wt% of different dispersed fillers to polyethylene (cellulose, flax straw, wood flour, etc.), the elongation at break value decreased by 90 ± 5%, which indicated the brittle fracture of materials [14–17]. The samples were fractured in the yield region before the plastic flow.

On the other hand, NR domains do not reinforce the PE matrix as opposed to most of the cellulosic fillers. In the case of addition 10 wt% of NR to PE, the tensile strength and modulus of elasticity went down by (20 ± 5)% compared to neat PE. Provided that NR content in the polyethylene matrix increases to 20 wt%, the further reduction occurred. The content of 20 wt% of NR in the composite is critical. In the case of the further increasing NR content, the tensile strength of composites stopped changing.

According to the DSC analysis, as the content of NR increased, PE melting onset temperature decreased. Apparently, the presence of NR prevents PE crystallization making the perfection of the crystallines lower. The degree of crystallinity of PE changed slightly. The range of PE crystalline particles was determined by width at height of melting peaks. When NR content varied, this value changed slightly.

The required mechanical and thermal properties of the PE/NR composites determined the availability of developing such composites for use in different consumption fields. Nevertheless, the main feature of the materials under investigation is biodegradability under environmental conditions. Therefore, a significant part of this study was devoted to this aspect. The biodegradability of neat NR and the composites based on PE and NR was studied by a water absorption test, a microbiological test with mold fungi and a full-scale soil test.

Water is widely distributed in the environment and is able to penetrate through the surface layers of material to diffuse deeper, causing both plasticizing and wedging action (Rebinder effect) [32]. Besides that, the bioavailability of NR in the PE matrix appears to depend on the polymeric matrix structure influencing the value of water absorption. Investigation of water absorption and kinetics of this process allows analyzing structural features and predicting changes while the biodegradation process of the composites.

NR was characterized by a high degree of water absorption (36 ± 0.2)%, which made it susceptible to soil microorganisms. For synthetic poly(isoprene) rubber SKI-3, the degree of equilibrium water absorption was lower (18 ± 0.2)%. This fact could be related to the difference in rubbers structure (SKI-3 consists of poly(trans-isoprene)) and accessory substances in NR [33].

According to the data of the water absorption test, the PE/NR composites had a higher water absorption degree compared to neat PE. Besides, the equilibrium water absorption for the composites increased with decreasing in PE content. In accordance with the kinetics of water absorption for the PE/NR composites, the rate of water absorption decreased in proportion to increasing NR content, which could be caused by high dispersion of NR domains in the PE matrix with increasing of the phases contact area.

The enhanced degree of water absorption facilitates penetrating vital activity products of microorganisms (acids and enzymes) into the materials that lead to hydrolysis of NR [34]. The products of NR hydrolysis, having lower molecular weight and higher diffusion coefficient through the polymer matrix, are able to leave the samples that in turn is resulting in the weight decrease of materials at exposure in soil. Moreover, PE in the PE/NR composites is exposed to the impact the absorbed water.

The analysis of materials bioresistance to mold fungi is one of the common model experiments for determining biodegradability of materials particularly polymeric composites. It was determined that fungal cultures of *T. harzianum* and *F. moniliforme* had the most intensive spreading on the PE/NR composites. So, these cultures could be indicated as the main biodestructors of PE/NR composites. The color change of the test samples was observed for cultures of *P. chrysogenum* and *C. globosum* that could be related to metabolites released by microorganisms (including pigments).

According to the results, most fungal stains had the common tendencies of growth depending on NR content in the composites. The composites with 30 wt% of NR were characterized by a slower growth rate compared to the composites with a lower content of NR. Apparently, this fact is connected with the accessibility of NR domains in the PE matrix. Since the composite of PE/NR = 90/10 contained large-dimension particles of NR (about 90 μm), NR domains located close to the sample's surface could be easily assimilated by fungi and used as a nutrient source. However, the remaining NR domains encapsulated in the PE matrix are not accessible to microorganisms. On the contrary, NR in the composites of PE/NR = 70/30 were fine-dispersed particles that could be a limiting factor of fungal growth at the first period of cultivation.

The biodegradability of PE/NR composites under environmental conditions was investigated by a full-scale soil test. By virtue of optical microscopy, biofouling of the composites by mold fungi was revealed. The release of microorganisms' metabolites, including melatonin type pigments, apparently leads to discoloration of the materials.

The mass loss is a basic criterion of the biodegradation rate under environmental conditions. The micromycete growth and the intensity of materials destruction evaluated by a weight loss after the soil tests largely depended on the NR content. Therefore, the samples containing 30 wt% of NR had the greatest weight loss and a significant change in the appearance in comparison with the other samples. The intensive weight reduction of these materials exposed in the soil is induced by a high biodegradability of materials in the environment. For composite samples with 10 and 20 wt% of NR, the weight reduction occurred during the first 1.5 months, after that the weight loss values remained unchanged. Probably, for these compositions, NR is encapsulated in the PE matrix; hence, microorganisms could assimilate only the surface NR particles. According to the results of a 90-day soil test, a mass loss of the PE/NR = 70/30 composite was about 7%.

After exposure to composites in both the aqueous and soil media, the structural changes of PE and NR were observed. It was determined that water including soil moisture promoted the recrystallization of PE crystallines. An increase in the degree of crystallinity of PE could be related to an enhanced degree of water absorption and an effect of NR particles on the material's structure. The phenomenon of PE recrystallization in the aqueous medium is connected with relaxation and the formation of a more ordered crystalline structure under the plasticizing effect of water. The structural reorganization of PE crystallines after exposure in water for 18 months was described in the work [35]. After exposure to the PE/NR composites in the moist soil, an additional peak was observed in the DSC curves at the melting of the exposed samples. This additional endothermic peak is probably responsible for the presence of bound moisture in these materials.

In addition, the changes in chemical composition were revealed for the PE/NR composites containing 20 and 30 wt% of NR exposed to the soil test. The decrease of IR absorption bands intensity of 836 cm^{-1} (C–H bending vibrations of sp^2 hybridized carbons in C(CH$_3$)=CH groups) and 1368 cm^{-1} (C–H bending vibrations in CH$_3$–groups) gave evidence of the utilization of NR by soil microorganisms. Like that, the rate of NR biodegradation was higher for composites with a higher content of NR.

The biodegradation of the developed composites could be described as a complex process including several stages. It is known that bioassimilation and subsequent degradation of polymeric materials start from adhesion and attachment of fungal spores to the substrate [36,37]. The work results demonstrated a high ability of the composites to biofoul by mold fungi that confirms susceptibility of the PE/NR composites to biodegradation by microorganisms.

After the biofouling of the sample's surface occurs, availability and usability as a nutrient source of the substrate play important roles at the further stage of biodegradation [38]. The usability of the filled composites by microorganisms is mainly dependent on the filler biodegradability, the diffusing properties of the polymeric matrix, the composite structure including the interfacial area between phases, as well as the degree of water absorption. Under the action of microorganism' enzymes and soil moisture on the composite materials, hydrolysis of the substances contained in the composites can occur [33]. According to the work results, natural rubber had an enhanced biodegradability (the period of full degradation of NR is about half of a year). Moreover, the PE/NR composites under investigation were characterized by an increased degree of water absorption that facilitates the penetration of vital products of microorganisms (acids and enzymes) into the materials, which leads to hydrolysis and oxidation [38,39]. In turn, the decomposition products having a low molecular weight and a higher diffusion coefficient can leave the samples, which leads to a decrease in the weight of materials when exposed in the soil.

It was revealed that the composites were characterized by an intensive weight reduction at the first three months in the soil medium (the mass loss of the PE/NR = 70/30 composite was about 7%). Provided that the FTIR analysis, this mass loss was related to the bioassimilation of NR domains by soil microorganisms. This fact is often observed for such types of biodegradable polymeric composites [14,16]. The further degradation is not limited by the natural rubber bioassimilation. Because of the NR utilization in the matrix of PE, the composites structure became porous with a great deal of different defects. For this reason, the uniform distribution of biodegradable filler (in this case, natural rubber) in the polymeric matrix is important for the degradation process. Moreover, the intensive growth of microbiota on the materials is accompanied by the formation of microorganisms' metabolites and the occurrence of mechanical stresses, which are the causes of destruction. In several papers [40–42], it was shown the biochemical destruction of poly(ethylene) under the action of certain enzymes of micromycetes and bacteria accompanied by a decrease in the molecular weight of PE.

Therefore, the main purpose of introducing natural additives into poly(olefins) is enhancing their bioavailability and biofouling, due to natural fillers are easier than poly(olefins) to be exposed to physicochemical and biological environmental factors. After the microorganisms have attached to the surface of the polymer material, biofouling and degradation of the material occur under the action of intracellular enzymes of microorganisms, which leads to the polymer depolymerization. Ultimately, the fragments of polyethylene macromolecules become short enough for assimilation by microorganisms and mineralization. According to the previous studies [43], microorganisms are able to assimilate paraffins with a molecular weight of less than 600.

The intensity of poly(ethylene) degradation in such types of composite materials is commonly evaluated after a longer period of degradation (1–2 years). The mass loss, the accumulation of oxidation products, and the decrease in the degree of crystallinity are used as criteria for assessing the destruction of polyethylene [16,44]. The further study of the degradation process of the developed composites based on poly(ethylene), natural rubber and technological additives is planned.

5. Conclusions

The development of composites based on poly(ethylene) and natural rubber allows modifying polyethylene structure and properties including biodegradability. Adding natural rubber makes polyethylene more susceptible to degradation agents, including moisture, corrosive chemical substances, oxidizing agents, and products of soil microorganisms' metabolism. As a result, the composites of polyethylene with natural rubber has a higher level of biodegradability under environmental conditions compared to neat polyethylene. In accordance with the obtained results, the biodegradability of polyethylene/natural rubber composites was revealed by the changes in materials weight, appearance and physical properties, poly(ethylene) structure as well as chemical composition. In addition, the composites based on polyethylene with natural rubber additives have satisfactory mechanical and technological properties that determine the suitability of such materials for application as packaging and agricultural films with advanced biodegradability.

Author Contributions: Conceptualization, A.P. and N.K.; methodology, N.K.; software, E.M.; validation, N.K. and E.M.; formal analysis, N.K.; investigation, I.V., M.I.C.G. and E.M.; resources, A.P.; data curation, E.M.; writing—original draft preparation, E.M. and I.V.; writing—review and editing, E.M. and M.I.C.G.; visualization, E.M. and I.V.; supervision, A.P.; project administration, A.P.; funding acquisition, A.P. All authors have read and agreed to the published version of the manuscript.

Funding: This research received no external funding.

Conflicts of Interest: The authors declare no conflict of interest. The funders had no role in the design of the study; in the collection, analyses, or interpretation of data; in the writing of the manuscript, or in the decision to publish the results.

References

1. Ebnesajjad, S. *Handbook of Biopolymers and Biodegradable Plastics, Properties, Processing, and Applications*; William Andrew Inc.: New York, NY, USA, 2012.

2. Karlsson, S.; Albertsson, A.C. Biodegradable polymers and environmental interaction. *Polym. Eng. Sci.* **1998**, *38*, 1251–1253.

3. Makio, H.; Fujita, T. Polyolefins—Challenges for the Future. In Proceedings of the 14th Asia Pacifil Confederation of Chemical Engineering Congress, Singapore, 22–24 February 2012; Research Publishing: Singapore, 2012.

4. Mittal, V.; Patwary, F. Polypropylene nanocomposites with oxo-degradable pro-oxidant: Mechanical, thermal, rheological, and photo-degradation performance. *Polym. Eng. Sci.* **2016**, *56*, 1229–1239. [CrossRef]

5. Montagna, L.S.; Catto, L.; Forte, M.; Chiellini, E.; Corti, A.; Morelli, A.; Campomanes Santana, R.M. Comparative assessment of degradation in aqueous medium of polypropylene films doped with transition metal free (experimental) and transition metal containing (commercial) pro-oxidant/pro-degradant additives after exposure to controlled UV radiation. *Polym. Degrad. Stabil.* **2015**, *120*, 186–192. [CrossRef]

6. Ol'khov, A.A.; Iordanskii, A.L.; Zaikov, G.E. Morphology and mechanical parameters of biocomposite based on LDPE-PHB. *Balk J. Tribol. Assoc.* **2014**, *20*, 101–110.

7. Sadi, R.K.; Fechine, G.J.M.; Demarquette, N.R. Effect of prior photodegradation on the biodegradation of polypropylene/poly(3-hydroxybutyrate) blends. *Polym. Eng. Sci.* **2013**, *53*, 2109–2122. [CrossRef]

8. Cruz-Navarro, D.S.; Espinosa-Valdemar, R.M.; Beltrán-Villavicencio, M.; Vázquez-Morillas, A. Degradation of Oxo-Degradable-Polyethylene and Polylactic Acid Films Embodied in the Substrate of the Edible Fungus Pleurotus ostreatus. *Nat. Resour. J.* **2014**, *5*, 949–957.

9. Madhu, G.; Bhunia, H.; Bajpai, P.K. Blends of high density polyethylene and poly(L-lactic acid): mechanical and thermal properties. *Polym. Eng. Sci.* **2014**, *54*, 2155–2160. [CrossRef]

10. Podzorova, M.V.; Tertyshnaya, Y.V.; Popov, A.A. Environmentally friendly films based on poly(3-hydroxybutyrate) and poly(lactic acid): A review. *Russ. J. Phys. Chem. B* **2014**, *8*, 726–731. [CrossRef]

11. Ten, E.; Jiang, L.; Zhang, J.; Wolcott, M.P. Mechanical performance of polyhydroxyalkanoate (PHA)-based biocomposites. In *Design and Mechanical Performance*; Misra, M., Pandey, J.K., Mohanty, A.K., Eds.; Elsevier Ltd.: Amsterdam, The Netherlands, 2015; pp. 39–52.

12. Hosseinnezhad, R.; Vozniak, I.; Morawiec, J.; Galeski, A.; Dutkiewicz, S. In situ generation of sustainable PLA-based nanocomposites by shear induced crystallization of nanofibrillar inclusions. *RSC Adv.* **2019**, *9*, 30370–30380. [CrossRef]

13. Ding, W.; Chen, Y.; Liu, Z.; Yang, S. In situ nano-fibrillation of microinjection molded poly(lactic acid)/poly(ε-caprolactone) blends and comparison with conventional injection molding. *RSC Adv.* **2015**, *5*, 92905–92917. [CrossRef]

14. Faruk, O.; Bledzki, A.K.; Fink, H.-P.; Sain, M. Biocomposites reinforced with natural fibers: 2000–2010. *Prog. Polym. Sci.* **2012**, *37*, 1552–1596. [CrossRef]

15. Mastalygina, E.E.; Shatalova, O.V.; Kolesnikova, N.N.; Popov, A.A.; Krivandin, A.V. Modification of isotactic polypropylene by additives of low-density polyethylene and powdered cellulose. *Inorg. Mater. Appl. Res.* **2016**, *7*, 58–65. [CrossRef]

16. Pantyukhov, P.; Kolesnikova, N.; Popov, A. Preparation, structure, and properties of biocomposites based on low-density polyethylene and lignocellulosic fillers. *Polym. Compos.* **2016**, *37*, 1461–1472. [CrossRef]

17. Ayswarya, E.P.; Vidya Francis, K.F.; Renju, V.S.; Thachil, E.T. Rice husk ash—A valuable reinforcement for high density polyethylene. *Mater. Des.* **2012**, *41*, 1–7. [CrossRef]

18. Youssef, A.M.; El-Gendy, A.; Kamel, S. Evaluation of corn husk fibers reinforced recycled low density polyethylene composites. *Mater. Chem. Phys.* **2015**, *26*, 152–167. [CrossRef]

19. Singh, N.P.; Aggarwal, L.; Gupta, V.K. Tensile and flexural behavior of hemp fiber reinforced virgin-recycled HDPE matrix composites. *Procedia Mater. Sci.* **2015**, *6*, 1696–1702. [CrossRef]

20. Martínez, E.S.M.; Méndez, M.A.A.; Solís, A.S.; Vieyra, H. Thermoplastic biodegradable material elaborated from unripe banana flour reinforced with metallocene catalyzed polyethylene. *Polym. Eng. Sci.* **2015**, *55*, 866–874.

21. Kaur, I.; Bhalla, T.C.; Deepika, N.; Gautam, N. Study of the biodegradation behavior of soy protein-grafted polyethylene by the soil burial method. *Appl. Polym. Sci.* **2009**, *111*, 2460–2469. [CrossRef]

22. Bode, H.B.; Kerkhoff, K.; Jendrossek, D. Bacterial Degradation of Natural and Synthetic Rubber. *Biomacromolecules* **2001**, *2*, 295–303. [CrossRef]

23. Steinbüchel, L.A. *Biodegradation of Natural and Synthetic Rubbers*; Wiley-VCH Verlag GmbH & Co. KGaA.: Münster, Germany, 2005.

24. Yáfiez-Flores, I.G.; Ramos-DeValle, L.F.; Rodriguez-Fernandez, O.S.; Sánchez-Valdez, S. Blends of Polyethylene-Polyisoprene Rubbers: Study of Flow Properties. *J. Polym. Eng.* **1997**, *17*, 295–310.

25. Mastalygina, E.E.; Varyan, I.A.; Kolesnikova, N.N.; Popov, A.A. Structural changes in the low-density polyethylene/natural rubber composites in the aqueous and soil media. *AIP Conf. Proc.* **2016**, *1736*, 020097.

26. Erman, B.; Mark, J.E.; Roland, C.M. *Science and Technology of Rubber (Third Edition)*; Elsevier Inc. Academic Press: Waltham, MA, USA, 2005.

27. Godowsky, J.K. *Thermophysical Research Methods Polymers (Teplofizicheskie Metody Issledovaniya Polimerov)*; Khimiya: Moscow, Russia, 1976.

28. Alef, K.; Nannipieri, P. *Methods in Applied Soil Microbiology and Biochemistry*; Academic Press: London, UK, 1995.

29. Chen, D.; Shao, H.; Yao, W.; Huang, B. Fourier transform infrared spectral analysis of polyisoprene of a different microstructure. *Int. J. Polym. Sci.* **2013**, *2013*, 1–5. [CrossRef]

30. Mastalygina, E.E.; Popov, A.A. Mechanical properties and stress-strain behaviour of binary and ternary composites based on polyolefins and vegetable fillers. *Solid State* **2017**, *265*, 221–226. [CrossRef]

31. Ting, S.S.; Achmad, N.K.; Ismail, H.; Santiagoo, R.; Noriman Zulkepli, N. Thermal Degradation of High-Density Polyethylene/Soya Spent Powder Blends. *J. Polym. Eng.* **2015**, *35*, 437–442. [CrossRef]

32. Troughton, M.J. *Handbook of Plastics Joining: A Practical Guide*; William Andrew Inc.: New York, NY, USA, 2008.

33. Mohan, K. Microbial deterioration and degradation of polymeric materials. *J. Biochem. Technol.* **2011**, *2*, 210–215.

34. Shah, A.A.; Hasan, F.; Shah, Z.; Kanwal, N.; Zeb, S. Biodegradation of natural and synthetic rubbers: A review. *Int. Biodeterior. Biodegrad.* **2013**, *83*, 145–157. [CrossRef]

35. Khvatov, A.V. Structure, properties and biodegradation of compositions based on polyethylene and natural additives. Ph.D. Thesis, Emanuel institute of Biochemical Physics, Russian Academy of Sciences, Moscow, Russia, 2009.

36. Gumargalieva, K.Z.; Kalinina, I.G.; Semenov, S.A.; Zaikov, G.E. Bio-damages of materials. Adhesion of microorganisms on materials surface. *Mol. Cryst. Liq. Cryst.* **2008**, *486*, 213–229. [CrossRef]

37. Whitehead, K.A.; Deisenroth, T.; Preuss, A.; Liauw, C.M.; Verran, J. The effect of surface properties on the strength of attachment of fungal spores using AFM perpendicular force measurements. *Colloids Surf. B Biointerfaces* **2011**, *82*, 483–489. [CrossRef]

38. Van der Zee, M. Structure-Biodegradability Relationships of 370 Polymeric Materials. Ph.D. Thesis, University Twente, Enschede, The Netherlands, 1997.

39. Ołdak, D.; Kaczmarek, H.; Buffeteau, T.; Sourisseau, C. Photo- and bio-degradation processes in polyethylene, cellulose and their blends studied by ATR-FTIR and raman spectroscopies. *J. Mater. Sci.* **2005**, *40*, 4189–4198. [CrossRef]

40. Arutchelvi, J.; Sudhakar, M.; Arkatkar, A.; Doble, M.; Bhaduri, S.; Uppara, P.V. Biodegradation of polyethylene and polypropylene. *Indian J. Biotechnol.* **2008**, *7*, 9–22.

41. Usha, R.; Sangeetha, T.; Palaniswamy, M. Screening of Polyethylene Degrading Microorganisms from Garbage Soil. *Libyan Agric. Res. Cent. J. Int.* **2011**, *2*, 200–204.

42. Pramila, R.; Ramesh, K. Biodegradation of low density polyethylene (LDPE) by fungi isolated from municipal landfill area. *J. Microbiol. Biotechnol. Res.* **2011**, *1*, 131–136.

43. Atlas, R.M. Microbial Degradation of Petroleum Hydrocarbons: an Environmental Perspective. *Microb. Rev.* **1981**, *45*, 180–209. [CrossRef]

44. Mastalygina, E.E.; Kolesnikova, N.N.; Popov, A.A.; Olkhov, A.A. Environmental degradation study of multilevel biocomposites based on polyolefins. *AIP Conf. Proc.* **2015**, *1683*, 020143.

 *polymers*

Article

Application of LCA Method for Assessment of Environmental Impacts of a Polylactide (PLA) Bottle Shaping

Patrycja Bałdowska-Witos [1,*], Weronika Kruszelnicka [1], Robert Kasner [1], Andrzej Tomporowski [1], Józef Flizikowski [1], Zbigniew Kłos [2], Katarzyna Piotrowska [3] and Katarzyna Markowska [4]

[1] Department of Technical Systems Engineering, Faculty of Mechanical Engineering, University of Science and Technology in Bydgoszcz, 85-796 Bydgoszcz, Poland; weronika.kruszelnicka@utp.edu.pl (W.K.); robert.kasner@gmail.com (R.K.); a.tomporowski@utp.edu.pl (A.T.); fliz@utp.edu.pl (J.F.)

[2] Institute of Machines and Motor Vehicles, Faculty of Transport Engineering, Poznan University of Technology, 60-965 Poznan, Poland; zbigniew.klos@put.poznan.pl

[3] Faculty of Mechanical Engineering, Lublin University of Technology, 20-618 Lublin, Poland; k.piotrowska@pollub.pl

[4] Department of Logistics and Transport Technologies, Faculty of Transport and Aviation Engineering, Silesian University of Technology, 40-019 Katowice, Poland; Katarzyna.Markowska@polsl.pl

* Correspondence: patrycja.baldowska-witos@utp.edu.pl

Received: 19 December 2019; Accepted: 2 February 2020; Published: 9 February 2020

Abstract: In recent years, there has been a significant increase in the consumption of single-use packaging. Their material diversity is a significant barrier to recycling, causing overloading of landfills. Increasing negative environmental aspects have highlighted the need to develop solutions to achieve a relatively high efficiency of the bottle shaping process with the lowest possible energy consumption. The aim of the project is to try to describe the impact of this process on the state, transformation and development of the natural environment. The work concerns current issues of the impact of packaging on the natural environment. The main goal was to conduct a life cycle analysis (LCA) of beverage bottles made of polylactide. The functional unit comprised a total of 1000 pieces of PLA bottles with a capacity of 1 L. The boundary of the adopted system included the steps from the delivery of the preforms to the production plant to their correct formation in the process of forming beverage bottles. Further stages of the production process were excluded from the system, such as beverage bottling, labeling, and storage and distribution. Processes related to transport and storage of raw material were also excluded. The LCA analysis was performed using the program of the Dutch company Pre Consultants called SimaPro 8.4.0. The ReCiPe 2016 method was chosen for the interpretation of the quantity of emitted substances into the natural environment. The test results were presented graphically on bar charts and subjected to verification and interpretation.

Keywords: PLA bottle; bio-based and biodegradable polymers; life cycle assessment; environmental impact; ReCiPe2016 method

1. Introduction

The commercialization of biodegradable materials under industrial composting in place of the non-degradable polymer materials currently used is a technologically and socially complex process [1]. Technological progress in the field of biodegradable polymers as well as consumer expectations motivate producers to undertake actions to gradually replace the fossil fuel-based sources with natural sources for polymer production [2,3].

Nowadays, environmentally conscious producers not only produce their products from partially or completely recycled materials, but they are also increasingly using biopolymers in addition to or instead of petroleum-based polymers. Like petroleum-based plastics, most biopolymers are used by the packaging industry [4]. However, due to their function, they have a very short life span (on average a few weeks), which is why they become a waste in a short time [5,6]. In 2016, 16.7 million tonnes of plastic packaging waste was collected, of which 40.8% was recycled, 38.8% was used to generate energy (combustion), and 20.4% went to landfills [7].

In May 2019, the Council of the European Union proposed new EU-wide rules for 10 single-use plastic products that are most often found in Europe's seas and beaches. Member States must, no later than two years after the entry into force of the Directive, prohibit the following single-use plastic products: plastic swabs, cutlery, plates, straws, stirrers and balloon sticks; all products made of oxo-plastic; and cups, food and beverage containers made of foam. In addition, 90% of single-use plastic bottles should be collected separately by 2029 [8].

In 2017, only about 2% of total plastic production was biopolymer, but its volume is growing every year [9]. With the entry into force of the directive, this increase is likely to be even greater. Poly (lactic acid) (PLA) is one of the most popular biodegradable biopolymers used in the packaging industry for the production of films, sheets, bottles and foams, and its growing use in industry intends to replace the currently used forms of polymeric polyethylene terephthalate (PET) [10,11]. Considering the above, it was considered necessary and possible to conduct an experiment to evaluate the cycle of the process of shaping biodegradable plastic bottles in an industrial composting plant [12,13].

The analysis of many works on the impact of the life cycle of products such as disposable bottles on the life cycle indicates that it is necessary to identify some sets of impacts and their hierarchy. At the same time, it is important to remember what environmental costs have already been incurred and what costs will probably be generated in the future. There is also a relationship between interactions and design solutions of the analyzed bottle, e.g., by using fillers or compatibilizers.

Using the life cycle analysis (LCA) method, companies declare a reduction of environmental impact to a greater extent than the competition, and the results of the assessment shape new production directions, taking into account such factors as: sources of materials, suitability for recycling and use of recycled materials, reduction of greenhouse gas emission rates. In the literature, you can often find the results of analyses of environmental impacts of various objects, production processes and packaging. The vast majority of these analyses are based on the concept of life cycle analysis. Analyzing the data of these analyses, it can be seen that the greatest environmental nuisance is the stage of operation of technical facilities. Among the results of analyses published by N. Horowitz, J. Frago, D. Mu (2018) [14] regarding the life cycle assessment of bottles made of PLA or PET, it is possible to determine potential levels of impact on the state of the environment. However, the presented results show only impact levels for the entire technological process of bottle production. In turn, the results presented by L. Chen, R. Pelton, T. Smith (2016) [15] illustrated the potential level of inflows for bottles produced in PET and PET obtained in 100% from the raw material subjected to the recycling process. Additionally, these studies did not refer to the estimation of the potential of levels of impacts of individual unit processes at the time of bottle formation. A similar approach was presented in the research of S. Papong, P. Malakul, R. Trungkavashirakun, P. Wenunun, T. Chomin, M. Nithitanakul, E. Sarobol (2014) [16], which focused on demonstrating the potential environmental impacts of the water bottle production process drinking from polylactic acid produced in Thailand. Analyzing the results of literature studies, no studies could be found regarding the assessment of potential impacts of the process of shaping beverage bottles [17,18]. Due to some limitations and a lack of data availability, most studies focus on analyzing cradle-to-grave processes. The reason for these limitations is primarily the varied means of operating machines and devices, which would require diversity and a multi-faceted approach to diagnose all impacts.

2. Materials and Methods

The methodology of environmental assessment of an object life cycle was developed in the 1990s and was formalized by introduction of norm of ISO 140040-43 series, which defines four stages of the assessment: selection of a functional unit, the system boundaries; analysis of the set of entries and exits; transformation of collected data into indexes of impact categories or damage categories and conclusions and verification of results [19].

The ReCiPe method, whose abbreviation stands for the institutions which were involved in its creation, that is: National Institute for Public Health and the Environment (RIVM) and Radboud University Nijmegen, CML (in Dutch: Centrum voor Milieuwetenschappen, in English: Institute of Environmental Sciences of the University of Leiden) and PRé, is the result of conversion of a long list of the life cycle inventory results into a limited number of indexes defined for two impact and damage category levels [20].

This method determines the middle and the final points, but also includes factors consistent with three cultural perspectives [21,22]. These perspectives make up a set of choices on issues such as time or expectations, which can provide the basis for appropriate management or future technological development in order to prevent from damaging the natural environment or reduce it as much as possible.

Contrary to other precisely described methods, such as Eco-Indicator 99, Environmental Priority Strategies (EPS) or IMPACT 2002+, indicator ReCiPe 2016 does not cover the potential future impacts for the category of damage assuming that such impacts have been taken into account in the analysis of inventory. Other differences can be observed in its approach to the category of damage which involves 'exhaustion of natural resources'.

The IMPACT 2002+ method accepts the amount of primary energy in megajoules (MJ) to be a unit, whereas an increase in fuel extraction costs is a unit of the ReCiPe method (in dollars) [23].

Moreover, ReCiPe 2016 uses an environmental mechanism which can be viewed as a series of results which when combined can cause a certain level of damage, for instance to human health or ecosystems. In result of the climate change, many substances increase their emissions which means that heat does not escape from the earth into space. In consequence, more energy is being trapped on earth and the temperature increases. This contributes to changes in the natural habitats of living organisms, which can lead to the extinction of some species of animals and plants [23]. Figure 1 shows that the longer it takes for the environmental mechanism to be formed, the higher the uncertainties. The temperature increase is more difficult to determine because there are many parallel positive and negative feedbacks [23,24].

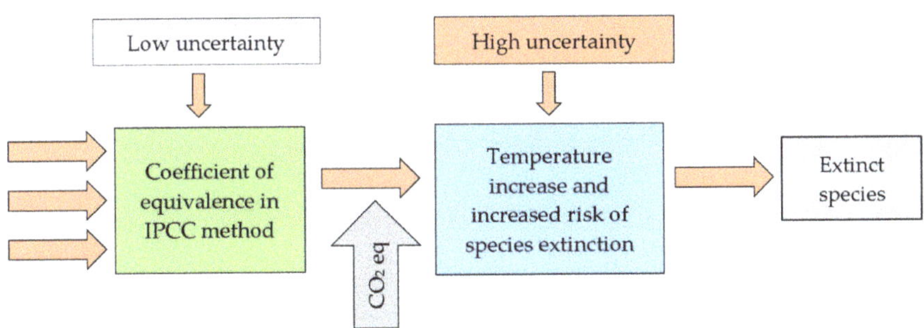

Figure 1. An example of a harmonized model of the final point regarding climate change connecting human health and damage to ecosystem.

The ReCiPe 2016 method involves performing calculations according to the specified eighteen indexes of the middle point and three indexes of the final points [25]. Since interpretation of a large number of indexes of the middle point is difficult, there is a motivation for calculation of indexes of the final point. Application of the final point level indexes is supposed to facilitate interpretation of the results because they represent impacts of the analyzed processes or products, only for three areas: human health, ecosystem quality, and climate changes [23].

2.1. Goal and a Functional Unit

The research goal is characterized by the research specificity, insight and scope and the types of data needed for the life cycle assessment. LCA is a tool for assessment of the product general impact on the natural environment "from the cradle to the grave". The choice of the process which is least harmful in terms of human health and natural environment is a priority. For this purpose, a detailed assessment of biodegradable polymer bottles shaping process was performed at the stage of characterizing. The major goal of this study is identification of the areas with potentially the highest harmful environmental impact. The basic assessment of the life cycle was limited to selected impact categories: global warming, formation of fine particles, use of water resources, water acidification and land use. Calculations were performed based on an Ecoinvent version 3.3, database library implemented in the SimaPro 8.4.1, program. One thousand bottles of 1 L volume were accepted to be a functional unit.

2.2. The System Boundary

Six separate unit operations consistent with the process accepted for production of bottles were assumed to be the system boundary. These include: preheating of the preform (PBH), heating the preform (PH), stretching and extension of the preform (PSE), pressure shaping of the preform (PPS), degasification (DB) cooling of the shaped bottle (CB) [23]. Adopting the same assumptions based on an identical division of unit operations as in the work [23] enables comparison of two processes of shaping bottles made of PET and PLA. The presented analyses of technological processes in the production of bottles for beverages [23] are characterized by diverse potential impact on the state of the natural environment and depend on the amount of electricity consumed in the production process. The high level of impact on human health is determined by the poly (ethylene terephthalate) used in the production process. To reduce the negative impact on the state of the natural environment, food industry producers should constantly seek substitutes for raw materials from exhaustible fossil resources for which the ideal replacement is biodegradable in an industrial composting facility. The right direction of development is to popularize biodegradable raw materials of natural origin. As a result, the technological operations of the accepted processes were burdened with the same simplifications which made it possible for each of the considered technological operation to accept the level of exclusion to be below 0.01% of the share in the whole life cycle and in all potential environmental impacts. The collected data cover the production, use of resources, energy consumption, emissions to the air, soil and water (Figure S1, Table S1).

The first stage of the research involved checking sufficiency and consistence of the accepted measurement data. Due to the character of the research which focused on assessment of the bottle shaping processes, whereas the elements related to the material delivery to the production plant and further stages of the product use were excluded (Figure 2).

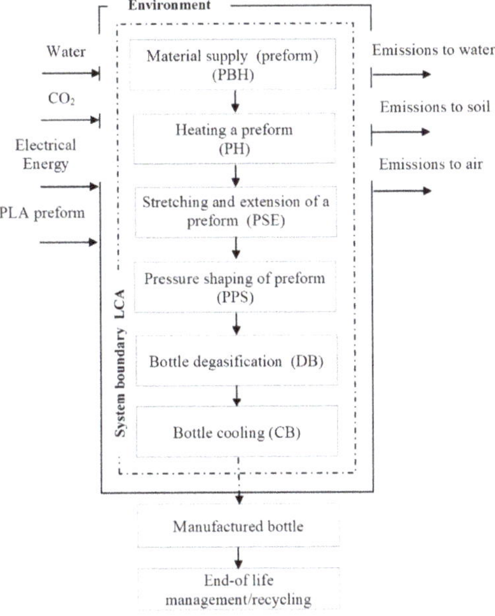

Figure 2. Boundaries of the accepted system [23].

3. Results

Taking into account the confidential character of the results presented in the study and company trade secrets, the values presented in Table 1 were changed by a coefficient ranging from 0.8 to 1.2 [20,24–26]. A tree of the processes depicting the flow of materials and energy during the whole life cycle of the analyzed object was developed. The tree, based on the above given assumptions of the beverage PLA bottle production process, is shown in Figure 3 while maintaining a 0.99% abscission level. This level covers all the process ranges including materials and media.

Table 1. Results LCI [23].

Technological Operations	
1. Material feed	
Mass of the one preform PLA	18.24 g
Total energy consumption by three motors of the carousel	0.368 kWh
2. Heating the preform	
Energy of infrared lamp (100 kW)	3.2 kWh
Energy of infrared lamps (200 kW)	6.4 kWh
Energy of supply chain	0.16 kWh
3. Stretching and extension of the preform	
Amount of electrical energy consumed	6.95 kWh
Amount of compressed air needed to blow the preform	0.0016 kg/m^3
4. Pressure shaping of the preforms	
Amount of electrical energy consumed	5.66 kWh
5. Degasification of the bottle	
Amount of electrical energy consumed	1.01 kWh
6. Bottle cooling	
Amount of electrical energy consumed	0.71 kWh
Water volume in a closed circulation	2.4 m^3

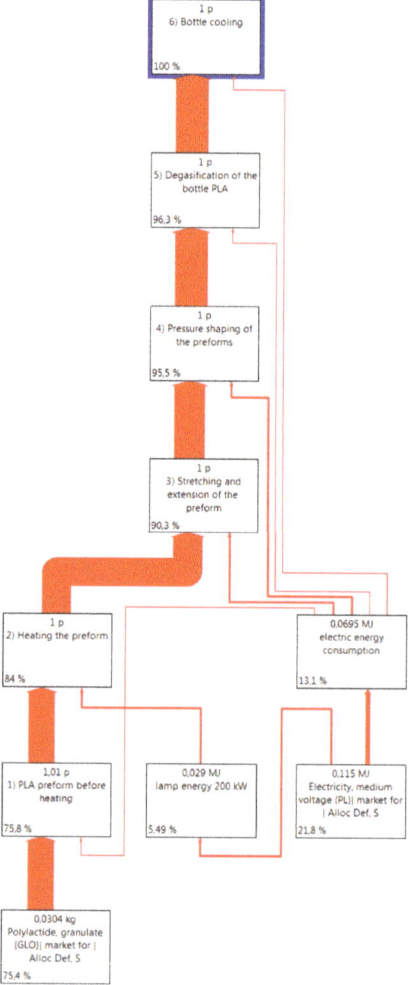

Figure 3. Tree depicts the processes involved in a PLA bottle shaping PLA.

4. Analysis of Test Results

The next stage of the analysis involved correct estimation of potential environmental impacts obtained at the stage of characterizing. The analysis results are included in Table 2. The results are presented in three basic units characteristic of the ReCiPe2016 method: Disability Adjusted Life Years (DALY)—life years of disability, Species Year (species.yr)—time-integrated loss of local species, Dollar ($)—additional costs connected with future extraction of fossil fuels and minerals [23–25,27–29].

Table 2. Results of characterizing environmental impacts of a bottle shaping process.

Impact Category	Unit	(CB)	(DB)	(PPS)	(PSE)	(PH)	(PBH)
Global warming, human health	DALY	4.5×10^{-10}	1.2×10^{-8}	8.9×10^{-9}	7.2×10^{-9}	1.2×10^{-9}	2.1×10^{-9}
Global warming. land ecosystems	species.yr	1.4×10^{-12}	3.5×10^{-11}	2.7×10^{-11}	2.2×10^{-11}	3.7×10^{-12}	6.3×10^{-12}
Global warming. fresh aquatic systems	species.yr	3.8×10^{-17}	9.5×10^{-16}	7.4×10^{-16}	5.9×10^{-16}	1.0×10^{-16}	1.7×10^{-16}
Startospheric ozone layer depletion	DALY	4.9×10^{-14}	1.3×10^{-12}	9.7×10^{-13}	7.9×10^{-13}	1.3×10^{-13}	3.8×10^{-13}
Ionizing radiation	DALY	6.1×10^{-15}	1.6×10^{-13}	1.2×10^{-13}	1.0×10^{-13}	1.6×10^{-14}	2.1×10^{-13}
Ozone formation. human health	DALY	7.4×10^{-13}	1.9×10^{-11}	1.4×10^{-11}	1.2×10^{-11}	2.0×10^{-12}	3.9×10^{-12}
Formation of fine particles	DALY	5.4×10^{-10}	1.2×10^{-8}	9.0×10^{-9}	7.4×10^{-9}	1.2×10^{-9}	2.2×10^{-9}
Formation of ozone land ecosystems	species.yr	1.1×10^{-13}	2.6×10^{-12}	2.1×10^{-12}	1.6×10^{-12}	2.8×10^{-13}	5.6×10^{-13}
Land acidification	species.yr	5.1×10^{-13}	1.3×10^{-11}	1.0×10^{-11}	8.1×10^{-12}	1.4×10^{-12}	2.1×10^{-12}
Eutrophication of fresh water	species.yr	3.5×10^{-14}	9.1×10^{-13}	6.9×10^{-13}	5.7×10^{-13}	9.5×10^{-14}	1.3×10^{-13}
Land ecotoxicity	species.yr	5.3×10^{-15}	1.4×10^{-13}	1.0×10^{-13}	8.4×10^{-14}	1.4×10^{-14}	4.8×10^{-14}
Fresh water ecotoxicity	species.yr	3.6×10^{-17}	9.0×10^{-16}	7.1×10^{-16}	5.6×10^{-16}	9.7×10^{-17}	1.0×10^{-15}
Marine water ecotoxicity	species.yr	2.8×10^{-17}	7.1×10^{-16}	5.5×10^{-16}	4.4×10^{-16}	7.5×10^{-17}	4.4×10^{-16}
Carcinogenic ecotoxicity for people	DALY	3.1×10^{-12}	7.3×10^{-11}	6.0×10^{-11}	4.5×10^{-11}	8.2×10^{-12}	2.6×10^{-10}
Non carcinogenic toxicity for people	DALY	3.6×10^{-12}	9.2×10^{-11}	7.0×10^{-11}	5.7×10^{-11}	9.6×10^{-12}	3.3×10^{-11}
Land use	species.yr	2.2×10^{-13}	5.8×10^{-12}	4.4×10^{-12}	3.6×10^{-12}	6.0×10^{-13}	8.9×10^{-13}
Deficiency of mineral resources	USD2013	5.1×10^{-08}	1.0×10^{-06}	8.5×10^{-07}	6.4×10^{-07}	1.2×10^{-07}	2.5×10^{-06}
Deficiency of fossil resources	USD2013	1.0×10^{-05}	1.4×10^{-04}	1.7×10^{-04}	1.0×10^{-04}	2.4×10^{-05}	1.7×10^{-05}
Use of water resources. human health	DALY	3.4×10^{-11}	8.7×10^{-10}	6.6×10^{-10}	5.4×10^{-10}	9.1×10^{-11}	6.8×10^{-09}
Use of water resources. land ecosystem	species.yr	2.1×10^{-13}	5.3×10^{-12}	4.0×10^{-12}	3.3×10^{-12}	5.5×10^{-13}	4.2×10^{-11}
Use of water resources. ecosystems	species.yr	9.2×10^{-18}	2.4×10^{-16}	1.8×10^{-16}	1.5×10^{-16}	2.5×10^{-17}	1.9×10^{-15}

Emission level with the highest potential environmental impact
Emission level with higher potential environmental impact
Emission level with high potential environmental impact
Emission level with low potential environmental impact
Emission level with lower potential environmental impact
Emission level with the potential lowest environmental impact

Greenhouse gases (GHG) that contribute to global warming [26], such as CO_2 and methane, are often released to the atmosphere by natural causes (that is volcanoes and breathing) of anthropogenic origin (that is, from chimneys, motor vehicles, stoves of apartment buildings, factories) [23]. The process of a properly shaped product degasification was reported to have the highest potential negative environmental impact on human health (1.16347×10^{-8} DALY), on land ecosystems (3.51106×10^{-11} species.yr) and fresh water ecosystems (9.5902×10^{-16} species.yr), with the material total life cycle contribution being about 78%, for all the three impact categories.

Carcinogenic toxicity for people and carcinogenic factors being non-carcinogenic or respiratory substances are the environmental effects. It has been proven that most carcinogenic substances PLA (2.59368×10^{-10} DALY) are released before the preform is supplied to a furnace for heating. However, the greatest potential effect on carcinogens for PET bottles was recorded for the entire production process with a value of (87.6%) [23]. The emissions of nickel and cadmium ions were found to have the most damaging effect in both of the considered processes. The highest potentially harmful impact was found for a shaped freshwater ecotoxicityPLA (3.0593×10^{-12} DALY) bottle during the process of cooling. This is caused by the fact that production of resin PET—oil derivative—is usually accompanied by a release of cancerogenic substances in the process of the material manufacturing (Figure S2, Table S2), (Figure S3, Table S3), (Figure S4, Table S4).

In the entire life cycle of beverage bottles, among all compounds in the analyzed category, the highest level of negative emissions to air is characterized by the radon isotope—^{222}Rn (Table 3, Figure 4). Most of this noble gas with high ionizing capacity is generated at the stage of cooling the 1.45×10^{-12} DALY bottle. The main source of emissions of this compound is the raw material and the amount of refrigerant used.

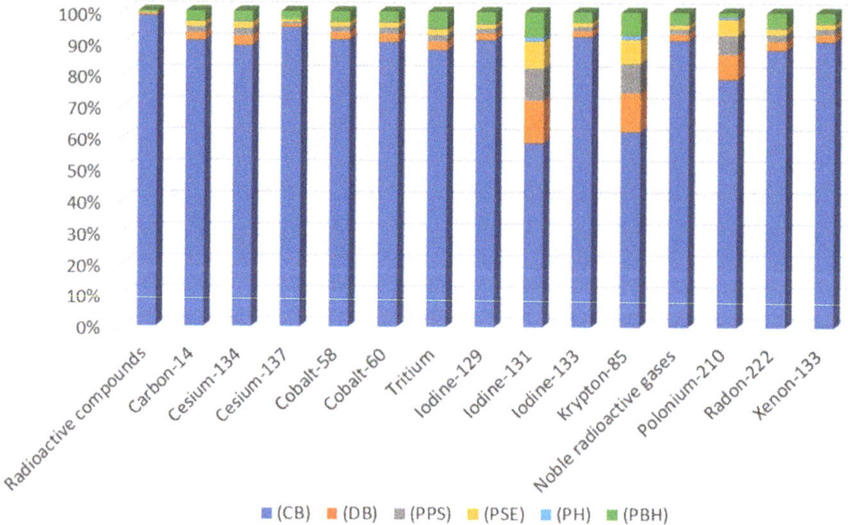

Figure 4. The results of characterization of the amount of ionizing radiation emissions into the air throughout the entire cycle of shaping bottles for beverages made of PLA, which is biodegradable in an industrial composting facility.

When analyzing the harmfulness of compounds emitted to water, the greatest pollution was recorded for Tritium at the stage of cooling the finished bottle (Table 4, Figure 5). The main source of environmental harm is the amount of refrigerant used and the raw material.

Table 3. Emission of ionizing radiation compounds into the air throughout the shaping cycle of PLA beverage bottles which is biodegradable in an industrial composting facility.

Substance	(CB)	(DB)	(PPS)	(PSE)	(PH)	(PBH)
Radioactive compounds	6.55×10^{-14}	2.69×10^{-16}	1.62×10^{-16}	1.66×10^{-16}	2.20×10^{-17}	2.95×10^{-16}
Carbon-14	3.50×10^{-12}	9.58×10^{-14}	7.17×10^{-14}	5.93×10^{-14}	9.78×10^{-15}	1.18×10^{-13}
Cesium-134	2.62×10^{-19}	9.24×10^{-21}	6.88×10^{-21}	5.72×10^{-21}	9.40×10^{-22}	9.23×10^{-21}
Cesium-137	9.91×10^{-15}	1.08×10^{-16}	7.70×10^{-17}	6.65×10^{-17}	1.05×10^{-17}	2.68×10^{-16}
Cobalt-58	2.59×10^{-20}	6.61×10^{-22}	4.89×10^{-22}	4.09×10^{-22}	6.68×10^{-32}	8.89×10^{-22}
Cobalt-60	7.57×10^{-18}	2.21×10^{-19}	1.64×10^{-19}	1.37×10^{-19}	2.24×10^{-20}	2.62×10^{-19}
Tritium	6.65×10^{-15}	2.09×10^{-16}	1.58×10^{-16}	1.30×10^{-16}	2.15×10^{-17}	3.96×10^{-16}
Iodine-129	1.02×10^{-14}	2.37×10^{-16}	1.75×10^{-16}	1.47×10^{-16}	2.39×10^{-17}	4.12×10^{-16}
Iodine-131	2.29×10^{-16}	5.34×10^{-17}	4.04×10^{-17}	3.31×10^{-17}	5.52×10^{-18}	3.10×10^{-17}
Iodine-133	4.22×10^{-20}	8.67×10^{-22}	6.38×10^{-22}	5.37×10^{-22}	8.71×10^{23}	1.43×10^{-21}
Krypton-85	3.02×10^{-18}	5.98×10^{-19}	4.53×10^{-19}	3.71×10^{-19}	6.18×10^{-20}	3.55×10^{-19}
Noble radioactive gases	7.39×10^{-14}	1.71×10^{-15}	1.26×10^{-15}	1.06×10^{-15}	1.73×10^{-16}	2.98×10^{-15}
Polonium-210	2.94×10^{-14}	3.01×10^{-15}	2.26×10^{-15}	1.86×10^{-15}	3.09×10^{-16}	4.86×10^{-16}
Radon-222	1.45×10^{-12}	4.81×10^{-14}	3.61×10^{-14}	2.98×10^{-14}	4.92×10^{-15}	7.71×10^{-14}
Xenon-133	2.75×10^{-16}	7.26×10^{-18}	5.38×10^{-18}	4.50×10^{-18}	7.35×10^{-19}	9.73×10^{-18}

Table 4. Emission of ionizing radiation compounds to water throughout the entire shaping cycle of PLA beverage bottles is biodegradable in an industrial composting facility.

Substance	(CB)	(DB)	(PPS)	(PSE)	(PH)	(PBH)
Iodine-131	8.21×10^{-16}	2.84×10^{-18}	1.69×10^{-18}	1.75×10^{-18}	2.31×10^{-19}	9.20×10^{-18}
Radioactive compounds	6.93×10^{-19}	1.60×10^{-20}	1.19×10^{-20}	9.92×10^{-21}	1.62×10^{-21}	2.79×10^{-20}
Antimony-124	6.88×10^{-15}	2.31×10^{-17}	1.37×10^{-17}	1.42×10^{-17}	1.86×10^{-18}	7.68×10^{-17}
Carbon-14	9.41×10^{-17}	2.92×10^{-19}	1.69×10^{-19}	1.80×10^{-19}	2.30×10^{-20}	1.01×10^{-18}
Tritium	3.52×10^{-14}	2.57×10^{-16}	1.77×10^{-16}	1.59×10^{-16}	2.41×10^{-17}	6.29×10^{-16}
Manganese-54	3.47×10^{-17}	1.06×10^{-18}	7.90×10^{-19}	6.58×10^{-19}	1.08×10^{-19}	1.28×10^{-18}
Strontium-90	1.60×10^{-14}	3.88×10^{-15}	2.94×10^{-15}	2.41×10^{-15}	4.02×10^{-16}	2.24×10^{-15}

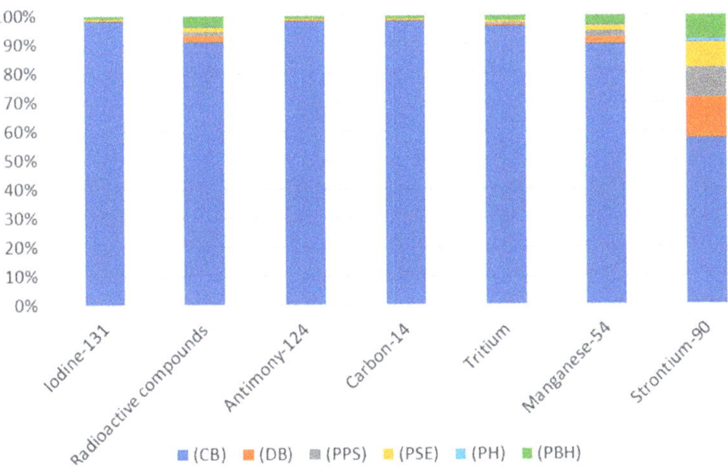

Figure 5. The results of characterizing the amount of ionizing radiation compounds emitted to water throughout the entire cycle of shaping bottles for beverages made of PLA which is biodegradable in an industrial composting facility.

The main sources of water pollution depend on the usability and innovation of the technological line, which has limited possibilities of reducing pollution as part of the production process. New installations have greater opportunities to adapt to environmental requirements through the use of technologies to prevent specific waste groups. The presented production process has a specific type of waste that can be determined knowing the characteristics of: construction materials, corrosion and erosion mechanisms, and consumables. Waste prevention includes: minimization of waste generation and recycling.

Non-carcinogenic toxicity for people had the highest potential impact for a bottle produced from PLA (9.21497×10^{-11} DALY). Emissions of non-cancerogenic compounds were found to be the highest at the stage of PLA bottle degasification, whereas the second highest value in terms of emission was the process of a bottle pressure shaping (7.04032×10^{-11} DALY) and the process of automatic stretching and extension of a pre heated preform PLA (5.71376×10^{-11} DALY). The total share of the PLA material in the process of technological shaping of a PLA bottle was merely (1.76×10^{-9} DALY); it exhibits almost four times lower potential environmental impact than a bottle produced from a non-biodegradable material. The source of electrical energy is largely responsible for emissions of non-cancerogenic substances and the amount of their emission gradually grows along with the progress of the production process. Degasification of a shaped PLA bottle is very fast and consumes a large amount of electric energy in a very short time, which leads to an increase in potential impacts on the noncarcinogenic substances.

Ultraviolet radiation can destroy organic matter [23]. Some compounds decompose the ozone layer, particularly derivatives of methane and ethane containing atoms of chlorine, chlorine and fluorine and bromine and fluorine [23]. In the case of PLA bottle shaping, the value (1.61×10^{-13} DALY) was reported while for the PET bottle shaping process a lower value was recorded (1.66×10^{-11} DALY) [23]. Similar values were found for unit processes for which the highest negative impact was reported for the bottle degasification process, whereas the lowest for the category of bottle cooling. The basic stage responsible for high value is the penultimate stage of bottle production, characterized by values of 97.81% of the impact for PLA bottles and 80.58% for a regular PET bottle [23].

Ecotoxicity of the aquatic and land environments is caused by poisonous toxic substances which are released to the natural environment [23]. Marine ecotoxicity has potentially the highest negative impact on the aquatic environment (Figure S5, Table S5) The highest similar emission levels were recorded for the degassing process of the shaped PLA bottle (7.1106×10^{-16} species.yr) and PET (6.97002×10^{-16} species.yr). Similar emission values were also found for both processes in which the PET bottle shaping process contributed to the ecotoxicity of the aquatic environment in 92.65%. While PLA bottles in 90.75% (Figure 4, Table 4) [23].

Freshwater ecotoxicity was featured by the highest potential value of emission defined for the process of PLA perform heating (1.04682×10^{-15} species.yr), second in terms of the emission of negative compounds to the aquatic environment was the same unit process describing the PET preform (1.01331×10^{-15} species.yr) [23].

The terrestrial ecotoxicity was highest affected by the PET bottle production process (4.95×10^{-12} species.yr) [23]. Among the six unit processes analyzed, the smallest negative impact of the shaped bottle cooling process was recorded for the PLA bottle (5.31222×10^{-15} species.yr). Comparing two bottle formation processes, it can be stated that the highest level of negative impact on terrestrial ecotoxicity was characterized by the raw material used for the PET bottle shaping process, as much as 92.97% [23], while the raw material used for the PLA bottle shaping process affected terrestrial ecotoxicity in 88.26% (Figure S6, Table S6).

Land environment acidification is caused by a decrease in the pH value. This phenomenon occurs as a result of disturbing the ecological balance of energy and matter exchange processes between elements of ecosystems. The reason for such changes is the presence of chemical substances. Regular land acidification occurs when gases such as CO_2 or SO_2, are absorbed by water and react to form acidic compounds on the surface of the earth [26–30]. The magnitude of the impact of the process of shaping bottles from biodegradable material (1.46×10^{-10} species.yr) was greater than that of shaping PET bottles (1.14×10^{-10} species.yr). The impact of individual raw materials in the bottle shaping process was 76.00% for PLA and 72.32% for PET, respectively. The smallest negative impact in the entire shaping process belonged to the cooling process of the finished product, while the largest degassing process of the bottle [23].

Land use was definitely higher for the production of corn-based bottles. This is confirmed by the fact that in the whole process of shaping the bottle, as much as 95.35% of the revenues of all categories is the raw material used for production. Characterizing the entire process of a PLA bottle shaping it needs to be noted that the highest potential influence on the land use was found for the process of a ready bottle degasification (1.73%), slightly lower influence (1.32%) was found for the process of the perform pressure shaping, nearly 1% impact was observed for the process of the preform stretching and extension in a mold, whereas, below 1%—for the preform processes applied before heating and cooling of a ready product.

The contribution of polylactide. being as much as 97.87% was found for the water eutrophication. This phenomenon is well, understood as cultivation of corn requires application of fertilizers which adversely affect the natural environment.

Ionizing radiation is a phenomenon which has always been present in the environment [31]. Its main source is radon emitted from the earth's crust [32]. The ubiquity of radioactive elements in nature makes humans a source of radiation as well [33]. The highest potentially negative environmental impact was involved in one of the sub-processes and depended on the amount of a semi product—PLA perform used in the process. The value of emissions was found to be (2.11×10^{-13} DALY). The PET preform 2.00518×10^{-13} DALY showed a slightly lower negative value. Relatively similar values were recorded for two raw materials used in the process of shaping PET bottles (88%) [23] and PLA (89%).

Particulate matter is generated during the process of bottle shaping by engines of the blowing machine. Fine particles referred to as PM (particulate matter), include those components of the bottle production process whose state when leaving the engine is different from gaseous. The most harmful substances accompanying formation of fine particles are PAH—polycyclic aromatic hydrocarbons [30]. Hence, based on the analysis results. the highest negative impact of the bottle shaping process on human health was found for the process of a PET bottle creation (1.17629×10^{-08} DALY) at the stage of degasification of a ready product. The highest value of negative emissions of particulate matter on human health was reported for the process of a PET bottle cooling (5.44468×10^{-10} DALY).

Water use has an influence on human health and the quality of land and aquatic ecosystems. Water is of key importance for the global industry. A bottle production plant uses significant amounts of raw material. that is, corn starch. The production material has an impact on the natural environment. For each of the three analyzed impact categories it was the process of PLA bottle formation that caused potentially the highest damage to human health and the quality of aquatic and land ecosystems (6.8291×10^{-9} DALY; 4.15283×10^{-11} species.yr; 1.85801×10^{-15} species.yr).

The PET bottle had the greatest potential impact on the mineral and fossil resource category. The process of shaping it showed the highest potential level of unfavorable influences in the process of collecting preforms for the heating furnace (2.71165×10^{-6} USD2013) for the category of mineral resource deficiency (Figure S7, Table S7). With the growing global demand for mineral resources, it is important to analyze whether the resources of geologically and technically available minerals in the earth's crust can meet the future needs of humanity. Widespread recycling. increasing material efficiency and demand management will certainly play an important role in satisfying future generations. A significant high value of potential inflows was recorded for the category of scarcity of fossil resources. The bottle with the greatest impact on this category was the PET bottle (8.58689×10^{-5} USD2013) [23] in the process of stretching and lengthening the preform, the process of collecting preforms for the heating furnace (7.00854×10^{-5} USD2013) was responsible for slightly less negative impact. This phenomenon is related to the fact that PET requires continuous extraction of fossil fuels. resulting in their depletion. This behavior confirmed the highest stage impact, which is bottle production, and therefore probably the PET bottle had the greatest impact in the category of mineral extraction. Bottle production had a slightly smaller impact in this category, probably due to the fact that extraction of resources is also necessary.

Out of all the discussed categories, those that were chosen were those whose summary life time impact involved in a biodegradable bottle shaping was 90%. Selected results of the characterizing of the environmental impacts for each stage of the bottle shaping process in terms of human health is presented in Figure 6.

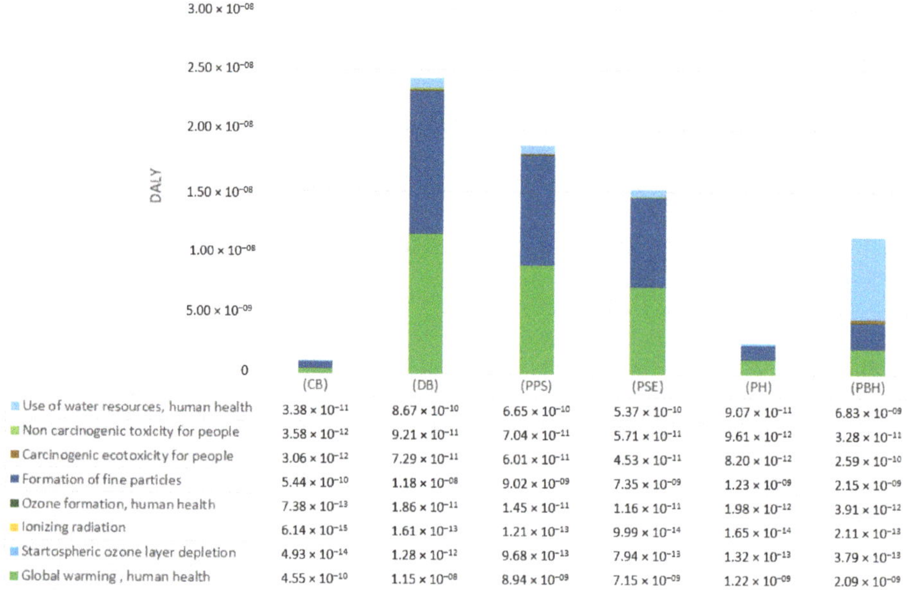

	(CB)	(DB)	(PPS)	(PSE)	(PH)	(PBH)
Use of water resources, human health	3.38×10^{-11}	8.67×10^{-10}	6.65×10^{-10}	5.37×10^{-10}	9.07×10^{-11}	6.83×10^{-09}
Non carcinogenic toxicity for people	3.58×10^{-12}	9.21×10^{-11}	7.04×10^{-11}	5.71×10^{-11}	9.61×10^{-12}	3.28×10^{-11}
Carcinogenic ecotoxicity for people	3.06×10^{-12}	7.29×10^{-11}	6.01×10^{-11}	4.53×10^{-11}	8.20×10^{-12}	2.59×10^{-10}
Formation of fine particles	5.44×10^{-10}	1.18×10^{-08}	9.02×10^{-09}	7.35×10^{-09}	1.23×10^{-09}	2.15×10^{-09}
Ozone formation, human health	7.38×10^{-13}	1.86×10^{-11}	1.45×10^{-11}	1.16×10^{-11}	1.98×10^{-12}	3.91×10^{-12}
Ionizing radiation	6.14×10^{-15}	1.61×10^{-13}	1.21×10^{-13}	9.99×10^{-14}	1.65×10^{-14}	2.11×10^{-13}
Startospheric ozone layer depletion	4.93×10^{-14}	1.28×10^{-12}	9.68×10^{-13}	7.94×10^{-13}	1.32×10^{-13}	3.79×10^{-13}
Global warming , human health	4.55×10^{-10}	1.15×10^{-08}	8.94×10^{-09}	7.15×10^{-09}	1.22×10^{-09}	2.09×10^{-09}

Figure 6. Results of characterization of environmental impacts for 8 impact categories in terms of human health.

The highest levels of negative impact on human health were found for: global warming (1.27×10^{-7} DALY), particulate matter formation (1.61×10^{-7} DALY), and use of water resources (2.93×10^{-8} DALY), which is caused by the use of a material—polylactide—as well as by emitting compounds to the natural environment during the process of bottle shaping.

The highest negative impact on human health was on the part of the global warming category emissions (Table 5, Figure 7) caused by emission of the compounds carbon dioxide (7.68×10^{-8} DALY) and nitric oxides (7.48×10^{-9} DALY) at the stage of end product cooling, which is connected with the process of bottle shaping and its given parameters.

Table 5. Results of characterization of environmental impacts for the category of global warming, (DALY).

Substance	(CB)	(DB)	(PPS)	(PSE)	(PH)	(PBH)
CO_2	7.68×10^{-8}	1.05×10^{-8}	8.14×10^{-9}	6.54×10^{-9}	1.11×10^{-9}	1.88×10^{-9}
Chloroform	3.72×10^{-14}	1.57×10^{-16}	1.23×10^{-16}	9.78×10^{-17}	1.68×10^{-17}	3.36×10^{-16}
Nitric oxide	7.48×10^{-09}	5.57×10^{-11}	4.2×10^{-11}	3.45×10^{-11}	5.74×10^{-12}	1.5×10^{-11}
Chlorinated hydrocarbons	3.72×10^{-14}	1.57×10^{-16}	1.23×10^{-16}	9.78×10^{-17}	1.68×10^{-17}	3.36×10^{-16}
Methane	7×10^{-15}	4.07×10^{-17}	3.61×10^{-17}	2.53×10^{-17}	4.92×10^{-18}	5.33×10^{-17}
Nitrogen fluoride	9.1×10^{-17}	9.28×10^{-19}	2.53×10^{-19}	5.64×10^{-19}	3.43×10^{-20}	5.01×10^{-19}
Sulfur hexafluoride	2.65×10^{-10}	3.56×10^{-11}	2.67×10^{-11}	2.2×10^{-11}	3.65×10^{-12}	7.3×10^{-12}

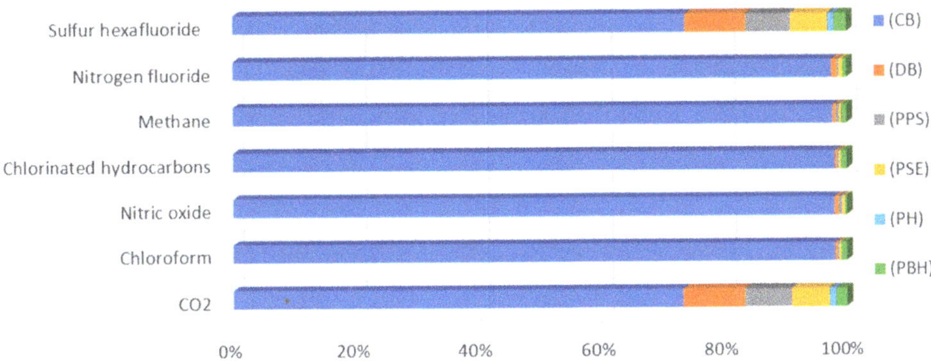

Figure 7. Results of characterization of environmental impacts for the category of global warming, (DALY).

The most detrimental effect on human health involved in the entire bottle shaping life cycle was found for the groups of particulate matter (Table 6, Figure 8), of which the largest impacts were caused by formation of fine particles < 2.5 um (7.13×10^{-8} DALY) and emission of sulfur dioxide (5.75×10^{-8} DALY) at the stage of the end product cooling. The least potentially harmful impact on human health and the environment was observed for emission of SO_3 (5.32×10^{-18} DALY) at the stage of the preform heating.

Table 6. Results of characterization of environmental impacts on category of particulate matter formation (DALY).

Substance	(CB)	(DB)	(PPS)	(PSE)	(PH)	(PBH)
Particulate matter formation < 2.5um	7.13×10^{-8}	2.1×10^{-9}	1.56×10^{-9}	1.3×10^{-9}	2.14×10^{-10}	6.13×10^{-10}
Sulfur dioxide	5.75×10^{-8}	9.76×10^{-9}	7.46×10^{-9}	6.05×10^{-9}	1.02×10^{-9}	1.54×10^{-9}
Sulfur oxides	3.08×10^{-13}	3.68×10^{-15}	2.67×10^{-15}	2.28×10^{-15}	3.64×10^{-16}	5.78×10^{-15}
Sulfur trioxide	2.1×10^{-13}	5.15×10^{-17}	3.9×10^{-17}	3.19×10^{-17}	5.32×10^{-18}	3.27×10^{-17}

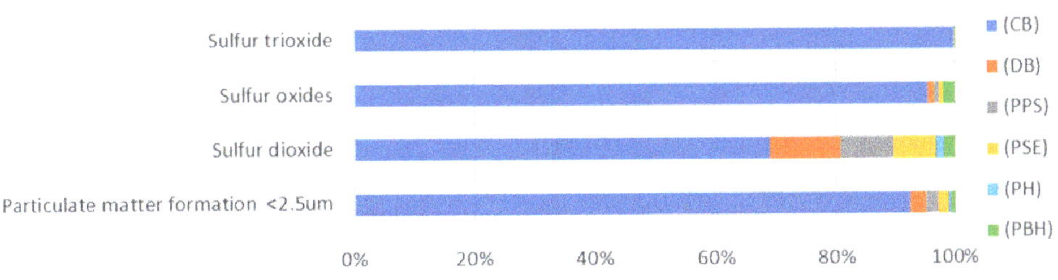

Figure 8. Results of environmental impact characterization for the category of particulate matter formation (DALY).

The results of environmental impact characterization for the category of water consumption, which is a medium (Table 7, Figure 9) used for the process of bottle production, show that the highest positive impact was found for the water used at the stage of a ready bottle degasification and cooling. This is due to the fact that the water used in the production process circulates in a closed system and its amount and control frequency is adjusted to the production size and volume and other respective parameters.

Table 7. Results of environmental impact characterization for the category of use of water resources (DALY).

Substance	(CB)	(DB)	(PPS)	(PSE)	(PH)	(PBH)
Water PL	-1.9×10^{-9}	-2.3×10^{-8}	-1.7×10^{-8}	-1.4×10^{-8}	-2.4×10^{-9}	-2.1×10^{-9}
Water in lakes PL	2.98×10^{-19}	3.16×10^{-18}	2.39×10^{-18}	1.96×10^{-18}	3.26×10^{-19}	2.88×10^{-19}
Water in rivers PL	6.85×10^{-16}	7.26×10^{-15}	5.49×10^{-15}	4.5×10^{-15}	7.5×10^{-16}	6.62×10^{-16}

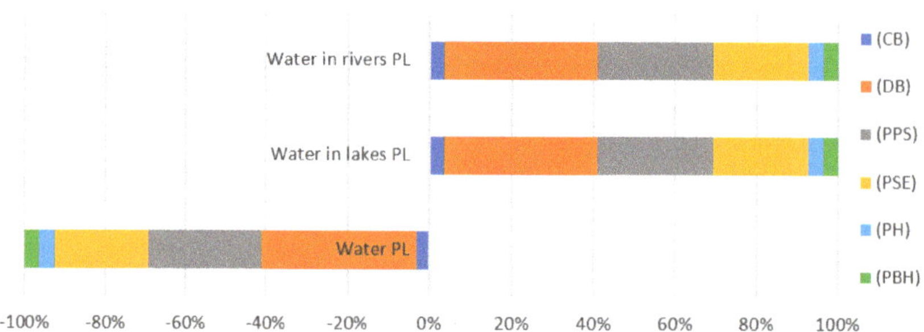

Figure 9. Results of environmental impact characterization for the category of water use (DALY).

Certain impact categories were particularly harmful to the natural environment throughout the life cycle of the bottle shaping process (Figure 10), i.e., global warming (1.04×10^{-14} species.yr), land acidification (1.47×10^{-10} species.yr), and land use (3.31×10^{-10} species.yr) contributing to climate changes. The lowest negative impacts were observed for the remaining categories, including land and marine toxicity, ozone layer creation, fresh water eutrophication and global warming affecting the fresh water ecosystems (Figure S8, Table S8), (Figure S9, Table S9), (Figure S10, Table S10), (Figure S11, Table S11), (Figure S12, Table S12).

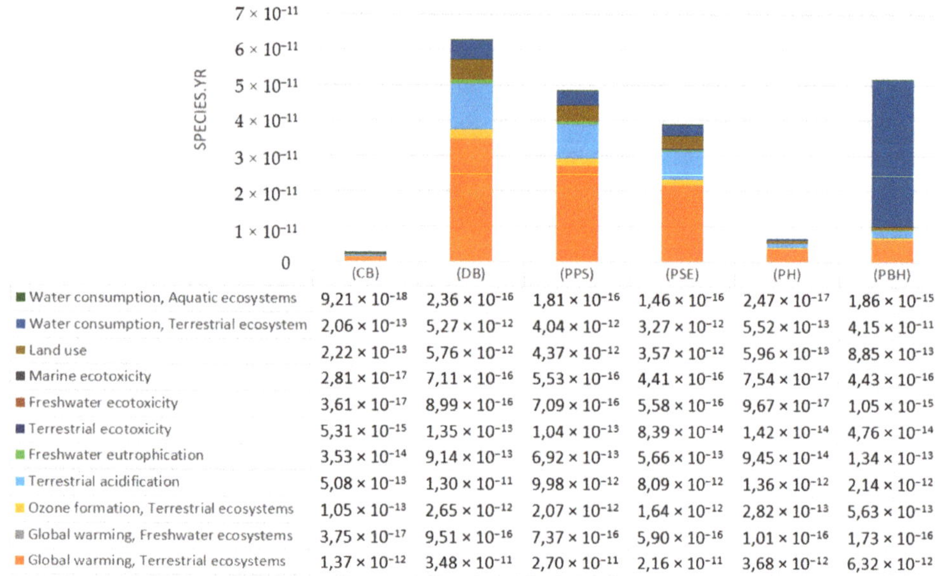

	(CB)	(DB)	(PPS)	(PSE)	(PH)	(PBH)
Water consumption, Aquatic ecosystems	$9,21 \times 10^{-18}$	$2,36 \times 10^{-16}$	$1,81 \times 10^{-16}$	$1,46 \times 10^{-16}$	$2,47 \times 10^{-17}$	$1,86 \times 10^{-15}$
Water consumption, Terrestrial ecosystem	$2,06 \times 10^{-13}$	$5,27 \times 10^{-12}$	$4,04 \times 10^{-12}$	$3,27 \times 10^{-12}$	$5,52 \times 10^{-13}$	$4,15 \times 10^{-11}$
Land use	$2,22 \times 10^{-13}$	$5,76 \times 10^{-12}$	$4,37 \times 10^{-12}$	$3,57 \times 10^{-12}$	$5,96 \times 10^{-13}$	$8,85 \times 10^{-13}$
Marine ecotoxicity	$2,81 \times 10^{-17}$	$7,11 \times 10^{-16}$	$5,53 \times 10^{-16}$	$4,41 \times 10^{-16}$	$7,54 \times 10^{-17}$	$4,43 \times 10^{-16}$
Freshwater ecotoxicity	$3,61 \times 10^{-17}$	$8,99 \times 10^{-16}$	$7,09 \times 10^{-16}$	$5,58 \times 10^{-16}$	$9,67 \times 10^{-17}$	$1,05 \times 10^{-15}$
Terrestrial ecotoxicity	$5,31 \times 10^{-15}$	$1,35 \times 10^{-13}$	$1,04 \times 10^{-13}$	$8,39 \times 10^{-14}$	$1,42 \times 10^{-14}$	$4,76 \times 10^{-14}$
Freshwater eutrophication	$3,53 \times 10^{-14}$	$9,14 \times 10^{-13}$	$6,92 \times 10^{-13}$	$5,66 \times 10^{-13}$	$9,45 \times 10^{-14}$	$1,34 \times 10^{-13}$
Terrestrial acidification	$5,08 \times 10^{-13}$	$1,30 \times 10^{-11}$	$9,98 \times 10^{-12}$	$8,09 \times 10^{-12}$	$1,36 \times 10^{-12}$	$2,14 \times 10^{-12}$
Ozone formation, Terrestrial ecosystems	$1,05 \times 10^{-13}$	$2,65 \times 10^{-12}$	$2,07 \times 10^{-12}$	$1,64 \times 10^{-12}$	$2,82 \times 10^{-13}$	$5,63 \times 10^{-13}$
Global warming, Freshwater ecosystems	$3,75 \times 10^{-17}$	$9,51 \times 10^{-16}$	$7,37 \times 10^{-16}$	$5,90 \times 10^{-16}$	$1,01 \times 10^{-16}$	$1,73 \times 10^{-16}$
Global warming, Terrestrial ecosystems	$1,37 \times 10^{-12}$	$3,48 \times 10^{-11}$	$2,70 \times 10^{-11}$	$2,16 \times 10^{-11}$	$3,68 \times 10^{-12}$	$6,32 \times 10^{-12}$

Figure 10. Results of environmental impact characterization for 11 categories of ecosystem shaping.

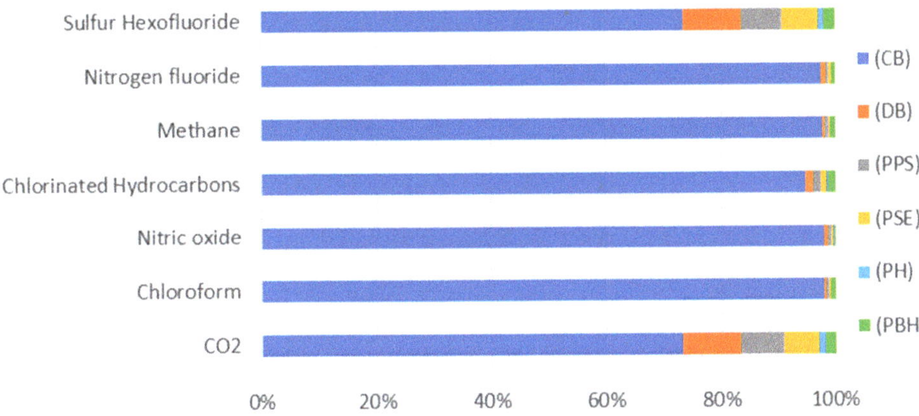

Figure 11. Results of environmental impact characterization for the category of global warming (species.yr).

In the case of the first category (global warming), featuring potentially the highest negative impact on the ecosystem (Table 8, Figure 11), just like in the category of human health, the same groups of components, although with different emission values, were observed: CO_2 (2.3162×10^{-10} species.yr), nitric oxide (2.2508×10^{-11} species.yr). A comparison of the same compounds but different damage categories shows that both CO_2 and NO_x featured higher negative effect on human health in the process of polylactide bottle shaping.

Table 8. Results of environmental impacts characterization for the category of global warming, land ecosystems (species.yr).

Substance	(CB)	(DB)	(PPS)	(PSE)	(PH)	(PBH)
CO_2	2.3162×10^{-10}	3.1827×10^{-11}	2.455×10^{-11}	1.9735×10^{-11}	3.3525×10^{-12}	5.6821×10^{-12}
Chloroform	1.1198×10^{-16}	4.7415×10^{-19}	3.7126×10^{-19}	2.9447×10^{-19}	5.0587×10^{-20}	1.0102×10^{-18}
Nitric oxide	2.2508×10^{-11}	1.6776×10^{-13}	1.266×10^{-13}	1.0395×10^{-13}	1.7281×10^{-14}	4.5046×10^{-14}
Chlorinated Hydrocarbons	5.5791×10^{-16}	9.1828×10^{-18}	7.561×10^{-18}	5.7239×10^{-18}	1.0271×10^{-18}	7.9708×10^{-18}
Methane	2.1078×10^{-17}	1.2247×10^{-19}	1.0875×10^{-19}	7.6347×10^{-20}	1.4835×10^{-20}	1.6051×10^{-19}
Nitrogen fluoride	2.7452×10^{-19}	2.8002×10^{-22}	7.6257×10^{-22}	1.702×10^{-21}	1.0352×10^{-22}	1.5108×10^{-21}
Sulfur Hexofluoride	7.9944×10^{-13}	1.073×10^{-13}	8.0621×10^{-14}	6.6478×10^{-14}	1.1006×10^{-14}	2.2029×10^{-14}

The stage of a ready bottle cooling (Table 9, Figure 12) featured the highest summary emissions of compounds into the atmosphere. At this stage, the highest levels of emissions were observed for sulfur dioxide (6.7×10^{-11} species.yr), sulfur oxides (3.59×10^{-16} species.yr), sulfuric acid (2.88×10^{-11} species.yr) and nitric oxide (1.6×10^{-11} species.yr), which are generated as a result of electric energy use in the process of a bottle manufacturing. At this stage. each reduction of the device working parameters to decrease electric energy consumption, contributes to atmospheric emission reduction.

Table 9. Results of characterization of environmental impacts for the land acidification category (species.yr).

Substance	(CB)	(DB)	(PPS)	(PSE)	(PH)	(PBH)
Ammonia	2.88×10^{-11}	1.18×10^{-13}	8.96×10^{-14}	7.33×10^{-14}	1.22×10^{-14}	2.63×10^{-14}
Nitrate	1.65×10^{-16}	1.43×10^{-18}	1.2×10^{-18}	8.92×10^{-19}	1.64×10^{-19}	2.18×10^{-18}
Nitric oxides	1.6×10^{-11}	1.56×10^{-12}	1.2×10^{-12}	9.65×10^{-13}	1.64×10^{-13}	3.2×10^{-13}
Sulfur dioxide	6.7×10^{-11}	1.14×10^{-11}	8.69×10^{-12}	7.05×10^{-12}	1.19×10^{-12}	1.79×10^{-12}
Sulfur oxides	3.59×10^{-16}	4.28×10^{-18}	3.11×10^{-18}	2.65×10^{-18}	4.24×10^{-19}	6.73×10^{-18}
Sulfur trioxide	2.46×10^{-16}	6.04×10^{-20}	4.57×10^{-20}	3.74×10^{-20}	6.23×10^{-21}	3.83×10^{-20}
Sulfuric acid	2.88×10^{-11}	1.18×10^{-13}	8.96×10^{-14}	7.33×10^{-13}	1.22×10^{-14}	2.63×10^{-14}

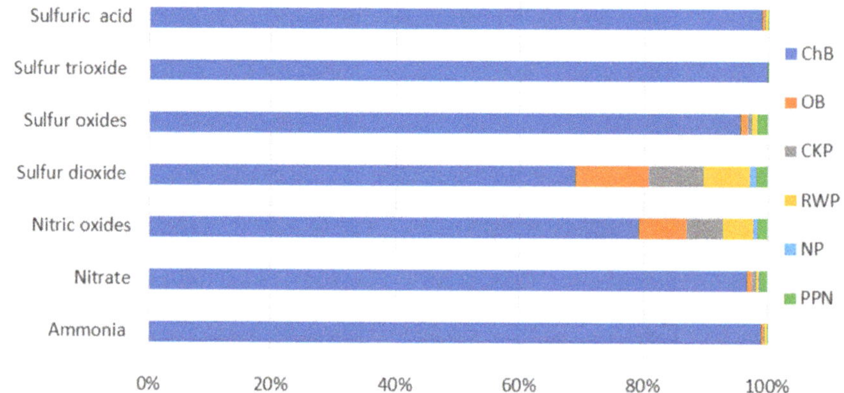

Figure 12. Results of environmental impact characterization for the category of land acidification (species.yr).

The results of characterizing of environmental impacts of the processes connected with land use involved in a biodegradable bottle production process life cycle are presented in Table 10 and Figure 13. Of all the considered processes of the analyzed impact category. the highest level of negative environmental impact was reported for the process of agricultural cultivation whose value at the last stage was (2.64×10^{-10} species.yr). Such a big difference in particular levels of emission is caused by increased use of land for cultivation of corn for the production of corn starch.

Table 10. Results of environmental impact characterizing for the category of land use (species.yr).

Substance	(CB)	(DB)	(PPS)	(PSE)	(PH)	(PBH)
Annual crops	2.64×10^{-10}	6.94×10^{-15}	5.26×10^{-15}	4.3×10^{-15}	7.18×10^{-16}	2.13×10^{-15}
Forest	5.93×10^{-14}	1.77×10^{-15}	1.18×10^{-15}	1.09×10^{-15}	1.61×10^{-16}	1.85×10^{-15}
Industrial areas	8.82×10^{-12}	5.71×10^{-14}	4.37×10^{-14}	3.54×10^{-14}	5.97×10^{-15}	3.61×10^{-14}
Places of mineral extraction	5.14×10^{-16}	2.94×10^{-18}	2.56×10^{-18}	1.83×10^{-18}	3.5×10^{-19}	1.41×10^{-18}
Graze lands	1.89×10^{-16}	2.59×10^{-18}	2.12×10^{-18}	1.62×10^{-18}	2.87×10^{-19}	1.57×10^{-18}
Forest transformations	7.94×10^{-13}	1.27×10^{-14}	9.61×10^{-15}	7.89×10^{-15}	1.31×10^{-15}	3.47×10^{-15}
Green land Transformations (unused)	2.35×10^{-15}	1.98×10^{-17}	1.5×10^{-17}	1.23×10^{-17}	2.04×10^{-18}	1.12×10^{-17}

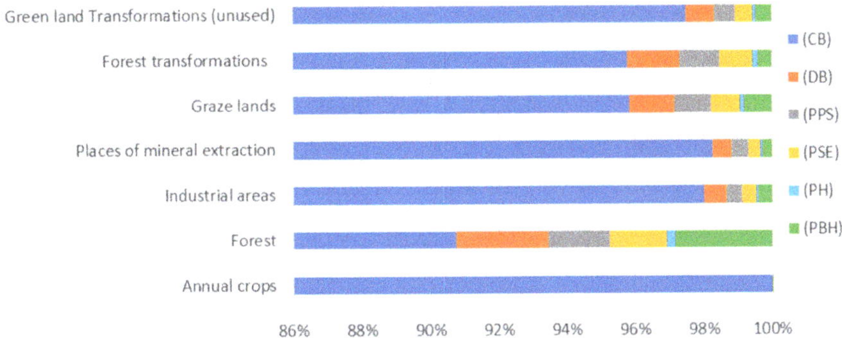

Figure 13. Results of environmental impact characterization for the category of land use (species.yr).

The last step of the analysis was to estimate the environmental impacts associated with exhaustion of fossil fuels (Table 11, Figure 14). According to the analysis results. the highest level of harmful environmental impact was found for the process of coal and natural gas extraction. at each stage of a bottle production. Emissions involved in the processes of oil extraction were observed to be positive at the stage of stretching of the preform and during degasification of the product.

Table 11. Results of characterization of environmental impacts for the category of fossil fuel deficiency (USD2013).

Substance	(CB)	(DB)	(PPS)	(PSE)	(PH)	(PBH)
Coal	0.000647	0.000143	0.000108	8.87×10^{-5}	1.48×10^{-5}	2.16×10^{-5}
Oil	0.001924	-2.5×10^{-5}	2.9×10^{-5}	-1.4×10^{-5}	3.95×10^{-6}	2.47×10^{-5}
Natural gas	0.0041	1.72×10^{-5}	3.6×10^{-5}	1.13×10^{-5}	4.92×10^{-6}	3.12×10^{-5}

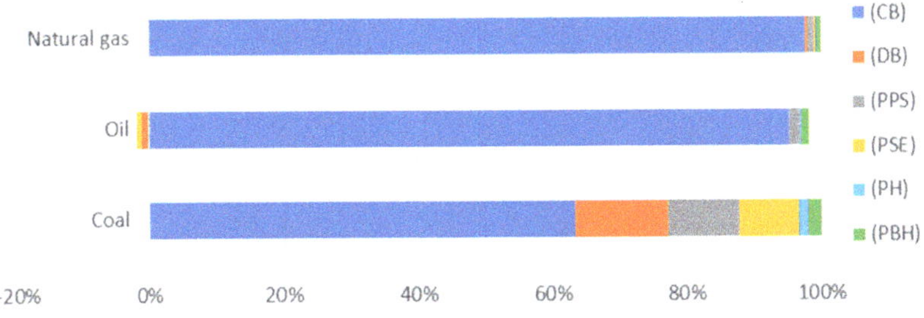

Figure 14. Results of environmental impact characterization for the category of fossil fuel depletion (USD2013).

5. Conclusions

The results presented in this study clearly indicate the emission of certain groups of components, that is, carbon dioxide. nitric oxides and sulfur oxides into the natural environment. This is caused by the use of raw materials, electrical energy and water throughout the life cycle of polylactide bottle shaping.

The results of this study show that the processes of end product degasification and cooling (in terms of used materials and electrical energy consumption) have the highest contribution in emissions of CO_2, nitric oxides, sulfur oxides and formation of fine particles of below 2.5 um.

In view of the obtained results, it needs to be said that in order to improve the natural environment, it is necessary to use methods for reduction of the perform mass as early as at the stage of the bottle shaping. The consumption of electrical energy should be reduced as well.

In result of the research it was assumed that for further analysis of the environmental impacts it is necessary to carry out an analysis of the quality of point data and uncertainty of results.

Supplementary Materials: The following are available online at http://www.mdpi.com/2073-4360/12/2/388/s1. Figure S1: Results of the characterization of environmental consequences for the category of scarcity of fossil resources obtained as a result of the PLA bottle shaping process (own study). Figure S2: Characterization results of environmental consequences for the category of ionizing radiation obtained as a result of the PLA bottle shaping process (own study). Figure S3: The results of the characterization of the environmental consequences for the carcinogenic toxicity category in humans obtained as a result of the PLA bottle shaping process (own study). Figure S4: Characterization results of environmental consequences for the category of non-creative toxicity to humans obtained as a result of the PLA bottle shaping process (own study). Figure S5: Characterization results of environmental consequences for the category of marine ecotoxicity obtained as a result of the PLA bottle shaping process (own study). Figure S6: The results of the characterization of the environmental consequences for the terrestrial ecotoxicity category obtained as a result of the PLA bottle shaping process (own study). Figure S7: The results of the characterization of environmental consequences for the category of mineral resource deficiency obtained as a result of the PLA bottle shaping process (own study). Figure S8: Results of the characterization of environmental consequences for the category of ozone formation affecting drow people obtained as a result of the process of shaping PLA bottles (own study). Figure S9: Results of characterization of environmental consequences for the ozone formation category affecting the ecosystem obtained as a result of the PLA bottle shaping process (own study). Figure S10: Results of the characterization of environmental consequences for the ozone layer depletion category obtained as a result of the PLA bottle shaping process (own study). Figure S11: Results of the characterization of the environmental consequences for the category of fresh water eutrophication obtained as a result of the PLA bottle shaping process (own study). Figure S12: Results of the characterization of the environmental consequences for the freshwater ecotoxicity category obtained as a result of the PLA bottle shaping process (own study). Table S1: Results of the characterization of environmental consequences for the category of scarcity of fossil resources obtained as a result of the PLA bottle shaping process (own study). Table S2: Characterization results of environmental consequences for the category of ionizing radiation obtained as a result of the PLA bottle shaping process (own study). Table S3: The results of the characterization of the environmental consequences for the carcinogenic toxicity category in humans obtained as a result of the PLA bottle shaping process (own study). Table S4: Characterization results of environmental consequences for the category of non-creative toxicity to humans obtained as a result of the PLA bottle shaping process (own study). Table S5: Characterization results of environmental consequences for the category of marine ecotoxicity obtained as a result of the PLA bottle shaping process (own study). Table S6: The results of the characterization of the environmental consequences for the terrestrial ecotoxicity category obtained as a result of the PLA bottle shaping process (own study). Table S7: The results of the characterization of environmental consequences for the category of mineral resource deficiency obtained as a result of the PLA bottle shaping process (own study). Table S8: Results of the characterization of environmental consequences for the category of ozone formation affecting drow people obtained as a result of the process of shaping PLA bottles (own study). Table S9: Results of characterization of environmental consequences for the ozone formation category affecting the ecosystem obtained as a result of the PLA bottle shaping process (own study). Table S10: Results of the characterization of environmental consequences for the ozone layer depletion category obtained as a result of the PLA bottle shaping process (own study). Table S11: Results of the characterization of the environmental consequences for the category of fresh water eutrophication obtained as a result of the PLA bottle shaping process (own study). Table S12: Results of the characterization of the environmental consequences for the freshwater ecotoxicity category obtained as a result of the PLA bottle shaping process (own study).

Author Contributions: Conceptualization, P.B.-W., Z.K.; methodology, Z.K., P.B.-W., K.M. and K.P.; software, K.P., P.B.-W., W.K.; validation, W.K., A.T., J.F., K.M. and R.K.; formal analysis, P.B.-W., R.K., Z.K., J.F., A.T. and W.K.; investigation, A.T., R.K. and P.B.-W.; resources, A.T., R.K., P.B.-W. and W.K.; data curation, A.T., J.F., P.B.-W. and Z.K.; writing—original draft preparation, P.B.-W., W.K., K.P. and R.K.; writing—review and editing, P.B.-W., W.K., A.T. and R.K.; visualization, W.K., R.K., A.T. and R.K.; supervision, Z.K., R.K., K.M., J.F., A.T. and K.P.; project administration, K.M. and P.B.-W. All authors have read and agreed to the published version of the manuscript.

Funding: This research received no external funding.

Conflicts of Interest: The authors declare no conflict of interest.

References

1. Castro-Aguirre, E.; Iñiguez-Franco, F.; Samsudinb, H.; Fang, X.; Auras, R. Poly(lactic acid)—Mass production. processing, industrial applications, and end of life. *Adv. Drug Deliv. Rev.* **2016**, *107*, 333–366. [CrossRef]
2. Murariu, M.; Dubois, P. PLA composites: From production to properties. *Adv. Drug Deliv. Rev.* **2016**, *107*, 17–46. [CrossRef]
3. Gere, D.; Czigany, T. Future trends of plastic bottle recycling: Compatibilization of PET and PLA. *Polym. Test.* **2019**, *81*, 106160. [CrossRef]
4. Da Silva Barrosa, K.; Zwolinski, P. Influence of the use/user profile in the LCA of 3d printed products. *Elsevier B.V.* **2016**, *50*, 318–323. [CrossRef]
5. Venkatachalama, V.; Spierlinga, S.; Hornb, R.; Endresa, H.J. LCA and Eco-design: Consequential and Attributional Approaches for Biobased Plastics. *Procedia CIRP* **2018**, *69*, 579–584. [CrossRef]
6. Lewandowska, A. Environmental life cycle assessment as a tool for identification and assessment of environmental aspects in environmental management systems (EMS) part 1: Methodology. *Int. J. Life Cycle Assess.* **2011**, *16*, 178–186. [CrossRef]
7. Kowalczuk, M. *Nowa generacja materiałów opakowaniowych z kompostowalnych tworzyw polimerowych. Materiały polimerowe nowej generacji z tworzywa polimerowego ulegajacego recyklingowi organicznemu. Materiały opakowaniowe z kompostowalnych tworzyw polimerowych*; Report 1; Centre of Polymer and Carbon Materials: Zabrze, Poland, 2012; pp. 3–14.
8. Borkowski, A. Recykling tworzyw sztucznych w Polsce. Czy można go zwiększyć? *PlasticsEurope* **2016**, *2017*, 1–12.
9. PlasticsEurope. Raport roczny 2016. 2017. Available online: https://www.plasticseurope.org/download_file/force/1265/521 (accessed on 20 November 2019).
10. Formela, K.; Hejna, A.; Zedler, L.; Przybysz, M.; Ryl, J.; Saeb, M.R.; Piszczyk, L. Structural. thermal and physico-mechanical properties of polyurethane/brewers' spent grain composite foams modified with ground tire rubber. *Ind. Crop. Prod.* **2017**, *108*, 844–852. [CrossRef]
11. Piasecka, I.; Tomporowski, A.; Piotrowska, K. Environmental analysis of post-use management of car tires. *Przem. Chem.* **2018**, *97*, 1649–1653.
12. Horowitz, N.; Frago, J.; Mu, D. Life cycle assessment of bottled water: A case study of Green2O products. *Waste Manag.* **2018**, *76*, 734–743. [CrossRef]
13. Kruszelnicka, W.; Bałdowska-Witos, P.; Kasner, R.; Flizikowski, J.; Tomporowski, A.; Rudnicki, J. Evaluation of emissivity and environmental safety of biomass grinders drive. *Przem. Chem.* **2019**, *98*, 1494–1498.
14. Piasecka, I.; Tomporowski, A. Analysis of Environmental and Energetical Possibilities of Sustainable Development of Wind and Photovoltaic Power Plants. *Problemy Ekorozwoju* **2018**, *13*, 125–130.
15. Chen, L.; Peltona, R.E.O.; Smith, T.M. Comparative life cycle assessment of fossil and bio-based polyethylene terephthalate (PET) bottles. *J. Clean. Prod.* **2016**, *137*, 667–676. [CrossRef]
16. Papong, S.; Malakul, P.; Trungkavashirakun, R.; Wenunun, P.; Chomin, T.; Nithitanakul, M.; Sarobol, E. Life Cycle Assessment of Poly(Lactic Acid) (PLA): Comparison Between Chemical Recycling. Mechanical Recycling and Composting. *J. Polym. Environ.* **2014**, *96*, 1221–1235.
17. Kuczenski, B.; Geyer, W. *Life Cycle Assessment of Polyethylene Terephthalate (PET) Beverage Bottles Consumed in the State of California*; The Department of Resources Recycling and Recovery: Sacramento, CA, USA, 2011; pp. 15–17.
18. Giridharreddy, K. Blow Mould Tool Design and Manufacturing Process for 1litre PET Bottle. *J. Mech. Civ. Eng.* **2013**, *8*, 12–21.
19. Bałdowska-Witos, P.; Kruszelnicka, W.; Kasner, R.; Tomporowski, A.; Flizikowski, J.; Mrozinski, A. Impact of the plastic bottle production on the natural environment. Part 2. Analysis of data uncertainty in the assessment of the life cycle of plastic beverage bottles using the Monte Carlo technique. *Przem. Chem.* **2019**, *98*, 1668–1672.
20. Piasecka, I.; Tomporowski, A.; Flizikowski, J.; Kruszelnicka, W.; Kasner, R.; Mrozinski, A. Life Cycle Analysis of Ecological Impacts of an Offshore and a Land-Based Wind Power Plant. *Appl. Sci.* **2019**, *9*, 231. [CrossRef]
21. Malinauskaite, J.; Jouhara, H.; Czajczyńska, D.; Stanchev, P.; Katsou, E.; Rostkowski, P.; Thorne, R.J.; Colon, J.; Ponsá, S.; Al-Mansour, F.; et al. Municipal Solid Waste Management and Waste-to-Energy in the Context of a Circular Economy and Energy Recycling in Europe. *Energy* **2017**, *141*, 2013–2044. [CrossRef]

22. Huijbregts, M.A.J.; Steinmann, Z.J.N.; Elshout, P.M.F.; Stam, G.; Verones, F.; Vieira, M.D.M.; Hollander, A.; Zijp, M.; Van Zelm, R. *Impact Pathways and Areas of Protection*; National Institute for Public Health and the Environment: Bilthofen, The Netherlands, 2016; pp. 16–17. Available online: https://www.rivm.nl/bibliotheek/rapporten/2016-0104.pdf (accessed on 28 October 2019).

23. Bałdowska-Witos, P.; Kruszelnicka, W.; Kasner, R.; Rudnicki, J.; Tomporowski, A.; Flizikowski, J. Impact of the plastic bottle production on the natural environment. Part 1. Application of the ReCiPe 2016 assessment method to identify environmental problems. *Przem. Chem.* **2019**, *98*, 1662–1667.

24. Kłos, Z.; Lewicki, R.; Koper, K. *Application of Environmental Characteristics of Materials in Sustainable Development*; Horvath, I., Mandorli, F., Rusak, Z., Eds.; Delft Univ Technology. Fac Indust Design Eng: Delft, The Netherlands, 2010.

25. Klos, Z. Ecobalancial assessment of chosen packaging processes in food industry. *Int. J. Life Cycle Assess.* **2002**, *7*, 309. [CrossRef]

26. Tomporowski, A.; Piasecka, I.; Flizikowski, J.; Kasner, R.; Kruszelnicka, W.; Mroziński, A.; Bieliński, K. Comparison Analysis of Blade Life Cycles of Land-Based and Offshore Wind Power Plants. *Pol. Marit. Res.* **2018**, *25*, 225–233. [CrossRef]

27. Kruszelnicka, W.; Bałdowska-Witos, P.; Tomporowski, A.; Piasecka, I.; Mroziński, A. Bilans energetyczny procesu spalania poużytkowych tworzyw polimerowych. *Apar. I Inżynieria Chem.* **2017**, *56*, 211–212.

28. Ogilvie, S.; Collins, M.; Aumônier, S. *Life Cycle Assessment of the Management Options for Waste Tyres*; Environment Agency: Bristol, UK, 2004.

29. Flizikowski, J.; Piasecka, I.; Kruszelnicka, W.; Tomporowski, A.; Mroziński, A. Destruction assessment of wind power plastics blade. *Polimery* **2018**, *63*, 381–386. [CrossRef]

30. Piotrowska, K.; Kruszelnicka, W.; Bałdowska-Witos, P.; Kasner, R.; Rudnicki, J.; Tomporowski, A.; Flizikowski, J.; Opielak, M. Assessment of the Environmental Impact of a Car Tire throughout Its Lifecycle Using the LCA Method. *Materials* **2019**, *12*, 4177. [CrossRef]

31. Lijewski, P.; Merkisz, J.; Fuc, P. Research of Exhaust Emissions from a Harvester Diesel Engine with the Use of Portable Emission Measurement System. *Croat. J. Eng.* **2013**, *34*, 113–122.

32. Tomporowski, A.; Flizikowski, J.; Kruszelnicka, W.; Piasecka, I.; Kasner, R.; Mroziński, A.; Kovalyshyn, S. Destructiveness of profits and outlays associated with operation of offshore wind electric power plant. Part 1 identification of a model and its components. *Pol. Marit. Res.* **2018**, *25*, 132–139. [CrossRef]

33. Serkowski, S.; Korol, J. Ocena środowiskowa technologii wytwarzania proppantu na podstawie analizy LCA—Analiza porównawcza. *Szkło I Ceram.* **2014**, *65*, 12–15.

Article

Water Vapor Permeability through Porous Polymeric Membranes with Various Hydrophilicity as Synthetic and Natural Barriers

Chalykh Anatoly [1],*, Zolotarev Pavel [1], Chalykh Tatiana [2], Rubtsov Alexei [2] and Zolotova Svetlana [2]

[1]   Frumkin Institute of Physical Chemistry and Electrochemistry of Russian Academy of Sciences, Leninsky Prospect, 31, bldg. 4, Moscow 119071, Russia; usecretar@phyche.ac.ru

[2]   Department of Commodity Science and Expertise, Plekhanov Russian University of Economics, Stremyanny per. 36, Moscow 117997, Russia; TCHalykh.TI@rea.ru (C.T.); ealexru@gmail.com (R.A.); goldoni@yandex.ru (Z.S.)

*   Correspondence: chalykh@mail.ru

Received: 26 December 2019; Accepted: 18 January 2020; Published: 1 February 2020

Abstract: The article is devoted to the analysis of sorption kinetics, permeability, and diffusion of water vapor in porous polymeric membranes of different hydrophilicities and through-porosities. The water transport measurement with a constant gradient of partial pressure allows the authors to obtain reliable characteristics for porous membranes, films, artificial leathers, and fabrics of various chemical natures (synthetic and bio-based) and phase structures. All the kinetic permeability curves were determined and effective diffusion coefficients, as well as their apparent activation energies, were calculated at the stationary and non-stationary stages of the mass transfer. The relationship between the sorption–diffusion characteristics of the polymer barriers and their vapor permeability is traced. Within the framework of a Zolotarev–Dubinin dual dispersive model, an analytical equation is obtained that relates permeability to diffusion coefficients of water vapor in the pore volume, polymer skeleton material using such characteristics as porosity and the solubility coefficient. It is proposed to use this equation to predict the sorption properties for barrier and porous materials of complex architecture specifically in food packaging.

Keywords: permeability; diffusion; sorption; porous membranes; hydrophilic and hydrophobic polymers

---

1. Introduction

Polymeric materials of different chemical natures and phase structures have such diverse functional purposes that the lifetime of products based on them varies drastically. It is natural that the products for long-term use, which are subject to high loads during operation, should be durable and resistant to environmental factors. Food packaging materials and containers, on the contrary, should be used no longer than the expiration period of the foodstuff. Otherwise, such materials, being discarded as solid domestic waste, pollute the environment. It is known that polymeric materials with short shelf life should be decomposable [1]. The best conditions for biodegradation of polymers are created in humid conditions, when moisture is immobilized in hydrophilic polymers, creating conditions for the development of microflora [2]. This process is significantly accelerated in porous polymeric membranes. Therefore, information on the interaction of moisture with polymers is of fundamental importance for solving the urgent problems of material science such as the choice of polymers for specific aims and the prediction of the behaviour of materials in the processes of saturation by water and biodegradation [3,4]. This article is devoted to the analysis of the process of sorption and diffusion of water vapor in porous polymers of different hydrophilicities.

From the standpoint of percolation theory, traditional analysis of the permeability of gas-filled polymers, and heterogeneous polymeric hybrid systems in general, can be divided into two states: Before the percolation transition, when the polymeric phase forms a continuous dispersion medium, and gas phase forms closed inclusions, and after the percolation transition, when the gas phase forms channels through the other phase. While the process phenomenology is sufficiently developed for the first state, much remains unclear for the second state [5,6]. First of all, what contribution to the transfer process is made by the matrix, its sorption and diffusion characteristics, tortuosity of transport channels, adsorption at the interface. The question of the methods of investigation of vapor permeability of these materials remains also open. Currently, the cup method has become the most widespread for the experimental determination of water vapor permeability through porous polymeric membranes. In this method, the value of flow I in the stationary area of the kinetic permeability curve is the main characteristic of the membrane, and it is usually identified with the permeability coefficient. This method is used successfully for standard permeability measurements of barrier materials.

On the other hand, the non-stationary part of the curve, within which the concentration distribution of the diffusant over the membrane cross-section is established, remains out of the attention of the researchers, although it is this very part that bears the greatest information on the features of water vapor diffusion through porous and monolithic polymeric materials [1,7]. Thus, the non-stationary area can be used to calculate the diffusion coefficient (D), to estimate its concentration and temporal dependencies, and to determine the solubility coefficients and Henry's constants (s). Unfortunately, the cup method in its modern version [8] does not allow the obtainment of information on the non-stationary part of the permeability curve since a gradient of the total pressure of the external environment is present in chromatographic and mass spectrometric cells along with the gradient of the partial pressure. Therefore, its use in porous membranes to study the mass transfer in isobaric–isothermal conditions in detail is problematic.

The purpose of this work was to study the kinetics of water vapor permeability through porous polymeric membranes of different hydrophilicity by continuous registration of the vapor diffusate amount, combined with the theoretical analysis of the mass transfer process in such systems.

2. Experimental

Monolithic and porous polymeric membranes with thicknesses of 150 μm and 500 μm made of collagen (C), polyesterurethane urea (PEU), polyamide 6.6 (PA), polyvinyl chloride (PVC) were used as the study objects. PEU membranes were obtained from the 4,4-diphenylmethane diisocyanate, polyetherdiols, and hydrazinehydrates dimethylformamide solution. Oligooxypropylenediol was used in the synthesis of PEU-1, statistical copolymer of propylene and ethylene oxides in the ratio of 1:1 for the synthesis PEU-2, and an ester of adipic acid and oxypropylenediol were used for PEU-3.

PEU of Sanpren brand (Japan) (M_W = 32.9 kDa, M_n = 22.1 kDa) was obtained from a solution in dimethylformamide of 4,4-diphenylmetandiisocyanate, polyesterdiols and hydrazine hydrates. Oligooxypropyleneddiol was used in the synthesis of PEU-1, and a statistical copolymer of propylene and ethylene oxides in the ratio of 1:1 was used for PEU-2, and for PEU-3—a complex ester of adipic acid and oxypropylenediol.

PVC monolithic films of C-70 brand (Nizhny Novgorod, Russia) (M_W = 107.2 kDa) were obtained by the casting of 5% wt% of polymer solution onto a glass substrate with subsequent drying of films to a constant weight.

Collagen membranes (M_W = 360 kDa) were obtained from acetic acid solution of the soluble part of animal dermis (Moscow, Russia). Fixed collagen film was washed from acetic acid residues in distilled water.

Porous PEU was obtained from 10 wt% solution in dimethylformamide by the condensation-induced structure formation method [9]. Water was used as a precipitator. Porous PVC films were obtained by its deposition from 5 wt% solutions in cyclohexanone via precipitation by ethanol, and PA films obtained from its solutions in ethanol via precipitation by water. The porous analogue of collagen was a sample of natural leather (raw and unpainted calfskin). The porosity of the samples and their specific surface were determined by measuring the density and sorption of the inert solvent, hexane. Characteristics of the studied samples are given in Table 1.

Table 1. Characteristics and kinetic constants of diffusion and permeability of porous polymeric membranes of different hydrophilicity at 298 K.

Material / Parameter	PVC	PEU-1	PEU-2	PEU-3	Collagen and Natural Leather
Porosity, %	38	55	58	65	44
Pore diameter, μm	8/12 ***	5/8	7/11	6/10	4/6
Water sorption, g/100 g	0.52	7.5	9.2	16.3	60
D_{Θ}, *10^{-7} cm^2/s	2	0.8	0.6	4	0.1
D_s, *10^{-7} cm^2/s	3.1	0.66	0.5	2.4	0.08
D_p, *10^{-2} cm^2/s *	~1	~1	~1	~1	~1
E_{Θ}, kJ/mol	51/52 **	48/46	54/49	56/54	61/63
E_S, kJ/mol	50/54	49/48	53/50	52/50	64/62
E_P, kJ/mol	54/29	51/25	57/18	58/21	63/26

* Calculated by ratio $D_P = P/\sigma$. σ is obtained from a real isotherm. ** In the numerator, the activation energy at a partial pressure difference of 44–60%, and in the denominator at a partial pressure difference of 44–90%. *** The range of pore size changes.

Isotherms of water vapor sorption were obtained on McBain's vacuum scales. The kinetics of water vapor permeability were studied using an experimental setup, the scheme of which is shown in Figure 1. The unit consists of a vapor source block (C), diffusion cell (B), which houses the membrane under study between the metal flanges, glass sorption column with quartz spiral, and a cup with water vapor absorber. In contrast to the previously created setups [8,9], the source block contains several glass thermostable cups mounted on a movable plastic disc.

Each source was filled with an aqueous salt solution before the experiment, in this case, $CaCl_2$. The composition of the solutions was chosen in such a way that it would be possible to change the gradient of partial pressure of steam from $(p/p_0)_1$ to $(p/p_0)_2$ by changing the sources, moving along the sorption isotherm. Here, $(p/p_0)_1$ and $(p/p_0)_2$ are relative humidity values above the source and the sink, respectively.

The disc is equipped with a special lifting device, with the help of which a tight connection was made between the cup and the metal flange of the cell. The distance from the membrane surface to the source is 2 cm, from the cup with absorber to the membrane surface is 2.5 cm. All units of the setup have independent temperature control, which allows us to carry out measurements both in isothermal and non-isothermal modes in a wide range of partial steam pressures.

The measurement method was as follows. The membrane sample after long conditioning was removed from the desiccator, in which the same absorber was used as a hydrostat (for example, K_2CO_3 provides a constant humidity of ≈44%), and then was installed between the flanges of the diffusion cell. The absorber was placed in the sorption column. The establishment of sorption equilibrium in the membrane-absorber system at the chosen temperature of the experiment was observed during the next 10–20 min. Then the block of sources (lower block) was brought into contact with the lower flange of the diffusion cell. From this moment on, the kinetics of the absorber weight change were measured. These measurements continued until a constant absorber weight change rate was reached, which corresponds to the establishment of a stationary state of the transfer process at the selected partial pressure of water vapor.

The design of the setup allowed to change the water vapor partial pressure difference measurements range via selection of the source (Figure 1b). The relative error in determining the kinetics of water vapor permeability at the stationary stage was 5%, and 8% at the non-stationary stage. Measurements were carried out in isothermal conditions of the process at temperatures of 298–230 K at various moisture differences between the source and absorber.

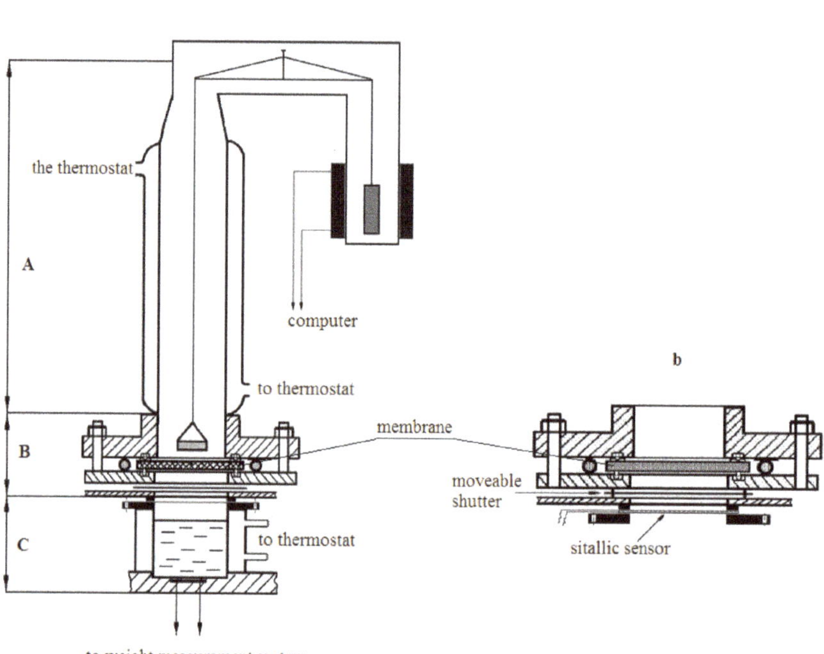

Figure 1. Experimental setup (**a**) and diffusion cell (**b**) for the study of the vapor permeability kinetics of polymeric membranes. Sitallic sensor is designed to measure the local humidity near the inner surface of the membrane. More detailed description of the setup is provided in the text.

3. Results and Discussion

Typical kinetic curves of permeability and sorption of water vapor by monolithic and porous membranes of different hydrophilicity are shown in Figures 2 and 3. It is seen that for both monolithic and porous membranes with through-porosity, two areas corresponding to the non-stationary and stationary stages of the process can be distinguished on the kinetic curves of permeability.

From the formal point of view, regardless of the phase state of the membrane and its porosity, the length of the non-stationary stage can be characterized by the time delay (Θ), which is numerically equal to the segment cut off on the time axis by the extrapolated linear part of the permeability curve (Figure 2).

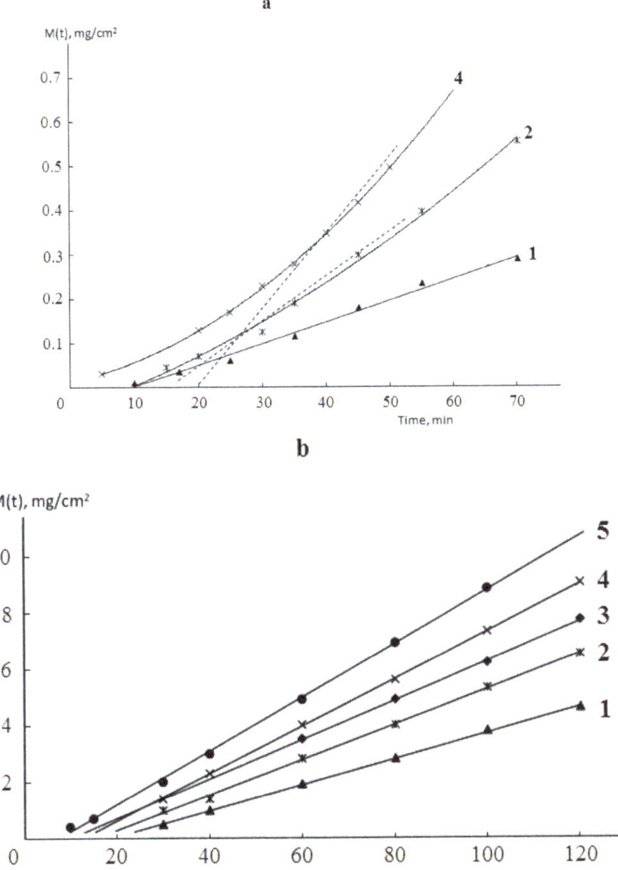

Figure 2. Kinetic curves of vapor permeability of monolithic (**a**) and porous (**b**) polymeric membranes obtained from polyvinyl chloride (PVC) (**1**), polyesterurethane urea (PEU)-1 (**2**), PEU-2 (**3**), PEU-3 (**4**), and collagen (**5**) at 298 K.

For porous membranes, Θ is always smaller than that for monolithic membranes and is significantly larger than that for the "source/air layer/absorber" system. For the latter, Θ is equal to 10–12 s, which, according to Barrer, gives the value of diffusion coefficient $D = 0.2$ cm^2/s, which coincides under normal conditions with the diffusion coefficient of water vapor in the air [10].

The following three facts are interesting. First, Θ for porous membranes is a function of hydrophilicity (Figure 4). In spite of the fact that the total porosity of the samples changes in quite a wide range (Table 1), it can be argued that Θ increases together with increasing of the sorption capacity of the polymer, from which the membrane is made. Changes in the humidity difference upon the measurement of vapor permeability shift the figurative point of the sample in one or another direction on the Θ-M (∞) curve. This effect is not observed for monolithic membranes.

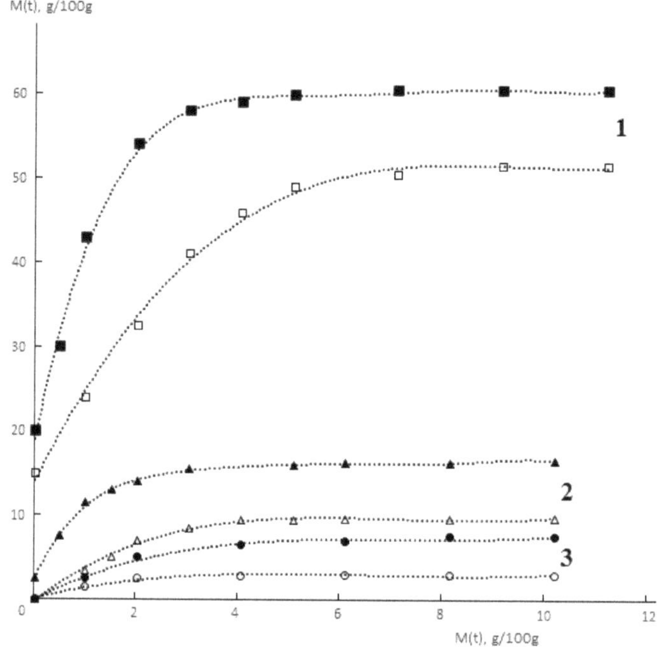

Figure 3. Kinetic curves of water vapor sorption by monolithic (**solid lines**) and porous (**dotted lines**) membranes made of collagen (**1**), PEU-3 (**2**), PEU-1 (**3**) at 298 K. Interval of relative water vapor pressure difference is 0.44–0.90.

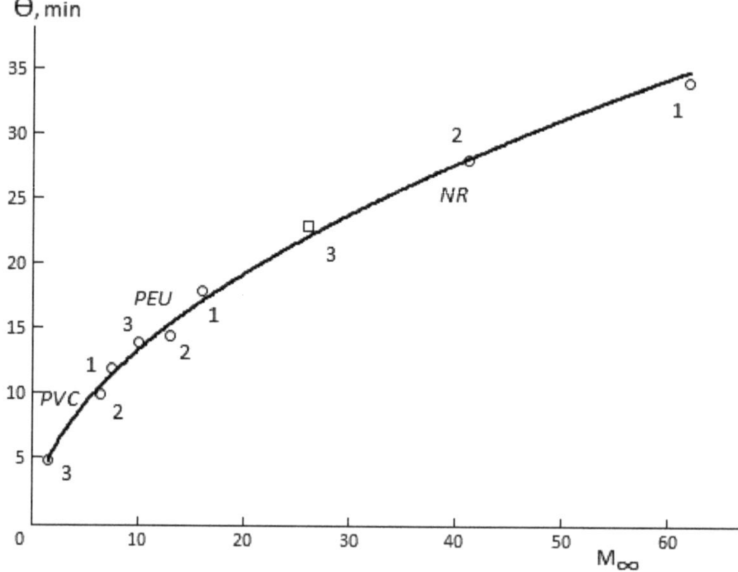

Figure 4. Dependence of the delay time (Θ) on the sorption capacity of the porous polymeric membranes ($M\infty$) at 298 K, relative humidity of 30% at various values of relative humidity differences: (**1**) 44–90 p/p_0; (**2**) 44–80 p/p_0; (**3**) 44–60 p/p_0. Polymer materials are denoted on the graph.

Secondly, there is a proportionality between Θ and the time of sorption equilibrium establishment in the membrane–water vapor system.

Thirdly, the ratios between the diffusion coefficients and temperature coefficients for porous and monolithic membranes, calculated from the stationary area of the permeability curve and the sorption kinetics, are unusual. Diffusion coefficients for monolithic membranes can be calculated by the following traditional equations [9–11]:

$$D_\Theta = l^2/6\,\Theta \tag{1}$$

$$D_s = (\pi/16)\,l^2\,(d\gamma/dt^{1/2})^2 \tag{2}$$

$$D_p = I/\sigma^{-1} \tag{3}$$

where l is the membrane thickness and γ is the relative degree of the membrane filling by diffusant.

While, the thus obtained diffusion coefficient values, as well as the apparent sorption and permeability activation energies E_Θ, E_s, E_p, (Table 1), are quite close to each other, the situation is completely different for the case of porous membranes.

The effective diffusion coefficients for porous membranes D_Θ' D_s' D_p', calculated using the same Equations (1)–(3), differ significantly among themselves (Table 1). For all humidity differences, as a rule, $(D_\Theta' = D_s') << D_p$, and for absolute values $D_\Theta' >> D_\Theta$. For hydrophilic membranes (collagen, PEU-2, and PEU-3), the difference between D_Θ' and D_p' is greater the greater the humidity difference. This effect is determined by the decrease of D_Θ' to a greater degree than by the increase of D_p' and I. Thus, upon transition from humidity difference of 44–60% to 44–90%, D_Θ' decreases by 2.5 times, and D_p' and I increase only by 30–35% due to the plasticization of polymer.

The temperature dependences of D_Θ' and D_s' for all membranes studied are described satisfactorily by the Arrhenius equation. At the same time, E_Θ' and E_s' are close to each other and, especially importantly, coincide with E_Θ and E_s for monolithic membranes. This means that the main contribution to the mass transfer at the non-stationary stage is made by the process of sorption saturation of the porous membrane frame by the diffusant.

However, the temperature dependence of permeability of porous membranes in the general case does not obey the Arrhenius equation. For PVC and PEU-1 membranes at all (p/p_o), and for collagen and PEU-3 membranes at $(p/p_o)_1 < 60\%$, the temperature dependences of I are described by the power function

$$I \cong k\,T^n \tag{4}$$

where n is an empirical constant, the value of which varies from 1.5 to 1.7, which indicates a significant contribution of free diffusion of water vapor through the pore space of the membranes to the stationary permeability. In the area of high humidity (more than 80%), the temperature dependence of water vapor permeability for collagen, PEU-2, and PEU-3 is described by a set of two functions

$$I \cong k_1\,T^n + k_2\,\exp\,(-E_\Theta/RT) \tag{5}$$

where k_1 and k_2 are empirical constants.

The experimental data obtained allow us to formulate the following ideas about the mechanism of vapor permeability of porous polymeric membranes. The process of moisture transfer in such membranes is a superposition of two flows: Phase transfer over the porous space of the membrane and diffusion transfer over the volume of the polymeric matrix (called activation diffusion according [10] or solid-state diffusion according to [9,11]). At $t < \Theta$, water vapor penetrates into the membrane through the system of through capillaries. Simultaneously with the diffusion through the free space of pores, vapor is adsorbed on the pore walls and absorbed–dissolved in the wall material. This process occurs before local sorption equilibrium is established in different parts of the membrane, and its rate is determined by the coefficient of water vapor diffusion into the polymer.

At t > Θ, some humidity gradient is set over the whole thickness of the sample. This moment corresponds to the beginning of the stationary stage of the process, in which both the pore space and the membrane material are involved simultaneously. Thus, until a constant moisture gradient is established over the cross-section of the membrane, most of the flow is absorbed by the material and does not participate in the direct transfer of water vapor to the outer surface of the sample. Once a constant gradient has been established, the flows through the porous space and polymer material become parallel, and the total flow is determined by their sum [10,11].

The theory of the mass transfer process in porous materials is described in detail in the following work's papers [12,13].

The averaged equations of mass transfer in the above described systems with a developed system of transport pores are as follows:

$$m\frac{\partial c}{\partial t} = mD_i\frac{\partial^2 c}{\partial x^2} - \kappa(\gamma c - a) \tag{6}$$

$$(1-m)\frac{\partial c}{\partial t} = (1-m)D_a\frac{\partial^2 a}{\partial x^2} + \kappa(\gamma c - a) \tag{7}$$

$$c(0,t) = c_0, \quad a(0,t) = \gamma c_0, \quad a(l,t) = c(l,t) = 0 \tag{8}$$

$$a(x,0) = c(x,0) = 0, \quad 0 \le x \le l \tag{9}$$

where m is the porosity of the membrane, l is its thickness, c and a are the local concentrations of the diffusant in the pores and the polymeric phase, D_i and D_a are the diffusion coefficients in the pores and the polymeric phase, respectively, k is the averaged constant of the mass transfer rate between the diffusant in the pores and the polymeric phase.

By introduction of dimensionless variables and parameters

$$\xi = \frac{x}{l} \quad \varepsilon = \frac{\tau_i}{\tau_a} \quad \beta = \kappa\tau_i/(1-m) \quad \alpha = \kappa\sigma\tau_i/m \quad \tau_i = \frac{l^2}{D_i} \quad \tau_a = \frac{l^2}{D_a} \quad \tau = \frac{t}{\tau_i} \tag{10}$$

we obtain the following equation for the amount of substance $M(t =) \int_0^t q(t)dt$, desorbed from the membrane at the moment t (i.e., M(t) absorbed by the sorbent) from the surface

$$M(t) = \frac{mD_i+(1-m)\sigma D_a}{l}c_0\times$$
$$\times\left\{t - t_1 - 2\tau_i \sum_{n=1}^{\infty} \frac{(-1)^n}{(\lambda_n^- - \lambda_n^+)}\left[\frac{\lambda_n^- + (\pi n)^2\Delta}{\lambda_n^+}e^{-\lambda_n^+\tau} - \frac{\lambda_n^+ + (\pi n)^2\Delta}{\lambda_n^-}e^{-\lambda_n^-\tau}\right]\right\} \tag{11}$$

where

$$t_i = \frac{\tau_i}{6}\frac{\beta+\alpha}{\beta+\alpha\varepsilon} \tag{12}$$

$$\lambda_n^\pm(=) \frac{[(\varepsilon+1)(\pi n)^2 + (\beta+\alpha)]}{2} \pm \sqrt{\frac{[(\varepsilon+1)(\pi n)^2 + (\beta+\alpha)]^2}{4} - (\pi n)^2[\varepsilon(\pi n)^2 + (\beta+\alpha\varepsilon)]} \tag{13}$$

Essential is the fact that at large times $t \to \infty$, the dependence of M on t is approximated by a straight line

$$M_{(t\to\infty)} = \frac{mD_i + (1-m)D_a}{l}c_0(t - t_1) \tag{14}$$

And upon extrapolation $t \to 0$, this line, in total accordance with the obtained experimental material, does not go to the origin of coordinates, but crosses the time axis at the values of

$$t_1 \equiv \Theta = \frac{l^2(m + \gamma(1-m))}{6(D_i m + \gamma(1-m)D_a)} \tag{15}$$

and ordinate axis at

$$M_{(t \to 0)} = \frac{l}{6}(m + \gamma(1 - m))c_0 \tag{16}$$

From Equations (12)–(16) it is clear that the obtained equations are more general than the Dynes–Barrer ratio [9]. They are very satisfactory in describing of the obtained set of experimental data. Thus, it follows from Equation (15) that the length of the non-stationary stage of the process of permeability is a function of the membrane porosity, its sorption capacity by diffusant, the ratio between the diffusion coefficients in the porous space and polymeric material. It is natural that in general case, upon consideration of $\left(\frac{\partial M}{\partial T}\right)_{t \to \infty}$ or $\left(\frac{\partial \Theta}{\partial T}\right)$ it is necessary to expect a quite complex temperature dependence of the mentioned process parameters:

$$\left(\frac{\partial I}{\partial T}\right) = const\left(\frac{\partial D_i}{\partial T}\right) + const\left(\frac{\partial D_a}{\partial T}\right) + const\left(\frac{\partial \gamma}{\partial T}\right) \tag{17}$$

which can be approximated by $I \approx T^{-n}$ or an exponential function only in the first approximation at certain relations between D_i, D_a, and σ.

Thus, the new possibility of calculating diffusion coefficients from kinetic sorption curves of monolithic samples and kinetic permeability curves of porous membranes, supplemented with information on porosity, is of great interest. For this purpose, it is possible to use either the ratio

$$D_i m = \frac{l}{c_0 \theta}(-M)_{t \to 0} - D_a \gamma^{(c)}(1 - m)$$

$$D_a = \frac{a_1 - a_2}{\left(\gamma_1^{(c)} - \gamma_2^{(c)}\right)(1 - m)} \tag{18}$$

$$D_i m = a_1 - \lambda_1^{(c)}\left(\frac{a_1 - a_2}{\gamma_1^{(c)} - \lambda_2^{(c)}}\right)$$

or, carrying out the measurements of I at different γ', i.e., at different humidity differences. In this case:

$$\begin{aligned} a_1 &= D_i m + \sigma_1 D_a(1 - m) \\ a_2 &= D_i m + \sigma_2 D_a(1 - m) \end{aligned} \tag{19}$$

where $a_i = I_i l(c_0)_i^{-1}$.

Calculations by these equations for all membranes have shown that the transfer of water vapor with a diffusion coefficient of $D_i = 0.18 - 0.21$ at 298 K occurs in their pore space, which is characteristic of the free diffusion of water vapor in air.

4. Conclusions

Thus, in the framework of the bidisperse Zolotarev–Dubininin model, an analytical equation, which connects permeability with water vapor diffusion coefficients in the pore space of the membrane, polymer frame material, porosity, and solubility coefficient, is derived. The obtained analytical ratios can be widely used to predict and calculate various situations arising from the operation of real porous polymeric membranes. The use of this equation is proposed to predict the sorption properties of barrier shells and porous materials of complex architectures, for example, in the analysis of mass transfer in disperse mixtures of polymers, block copolymers, gradient barrier layers. It should be noted that the proposed analytical equations allow one to also solve the inverse problem to estimate structural and defective characteristics of a given polymer membrane [14,15].

Author Contributions: Conceptualization, C.A. and Z.P.; methodology, R.A. and C.T.; validation, R.A.; formal analysis, Z.S.; investigation, C.A., R.A., Z.S.; resources, Z.S.; writing—original draft preparation, C.A. and C.T.; writing—review and editing, Z.P.; visualization, R.A.; supervision, C.A.; All authors have read and agreed to the published version of the manuscript

Funding: Presidium of Russian Academy of Sciences P22 (2019).

Conflicts of Interest: The authors declare no conflict of interest.

References

1. Gunatillake, P.; Mayadunne, R.; Adhikari, R. Recent developments in biodegradable synthetic polymers. *Biotechnol. Ann. Rev.* **2006**, *12*, 301–347. [CrossRef]
2. Nair, L.S.; Laurencin, C.T. Biodegradable polymers as biomaterials. *Prog. Polym. Sci.* **2007**, *32*, 762–798. [CrossRef]
3. Sangroniz, A.; Sarasua, J.R.; Iriarte, M.; Etxeberria, A. Survey on transport properties of vapours and liquids on biodegradable polymers. *Eur. Polym. J.* **2019**, *120*, 109232. [CrossRef]
4. Moustafa, H.; Youssef, A.M.; Darwish, N.A.; Abou-Kandil, A.I. Eco-friendly polymer composites for green packaging: Future vision and challenges. *Compos. Part B Eng.* **2019**, *172*, 16–25. [CrossRef]
5. Petropoulos, J.H. Mechanisms and Theories for Sorption and Diffusion of Gases in Polymers. In *Polymer Gas Separation Membranes*; Paul, D.R., Yampolskii, Y.P., Eds.; CRC Press: London, UK; Tokyo, Japan, 1994; pp. 17–83.
6. Bekman, I.N. *The Mathematics of Diffusion*; ONTO Print: Moscow, Russia, 2016; p. 400.
7. Iordanskii, A.L.; Olkhov, A.A.; Karpova, S.G.; Kucherenko, E.L.; Kosenko, R.Y.; Chalykh, A.E.; Berlin, A.A. Influence of the structure and morphology of ultrathin poly(3-hydroxybutyrate) fibers on the diffusion kinetics and transport of drugs. *Polym. Sci. Ser. A* **2017**, *59*, 352–362. [CrossRef]
8. Kellert, H.J. Determination of water vapour permeability of leather—A report of the VGCT Commision for leather testing and evaluation. *J. Soc. Leather Technol. Chem.* **2004**, *88*, 63–65. GOST ISO 14268-2011/ISO 14268:2002.
9. Chalykh, A.E.; Zlobin, V.B. Modern concepts of diffusion in polymer systems. *Rus. Chem. Rev.* **1988**, *57*, 903. [CrossRef]
10. Reid, R.S.; Sherwood, T.K. *The Properties of Gases and Liquids: Their Estimation and Correlation*; McGraw-Hill Book Co, INC.: New York, NY, USA, 1958; p. 336.
11. Malkin, A.Y.; Askadsky, A.A.; Kovriga, V.V.; Chalykh, A.E. *Experimental Methods of Polymer Phisycs*; Mir Publishers: Moscow, Russia, 1983; 520p.
12. Ohya, H.; Kudryavtsev, V.V.; Semenova, S.I. *Polyimide Membranes*; Kodansha: Tokyo, Japan, 1996; 314p.
13. Siracusa, V.; Ingrao, C.; Karpova, S.G.; Olkhov, A.A.; Iordanskii, A.L. Gas transport and characterization of poly(3 hydroxybutyrate) films. *Eur. Polym. J.* **2017**, *91*, 149–161. [CrossRef]
14. Chalykh, T.I.; Zolotarev, P.P.; Chalykh, A.E. *The Structure and Dynamics of Molecular Systems*; Kazan—Yoshkar-Ola: Yoshkar-Ola, Russia, 1998; p. 142.
15. Chalykh, T.I. Structure and Structure and Water-Exchange Properties of Porous Polymer Materials. Ph.D. Thesis, MITHT Moscow Institute of Fine Chemical Technology, Moscow, Russia, 2000; 306p.

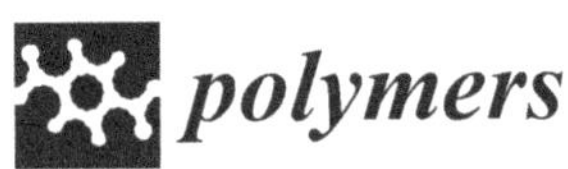 

Article

Research of the Influence of the Ultrasonic Treatment on the Melts of the Polymeric Compositions for the Creation of Packaging Materials with Antimicrobial Properties and Biodegrability

Irina Kirsh [1], Yuliya Frolova [2,*], Olga Bannikova [1], Olga Beznaeva [1], Isabella Tveritnikova [1], Dmitry Myalenko [3], Valentina Romanova [1] and Daria Zagrebina [1]

[1] Scientific and Educational Center Advanced Packaging Materials and Recycling Technologies, Center of the Collective Use, Moscow State University of Food Production, 125080 Moscow, Russia; irina-kirsh@yandex.ru (I.K.); bannikovaoa@mgupp.ru (O.B.); olgazaikina@mail.ru (O.B.); iza-1995@bk.ru (I.T.); bwal1307@mgupp.ru (V.R.); zagrebinadm@mgupp.ru (D.Z.)

[2] Laboratory of Food Biotechnology and Specialized Products, Federal Research Center of Nutrition and Biotechnology, 109240 Moscow, Russia

[3] Russian Research Institute of Dairy Industry, 115093 Moscow, Russia; myalenkod@list.ru

* Correspondence: Y.operarius@yandex.ru

Received: 31 December 2019; Accepted: 27 January 2020; Published: 30 January 2020

Abstract: Ensuring the microbiological safety of food products is a problem of current interest. The use of antimicrobial packaging materials is a way of solving the problem. When developing packaging materials, it is advisable to use a modern approach based on the creation of biodegradable materials. The difficulty in the selection of the polymer compositions' components lies in solving the dilemma of the joint introduction and processing of antimicrobial and biodegradable agents. The studies of the ultrasound treatment on the melts of polymer mixtures showed an increase in the dispersion process of the components of the mixture. In this regard, this work aimed to study the effect of the ultrasonic treatment on the melts of polymer compositions containing thermoplastic starch and birch bark extract (BBE). In the work, the properties of PE-based packaging materials with various BBE concentrations obtained with ultrasonic treatment of melts on a laboratory extruder were studied. Biodegradable polymer compositions containing thermoplastic starch and BBE, obtained with the use of the ultrasonic treatment during extrusion, were investigated. The methods for studying rheological, physic-mechanical, antimicrobial properties and sanitary chemical indicators of materials were used in the article. It was found that ultrasonic treatment increases the melt flow and contributes to the production of materials with the uniform distribution of additives. The BBE content from 1.0% and higher in the contents of the material provides antimicrobial properties. When studying the permeability of oxygen and water vapor of the polymer compositions based on PE and BBE, it was found that the introduction of a filler increases vapor permeability by about 8–12% compared with control samples. The optimal concentration of BBE in polyethylene compositions containing thermoplastic starch was determined. The extension of the shelf life of the food product during storage in the developed material was established.

Keywords: food packaging; antimicrobial properties; polyethylene; birch bark extract; ultrasound; thermoplastic starch; biodegradation

1. Introduction

Ensuring the quality and safety of food products is one of main problems in the food industry [1]. Losses of unpackaged food products associated with spoilage can reach up to 50% [2]. Improving production and processing technologies, as well as correctly selecting packaging material, can significantly reduce product losses [1]. The safety of food products, of both animal and plant origin, is determined primarily by microbiological indicators, taking into account the quantitative and qualitative content of contaminants of chemical, biological and microbiological nature [3]. Loss of food integrity is the result of chemical and microbiological spoilage, which are the main causes of spoilage of many products during production, transportation, processing, storage and marketing. These processes are directly connected with the loss of the food products' quality [1,4]. According to statistics, 80% of food poisoning is due to the presence of sanitary-indicative, pathogenic microorganisms, yeast, molds and their toxins in the products. When microorganisms get to the surface of a food product, due to the favorable environment, they develop, worsening the appearance, reducing taste, chemically reacting with the components, causing change in proteins and lipids, producing toxic substances that cause food poisoning and also creating favorable conditions for the growth of bacteria [5,6]. In this regard, one of the current trends in the food industry is the creation of packaging materials with antimicrobial properties.

Over the past decades, many packaging materials based on polymers and coatings with antimicrobial properties have been developed [1,7–12]. Substances of natural [8,13,14] and synthetic [15,16] origin, including nanoparticles, are used as antimicrobial agents [12,17]. Polyolefin-based materials (polyethylene, polypropylene) are often used as barrier packaging. However, polyethylene and polypropylene films do not have inherent antimicrobial properties [12]. Various approaches are used to impart antimicrobial properties to polymeric materials based on polyolefins: surface treatment of the polymeric material with antimicrobial additives (spraying onto the surface) followed by fixing the antimicrobial agent on the packaging surface [12] or the introduction of antimicrobial components directly into the polymer matrix [13,15,16,18–20]. Using the approach of the surface treatment of packaging materials with antimicrobial agents has a significant drawback in that there is a risk of the migration of the antimicrobial component into the food product, which can change its organoleptic characteristics. The advantage of this method is the possibility of using thermally unstable antimicrobial additives. When using this type of packaging material, it is necessary to strictly control the concentration and possible conditions of the migration of the antimicrobial additive into the product. Obtaining packaging materials using the second approach consists of directly mixing the additive and the polymer before loading into the extruder, followed by obtaining the material. The disadvantage of this method is the possibility of using only heat-resistant antimicrobial additives. Also, when mixing the additive and the polymer before loading into the extruder, there is the possibility of uneven distribution of the antimicrobial additive in the polymer matrix of the finished material.

A promising antimicrobial additive, able to modify the properties of polyolefins, is birch bark extract (BBE). BBE ($C_{36}H_{60}O_3$) is the multicomponent mixture containing betulinol, lupeol, lupenon, uveol, betulinol acetate, allobetulin, isobetulinol, oleanolic acid and other substances [21].

BBE is known for its antibacterial, antiviral, anti-inflammatory and antimutagenic properties, as well as its resistance to molds and bacteria [21,22]. BBE is resistant to oxygen and sunlight and is nontoxic, which allows it to be used in polymeric materials in contact with food. To intensify the production process of polymeric materials with a given set of properties, including those with antimicrobial properties, it is advisable to use superconcentrates (or masterbatches) [23,24]. This technology involves obtaining packaging material with a modifier in two stages: obtaining superconcentrate, followed by mixing it with a polymer base and obtaining a film. This approach allows obtaining polymer materials with the more uniform distribution of additives in the polymer matrix. However, at high concentrations of antimicrobial additives, the probability of its agglomeration increases and, as a consequence, the uneven distribution is possible. The formation of structures with the uniform distribution of component composition can be achieved by additional mechanical mixing, the use of dispersants, etc. However, the use of ultrasonic treatment of polymer melts is gaining practical and scientific interest [25,26]. The work performed to study the effect of the ultrasound treatment on the melts of polymer compositions containing agricultural wastes showed the increase in the uniform distribution of the mixture components even when the filler content is greater than 30%. In addition, it was found that ultrasonic treatment of melts of polymer compositions leads to the acceleration of the destruction of the polymer matrix due to the increase in oxygen-containing groups in polyethylene and the increase in water absorption [27]. Therefore, to improve the dispersion process of the components of polymer compositions, a laboratory extruder with ultrasonic treatment of polymer melts, developed at the university, was used. The aim of the present work was to study the effect of the ultrasonic treatment of melts of polymer compositions in order to create packaging materials with antimicrobial properties that ensure prolonged shelf life of packaged food products and have the ability to accelerate biodegradation.

2. Materials and Methods

2.1. Materials

In the experimental work birch bark extract (BBE) (Birch World, St. Petersburg, Russia), which is the multicomponent mixture containing betulinol, lupeol, lupenon, uveol and other substances, was used. It is registered as a biologically active additive in the state register (certificate No. 77.99.23.3.U.3440.4.08 of April 29, 2008). Other materials used were high-pressure polyethylene (PE) (Kazanorgsintez, Kazan, Russia), grade Kazpelen 15813-020, the main characteristics of which are presented in Table 1; thermostabilizer Irganox 1010 (pentaerythritol tetrakis(3-(3,5-di-tert-butyl-4-hydroxyphenyl)propionate)) (BASF, The Chemical Company, Basel, Switzerland); corn starch with a particle size of 50 μm and a share of fraction up to 30 μm of at least 50%. Nutrient media for the microbiological research was as follows: dry nutrient medium Sabouraud (GNTsPMB, Obolensk Russia), dry nutrient medium Czapek-Dox (GNTsPMB, Obolensk, Russia), meat-peptone agar (GNTsPMB, Obolensk, Russia). All other reagents used to analyze the properties of the materials were of analytical purity.

Table 1. The main characteristics of polyethylene.

Parameter	Value
Average molecular weight	78,600
Melting point, mp, °C	105–108
Density, ρ, g/cm^3, ρ	0.919–0.925
Melt flow rate, g/10 min	2–2.5
Elongation at break, ε_p, %	600–800
Breaking stress, σ_p, MPa	11–15

2.2. The Technology of Polymer Materials

Polymer compositions based on polyethylene, thermoplastic starch and birch bark extract (BBE) were used. Thermoplastic starch was obtained in a laboratory mixer by mixing starch, plasticizers and additives to improve processing of the compositions. In polymer composite materials (PCM), the amounts of thermoplastic starch were 20%, 40% and 60%. The amount of BBE in PCM ranged from 1% to 12%.

Obtaining samples of films based on PE with BBE was carried out in two stages:

1. Obtaining PE with BBE in the form of granules with diameter 3 mm and length 6 mm.

2. Obtaining film material (film thickness 62 ± 2 μm) from granules of PE with BBE and thermoplastic starch.

To obtain granules and film material, the laboratory extruder with the screw diameter of 16 mm and with ultrasonic (US) treatment of the melt, developed at the university, was used. Ultrasonic treatment of the melt of the polymer compositions was conducted at 22.4 kHz. Processing modes are presented in Table 2.

Table 2. Temperature conditions for the processing of the polyethylene compositions in the extruder.

Temperature in the Extruder by Zone, °C				Polymer Compositions [2]
Zone 1 [1]	Zone 2	Zone 3	Zone 4	
120	160	180	190	PE with BBE
120	150	150	160	PE with BBE and starch

[1] Zones 1–3 represent zones in the extruder; zone 4 is the extrusion head. The ultrasonic processing frequency was 22.4 kHz, the ultrasonic power was 300 W, and the screw rotational speed was 90 rpm. [2] PE with BBE, polymeric composition based on polyethylene with birch bark extract; PE with BBE and starch, polymeric composition based on polyethylene with birch bark extract and thermoplastic starch.

The use of the ultrasonic treatment allowed the production of materials with the temperatures of extruder zones 3 and 4 being 10 °C lower than without the ultrasonic treatment.

The concentration of BBE in PE varied between 0.5%, 1.0%, 2.0% and 5.0%. Irganox 1010 (BASF, The Chemical Company, Basel, Switzerland) was introduced as thermostabilizer in the amount of 1.0%. The content of the compositions is presented in Table 3.

Table 3. Component composition of polymeric compositions based on polyethylene (PE) with birch bark extract (BBE).

Polymer Composition	The Content of the Compositions, %		
	Polyethylene	Birch Bark Extract	Irganox 1010
PE 0.5% BBE	98.5	0.5	1
PE 1.0% BBE	98	1	1
PE 2.0% BBE	97	2	1
PE 5.0% BBE	94	5	1
PE 0% BBE	100	0	0

Materials, based on PE and BBE without ultrasonic treatment and pure PE, not containing BBE, were used as control samples. The concentration of BBE in PE with thermoplastic starch varied between 0%, 2.0%, 5.0%, 8.0% and 12.0%. Irganox 1010 was introduced as thermostabilizer in the amount of 1%. The amounts of thermoplastic starch were 20%, 40% and 60%. The content of the compositions is presented in Table 4.

Table 4. The component composition of the polymeric compositions based on polyethylene (PE) with birch bark extract (BBE) and thermoplastic starch.

Polymer Composition	The Content of the Compositions, %		
	Polyethylene with Irganox 1010	Birch Bark Extract	Thermoplastic Starch
PC 20 BBE 0	80	0	20
PC 40 BBE 0	60	0	40
PC 60 BBE 0	40	0	60
PC 20 BBE 2	78	2	20
PC 40 BBE 2	58	2	40
PC 60 BBE 2	38	2	60
PC 20 BBE 5	75	5	20
PC 40 BBE 5	55	5	40
PC 60 BBE 5	35	5	60
PC 20 BBE 8	72	8	20
PC 40 BBE 8	52	8	40
PC 60 BBE 8	32	8	60
PC 20 BBE 12	68	12	20
PC 40 BBE 12	48	12	40
PC 60 BBE 12	28	12	60

Compositions obtained without the influence of ultrasound on their melts were selected as control samples.

2.3. Characteristic of Polymer Materials

2.3.1. Appearance and Microstructure

The visual assessment of the developed films was carried out in order to determine the color of the outer and inner surfaces. The assessment of the surface of the sample aimed to detect cracks, sagging, bumps, roughness. Microphotographs of the surface were obtained using the Bresser digital light microscope (Bresser GmbH, Germany). The structural-morphological properties of the samples were studied using the scanning electronic microscope, JSM-U3 (JEOL, Tokyo, Japan).

2.3.2. Rheological Research

Studies of polymer melts were carried out by the standard method of capillary viscometry [28]. This experiment was carried out five times.

2.3.3. Determination of Physico-Mechanical Properties

Studies of polymeric materials samples were carried out on a universal testing machine BM-50 (Moscow, Russia) with fixation of and elongation at break. The tensile speed of the materials was 10 mm/s. This experiment was carried out five times.

2.3.4. Determination of Antimicrobial Properties

Determination of Antibacterial Properties: To determine the antibacterial properties of polymeric materials the culture of microorganisms *Escherichia coli* M 17 (*E. coli*) and *Candida albicans* (*C. albicans*) was used. The microorganism culture was grown in 50 mL glass tubes with beveled nutrient agar for 24 h. As the nutrient medium, meat-peptone agar was used. The grown culture was suspended in physiological saline with the turbidity of 0.5 according to the McFarland standard (1.5×10^8 CFU/mL) and was used for 15 min. Then, 22 ± 1 mL of meat-peptone agar was poured into Petri dishes and left until the agar solidified. Inoculation of the culture on the surface of the agar was carried out by distributing 0.2 mL of the suspension with a glass microbiological spatula of Drigalski. Next, disks of polymer material with the diameter of 20 mm were laid on the inoculated surface. Preliminarily, the polymeric material was disinfected by the double immersion in pure ethanol and the solvent vapor was allowed to evaporate completely before use. Petri dishes with the studied samples were placed in the TS-1/20 SPU thermostat (Smolensk SKTB SPU, Smolensk, Russia) at the temperature of 37 ± 1 °C for 48 h. After 24 h, the intermediate inspection of the plates was performed. The development of microorganisms on the materials' surface and the presence of the inhibition growth zone were visually assessed. The experiment was conducted three times.

Determination of Fungicidal Properties: To determine the resistance of polymeric materials to mold, the following cultures were used: *Penicillium commune* F-4486 (*P. commune*) and *Aspergillus niger* strain 82 (*A. niger*). The studies were carried out in two ways. According to the first method, *P. commune* mold was grown in 50 mL glass tubes with beveled nutrient agar for 7 days. Sabouraud was used as the nutrient medium. The grown culture was suspended in physiological saline. The spore concentration in the solution was $(2 \pm 1) \times 10^6$ CFU/mL. Then, the experiment was carried out according to the method described in Section 2.3.4. Petri dishes with the studied samples were placed in the TS-1/20 SPU thermostat (Russia) at the temperature of 28 ± 1 °C for 72 h. After 48 h, the intermediate inspection of the plates was performed. The development of microorganisms on the materials' surface and the presence of the inhibition growth zone were visually assessed. The experiment was carried out three times. The second method consisted of keeping the polymer materials contaminated with the spores of *A. niger* mold in the absence of mineral and organic pollutants at high humidity. A fungus culture of *A. niger* was grown in test tubes with the beveled nutrient medium (Czapek-Dox medium with agar) for 14 days. The spore suspension was prepared in sterile distilled water. The spore concentration was $(1.5 \pm 0.5) \times 10^6$ CFU/mL. Disks of polymer material with the diameter of 20 mm were cleaned before use according to the method described in Section 2.3.4. Prepared film samples were placed one per Petri dish. Inoculation of the samples was carried out by spraying the suspension of the mold spores on the surface of the film, preventing droplets from merging. The inoculated samples were kept in a box for the drying of drops at the temperature of 23 ± 1 °C for no more than 60 min. Petri dishes were closed and placed in desiccators, on the bottom of which was poured distilled water to create conditions of high humidity (greater than 90%). The desiccators were closed and kept in a dark room at the temperature of 29 ± 1 °C. The test duration was 14 days. After 7 days, the covers of the desiccators were opened for 3 min for the oxygen access. The surface was evaluated after incubation by light microscopy at the magnifications of 100× and 400×. The experiment was carried out three times.

2.3.5. Sanitary Chemical Research

The organoleptic evaluation was performed in the universal model medium (distilled water). The purified polymeric material was immersed in the model medium in the ratio 1:1 and kept at temperatures of 20, 40 and 60 °C for 7, 14, 21 and 28 days. During the organoleptic assessment, the presence of turbidity, sediment and extraneous odor of aqueous extracts was determined. The turbidity of the extracts was characterized descriptively as either lack of turbidity, weak opalescence, opalescence, strong opalescence, weak turbidity, noticeable turbidity or strong turbidity. Sediment was characterized by the absence of sediment as either negligible, insignificant, noticeable and large. Its properties were noted as crystalline, amorphous, etc. Its color was also noted. The smell of water extracts was expressed descriptively as lacking smell, phenolic, aromatic, extraneous indefinite, etc. Odor intensity was expressed on a 5-point scale, where 0 indicated no tangible smell and 5 indicated an obviously strong smell that caused a persistent negative feeling. The experiment was carried out three times.

Quantitative assessment of the migration of low-molecular-weight substances from polymeric materials to model media (2.0% citric acid; 3.0% lactic acid; 0.3% solution of sodium chloride) was carried out by the standard gas chromatographic method with the use of high-performance modular liquid chromatography (Agilent 1200, Agilent Technologies Inc, USA).

2.3.6. Determination of Permeability of Packaging Materials

To determine the vapor permeability of the packaging materials, the method of determining the amount of water vapor through polymeric materials by changing the mass of the container with distilled water and a sample over 24 h at the temperatures of 23 and 38 °C and the humidity of 90–98% (ASTM D1653) was used [29]. The appliance W3/030 Labthink (Labthink Instruments Co. Ltd., China (CPR)) was used.

To determine oxygen permeability of the packaging materials, the equilibrium pressure method at the temperature 23 °C (GB/T 1038-2000) was used [30]. The appliance PERME OX2/231 Labthink (Labthink Instruments Co. Ltd., China (CPR)) was used.

2.3.7. Biodegradability Assessment of the Materials

The samples were placed in containers with biohumus at the temperature 23 ± 2 °C and the humidity 60% $\pm$ 5% [31]. Samples of polymeric materials were cut out with a size of 10 × 10 cm and placed in a container with soil at a humidity of at least 50% of its maximum moisture capacity. A layer of soil 1.5 ± 0.5 cm thick was poured on top of the samples and loosely closed containers were placed in a chamber at a temperature of 23 ± 2 °C and humidity of 60% $\pm$ 5%. The temperature and humidity levels were monitored throughout the composting process. Composting time for polymeric materials was 6 months. The degree of biodegradation of the materials was determined by the change in physico-chemical properties during composting.

2.4. Packed Product Storage Studies

Studies of the storage of products in polymeric materials were carried out on chilled parts of broiler carcasses. Fresh carcass parts were bought in a local supermarket. The weight of each part of the carcass was 120 ± 30 g. Fresh carcass parts were hermetically sealed in the polymeric material under sterile conditions. The control samples were carcass parts of the same batch, packaged under the same conditions in PE without BBE. The packaged products were stored for 5 days at the temperatures from 0 to 2 °C. The content of the quantity of mesophilic aerobic and facultative anaerobic microorganisms in the product were controlled as a standardized indicator of microbiological spoilage. Control points were 0, 2, 3, 4 and 5 days. ISO 4833:2003 conventional method for determining the quantity of mesophilic aerobic and facultative anaerobic microorganisms (QMAFAnM) was used [32].

To determine the optimal concentration of BBE in PE compositions containing thermoplastic starch, the relative increase in shelf life was used as a criterion, which was calculated by the following formula:

$$W = \frac{ti - tk}{tk} \times 100\% \tag{1}$$

where W is the relative increase in shelf life; ti is product storage time in PCM-based films; tk is the storage time in films based on PE without additives.

2.5. Statistics

Statistical processing of the results was performed using the IBM SPSS Statistics Program 20 (SPSS Inc. USA).

3. Results and Discussion

3.1. Study of the Effect of the Ultrasonic Treatment on the Properties of Polyethylene Compositions Modified by BBE

3.1.1. Appearance

The appearance of the obtained polymer films is presented in Figure 1.

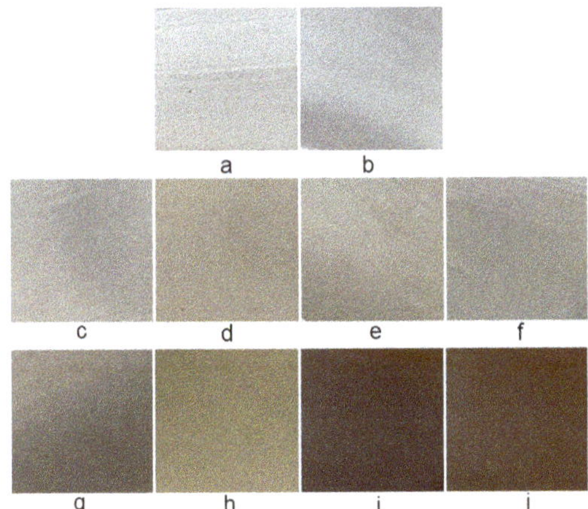

Figure 1. External appearance of the obtained polymeric materials, containing 0%, 0.5%, 1.0%, 2.0% and 5.0% of birch bark extract (BBE) and received with the use of ultrasonic treatment (with US) and without the use of ultrasonic treatment (without US): (**a**) 0% BBE without US; (**b**) 0% BBE with US; (**c**) 0.5% BBE without US; (**d**) 0.5% BBE with US; (**e**) 1.0% BBE without US; (**f**) 1.0% BBE with US; (**g**) 2.0% BBE without US; (**h**) 2.0% BBE with US; (**i**) 5.0% BBE without US; (**j**) 5.0% BBE with US.

The inclusion of BBE in the polymer matrix of polyethylene leads to the color change of the materials obtained. At low concentrations (1–2% BBE), the films had slight yellowness (Figure 1e–h), and at high concentrations (5% BBE), the films turned brown (Figure 1i–j). The films containing BBE had a rough but smooth surface. The samples of films obtained without the treatment of ultrasonic melt were distinguished by the presence of the agglomerated filler; with the increase of the BBE content, the number of agglomerates increased. The agglomerate formation was confirmed by electron microscopy images (Figures 2 and 3).

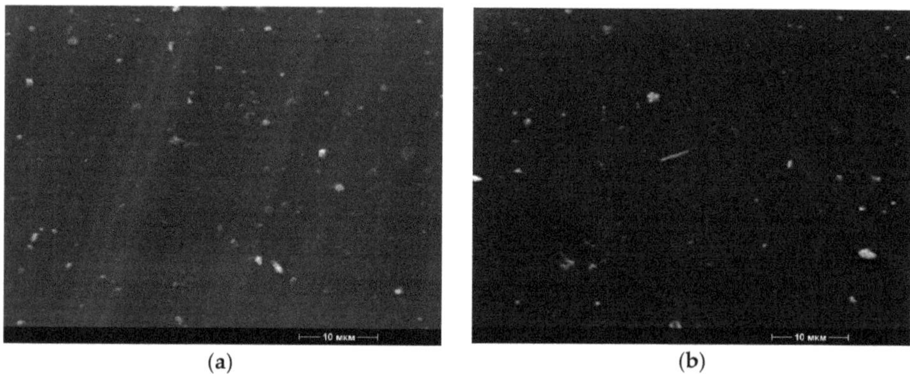

(a)   (b)

Figure 2. Microphotographs of the surface structure of the materials (5000× magnification) based on polyethylene with 2.0% content of birch bark extract, received with the use of ultrasonic treatment (**a**) and without the use of ultrasonic treatment (**b**).

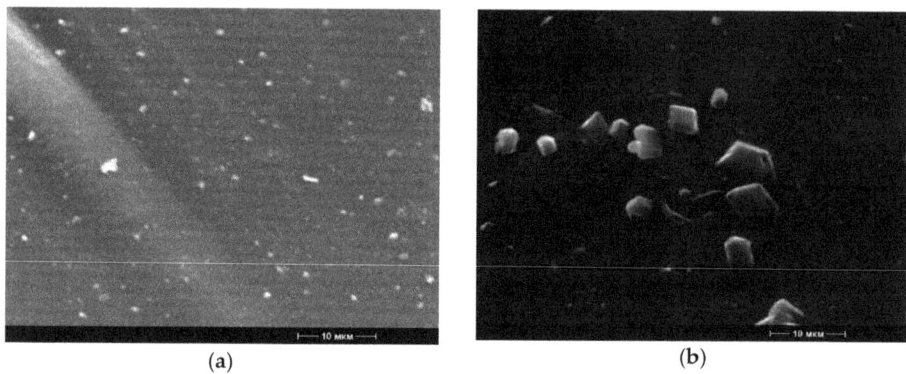

(a)   (b)

Figure 3. Microphotographs of the surface structure of the materials (5000× magnification) based on polyethylene with 5.0% content of birch bark extract, received with the use of ultrasonic treatment (**a**) and without the use of ultrasonic treatment (**b**).

The use of the ultrasonic treatment of the polymer melt in the production of materials with BBE made it possible to obtain more transparent samples with the more uniform distribution of the additive in the material.

3.1.2. Rheological Properties

Figure 4 shows the dependence of the melt flow rate on the concentration of BBE and ultrasound.

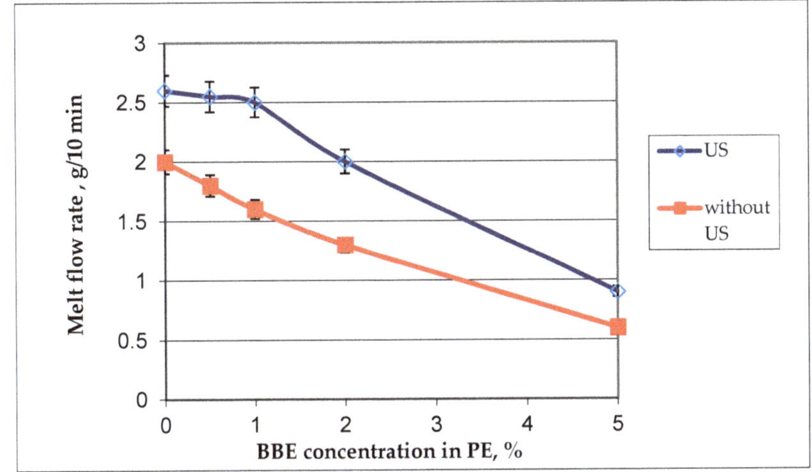

Figure 4. Graphic dependance of the melt flow rate on the birch bark extract (BBE) content in polyethylene (PE) and on ulrasonic (US) treatment.

Processing of ultrasonic melts leads to the increase in the MFR of pure polyethylene and compositions with BBE. The increase in the MFR value under the influence of the ultrasonic treatment is associated with the peculiarities of the effect of the ultrasonic treatment on polymer melts, as presented in [25]. The increase in the content of BBE in the composition of PE leads to the decrease in the MFR, this is due to the fact that the BBE introduced into the polymer behaves as the "filler-polymer".

3.1.3. Physico-Mechanical Properties

The influence of the BBE concentration and ultrasound treatment on the physico-mechanical properties was evaluated by two indicators: breaking stress and elongation at break. The results are presented in Figures 5 and 6.

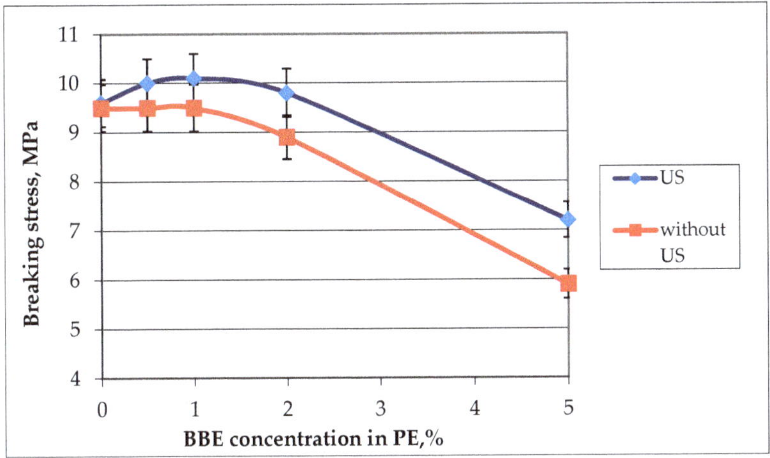

Figure 5. Graphic dependence of the breaking stress on the birch bark extract (BBE) content in polyethylene (PE) and on ultrasonic (US) treatment.

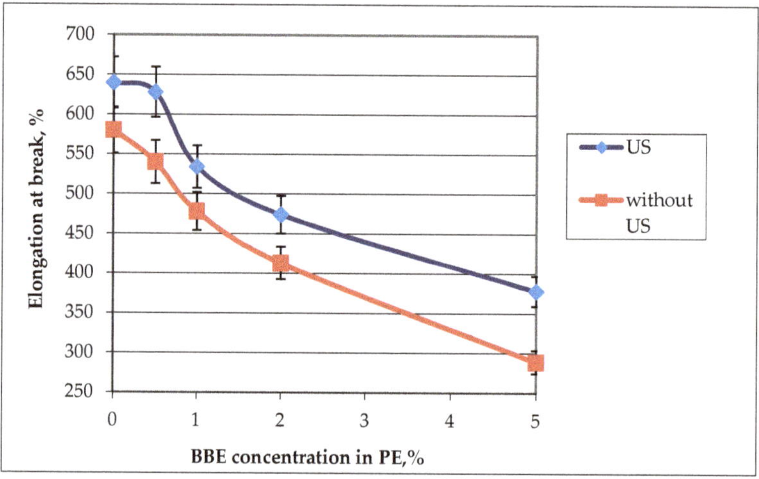

Figure 6. Graphic dependence of the elongation at break on the birch bark extract (BBE) content in polyethylene (PE) and on ultrasonic (US) treatment.

It was found that with the increase of the concentration of BBE in the composition of the polymeric material based on PE, the breaking stress and elongation at break also increase, which is connected with the distribution of the BBE in the structure of PE and the violation of the integrity of its primary structure. Ultrasonic treatment of melts in the production of polymeric materials leads to the increase in breaking stress and elongation at break by 1.5 times in comparison with the control samples. This is due to the fact that, during the ultrasonic treatment, the distribution of the additive is uniform (Figures 1–3). The low content of BBE (up to 1%) in the composition of the polymeric material has practically no effect on the physico-mechanical properties.

3.1.4. Antimicrobial Properties

The results of the evaluation of the antimicrobial properties of the studied materials samples in relation to *E. coli*, *C. albicans* and *P. commune* are presented in Table 5.

Table 5. Results of the visual assessment of the surface of polymeric materials based on polyethylene (PE) with birch bark extract (BBE) received with (with US) and without the use of ultrasonic (without US) treatment, inoculated for 24–48 h.

Polymer Composition	Visual Assessment		
	E. coli	*C. albicans*	*P. commune*
PE 0% BBE without US	Surface growth	Surface growth	Surface growth
PE 0% BBE with US	Surface growth	Surface growth	Surface growth
PE 0.5% BBE without US	Slow surface growth	Slow surface growth	Slow surface growth
PE 0.5% BBE with US	Slow surface growth	Slow surface growth	Slow surface growth
PE 1.0% BBE without US	Lack of surface growth	Lack of surface growth	Lack of surface growth
PE 1.0% BBE with US	Lack of surface growth	Lack of surface growth	Lack of surface growth
PE 2.0% BBE without US	Lack of surface growth	Lack of surface growth	Lack of surface growth
PE 2.0% BBE with US	Lack of surface growth	Lack of surface growth	Lack of surface growth
PE 5.0% BBE without US	Zone of inhibition 1.8 ± 0.2 mm	Zone of inhibition 1.5 ± 0.1 mm	Zone of inhibition 3.0 ± 0.5 mm
PE 5.0% BBE with US	Zone of inhibition 1.7 ± 0.2 mm	Zone of inhibition 1.5 ± 0.2 mm	Zone of inhibition 3.1 ± 0.5 mm

Based on the data obtained, it was found that ultrasonic treatment of melts during the development of materials does not affect their antimicrobial properties. The dependence of antimicrobial activity on the concentration of BBE in the polymer matrix was revealed. For *C. albicans*, after 24 h of exposure, the growth of cultures was observed on the surface of the control samples containing no BBE (Figure 7a).

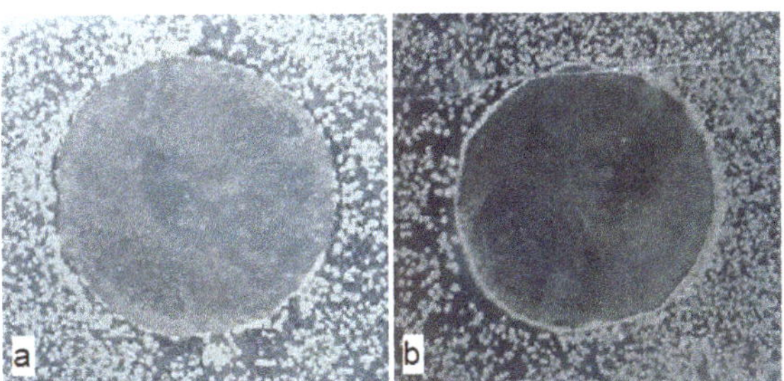

Figure 7. External appearance of polyethylene without birch bark extract (**a**) and polyethylene with 1.0% content of birch bark extract (**b**) and the use of ultrasonic treatment after the incubation under the influence of *C. albicans* for 24 h.

The pattern of the antimicrobial properties of the materials with BBE compared with the control samples (example, Figure 7) was observed for all cultures of microorganisms: *E. coli*, *C. albicans* and *P. commune*. At the BBE concentration of 0.5% for 24 h, the growth inhibition was observed on the surface of the materials, but after 48 h there was the growth on the surface. The development of *P. commune* occurred with a certain delay compared to *E. coli and C. albicans*, which is connected with the peculiarities of the growth rate of microorganisms, in this regard, the development of molds was evaluated after 48 h. Materials containing 1% and 2% BBE are characterized by the absence of growth of microorganisms on their surface and under the material for 168 h of exposure. In this regard, materials containing 1% and 2% BBE have bacteriostatic and fungistatic properties. The increase in the BBE content to 5% leads to the appearance of the inhibition zone (3.0 ± 0.5 mm), which may indicate the migration of BBE into the agar medium.

In addition, the stability of materials with BBE exposure under conditions of high humidity (greater than 90%) at 29 ± 1 °C was evaluated against *A. niger* according to the degree of fungus development on the film surface, using a microscope (Figure 8).

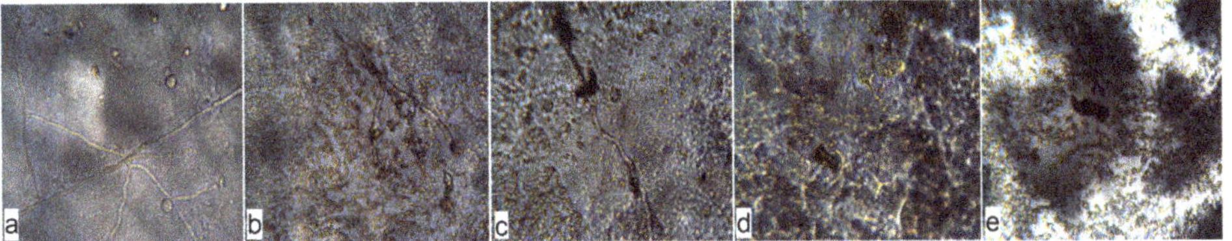

Figure 8. Microphotographs of the surface of the inoculated polyethylene materials with birch bark extract and with the use of ultrasonic treatment on the 14th day of *A. niger* exposure (100× magnification): (**a**) PE 0% BBE ultrasound; (**b**) PE 0.5% BBE, ultrasound; (**c**) PE 1.0% BBE, ultrasound; (**d**) PE 2.0% BBE, ultrasound; (**e**) PE 5.0% BBE, ultrasound.

On the surface of the materials based on PE without BBE, after 14 days of exposure under conditions of forced surface inoculation with *A. niger* fungus at a temperature of 29 ± 1 °C and humidity of greater than 90%, the formation of mycelium (Figure 8a) and conidia with conidiospores (Figure 9a) was observed. On the surface of PE materials with 0.5% BBE, partial destruction of mycelium was observed after 14 days of exposure (Figures 8b and 9b), which indicates the death of the culture. On microphotographs of the surface of the material with BBE concentration of 2.0%, complete destruction of the mycelium was observed (Figures 8d and 9c). Based on the results of the growth of *A. niger* culture on the surface of materials containing BBE, it was found that on the surface of materials containing 2.0% or more BBE, under conditions of forced inoculation, the culture is destroyed, which indicates the mold resistance of materials with BBE. On the basis of rheological, physico-mechanical and antimicrobial studies of materials based on PE with BBE obtained with and without ultrasonic treatment, it is recommended to use material containing 2.0% BBE for packaging to ensure microbiological safety of food products.

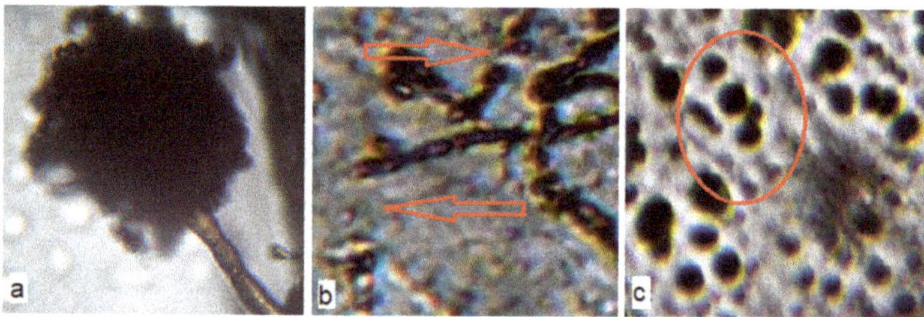

Figure 9. Microphotographs of the surface of the inoculated polyethylene (PE) materials with birch bark extract (BBE) and with the use of ultrasonic treatment on the 14th day of the exposure of *A. niger* (400× magnification): (**a**) PE 0% BBE, ultrasound; (**b**) PE 0.5% BBE, ultrasound; (**c**) PE 2.0% BBE, ultrasound.

3.1.5. Sanitary Chemical Research

Materials intended for contact with food products were subjected to sanitary chemical studies for the migration of low-molecular-weight substances.

Organoleptic studies at temperatures of 20, 40 and 60 °C for 7, 14, 21 and 28 days showed that, in the analyzed extracts from all the materials, there was no sediment, turbidity or color change in the extracts. However, there was the increase in the intensity of odor from materials with the increase in BBE concentration and temperature to 60 °C, while the average value was no more than 1 point, which is allowed for the materials contacting food products.

The results of the study of the migration of low-molecular-weight substances from the materials are presented in Table 6.

Table 6. Results of sanitary chemical research of extarcts of various model environments of materials based on polyethylene with 2.0% content of birch bark extract and the use of ultrasonic treatment.

Name of Indicator, mg/dm³	Permissible Norm	Citric Acid 2.0%	Lactic Acid 3.0%	Solution NaCl 0.3%
Acetaldehyde	≤0.2	<0.05	<0.05	<0.05
Ethyl acetate	≤0.1	<0.05	<0.05	<0.05
Hexane	≤0.1	<0.05	<0.05	<0.05
Heptane	≤0.1	<0.05	<0.05	<0.05
Acetone	≤0.1	<0.05	<0.05	<0.05
Formaldehyde	≤0.1	<0.025	<0.025	<0.025
Methyl alcohol	≤0.2	0.08	0.12	0.17
Butyl alcohol	≤0.5	<0.05	<0.05	<0.05
Isobutyl alcohol	≤0.5	<0.05	<0.05	<0.05
Propyl alcohol	≤0.1	<0.05	<0.05	<0.05
Isopropyl alcohol	≤0.1	<0.05	<0.05	<0.05

When using a model medium of 3.0% lactic acid solution, the increase in the migration of methyl alcohol and formaldehyde is observed, although its value does not exceed permissible norms (Table 6). A study of the migration of low-molecular-weight components from developed materials with BBE shows that migrating substances do not exceed permissible norms. It was found that ultrasonic treatment of the melt and the content of the additive at the concentration of up to 5% do not affect the formation of low-molecular-weight substances that can migrate into the food product upon contact.

Based on the carried out rheological, physico-mechanical, antimicrobial and sanitary chemical studies, it is proposed to use 2% BBE with ultrasound treatment as the packaging material for contact with food products. This material has sufficient physico-mechanical properties, has antimicrobial properties and complies with sanitary chemical standards. The use of the ultrasound treatment makes it possible to obtain material with uniformly distributed additive, and the content of 2% BBE is justified by the economic factor.

3.1.6. Study of the Permeability of the Packaging Materials

When studying the vapor permeability of polymeric compositions based on PE and BBE, it was found that the introduction of a filler increases the vapor permeability by about 8–12% in comparison with the control samples. For example, the vapor permeability of PE without BBE is 0.18–0.21 g/m^2. When the BBE content is 5% in the polyethylene composition, the vapor permeability is about 0.23–0.25 g/m^2.

When studying the oxygen permeability of the samples, it was found that oxygen permeability coefficients for the compositions based on PE and BBE 5% are 1.42×10^{-10} cm^3 cm/m^2 s Pa. For the PE film without BBE, the oxygen permeability coefficient is 1.2×10^{-10} cm^3 cm/m^2 s Pa.

3.2. Packed Product Storage Studies

The results of storing chilled meat are presented using broiler chicken carcasses' parts as an example. Packed product samples were stored at 0 ± 2 °C. The quantity of mesophilic aerobic and facultative anaerobic microorganisms (QMAFAnM) was determined by the accelerated method of seeding on test plates—petrifilms containing dehydrated nutrient gel-like chromogenic substrate.

In the process of storage, the quantity of mesophilic aerobic and facultative anaerobic microorganisms increases in the packed parts of poultry carcasses (Table 7). The acceptable content of QMAFAnM in chilled poultry is 1×10^3 CFU/g. It was found that during the storage of poultry carcasses at the temperature of 0 ± 2 °C for 4 days in a packaging material based on pure polyethylene (PE), the QMAFAnM index approaches a critical value and significantly exceeds it by 5 days. The use of such products for food is considered unsafe for health. During the storage of poultry carcasses at the temperature of 0 ± 2 °C, the QMAFAnM index approaches the critical value on the 5th day. The rate of accumulation of the quantity of mesophilic aerobic and facultative anaerobic microorganisms is higher in a product packaged in pure polyethylene compared to a product packaged in plastic film with birch bark extract. This is due to the fact that upon the contact of the packaging material containing BBE with the product, partial suppression of the growth of microorganisms located on its surface occurs.

Table 7. Change in the content of the quantity of mesophilic aerobic and facultative anaerobic microorganisms, CFU/g, in parts of poultry carcasses during storage.

Type of Packaging	Duration of Storage, days				
	0	2	3	4	5
PE 0% BBE US	$<10^2$	$(5.75 \pm 0.30) \times 10^2$	$(7.40 \pm 0.28) \times 10^2$	$(0.95 \pm 0.20) \times 10^3$	$(2.11 \pm 0.14) \times 10^3$
PE 2% BBE US	$<10^2$	$(3.10 \pm 0.22) \times 10^2$	$(5.50 \pm 0.20) \times 10^2$	$(0.72 \pm 0.18) \times 10^3$	$(0.90 \pm 0.21) \times 10^3$

The data obtained show that the use of the developed PE BBE material in the packaging technology of chilled meat (poultry meat in particular) leads to the increase in shelf life.

3.3. Investigation of the Effect of Ultrasonic Treatment on the Properties of Polyethylene Compositions Based on Birch Bark Extract and Thermoplastic Starch

At this stage of the work, the development of polymer composite materials (PCM) with antimicrobial properties and biodegradability was carried out. In this work, compositions based on PE, thermoplastic starch and BBE were obtained. In PCM, the amounts of thermoplastic starch were 20%, 40% and 60%. The amount of BBE in PCM ranged from 0% to 12%.

The obtained film samples were investigated by rheological and physico-mechanical properties (Tables 8 and 9).

Table 8. Dependance of the melt flow rate index (MRF) on the composition of polymeric compositions material, received with (with US) and without the use of ultrasonic treatment (without US).

Polymer Composition	Average Values MRF, g/10 min	
	with US	without US
PC 20 BBE 0	1.7	1.2
PC 40 BBE 0	1.2	0.7
PC 60 BBE 0	0.9	0.6
PC 20 BBE 2	1.3	1.0
PC 40 BBE 2	0.9	0.8
PC 60 BBE 2	0.7	0.4
PC 20 BBE 5	1.1	0.8
PC 40 BBE 5	0.8	0.4
PC 60 BBE 5	0.5	0.2
PC 20 BBE 8	0.9	0.5
PC 40 BBE 8	0.4	0.2
PC 60 BBE 8	0.3	0.1
PC 20 BBE 12	0.8	0.5
PC 40 BBE 12	0.4	0.2
PC 60 BBE 12	0.2	0.05

Table 9. Physico-mechanical properties of composite polymeric films (PC) according to the content of birch bark extract (BBE) and the use of ultrasonic treatment (US).

Polymer Composition	Breaking Stress, MPa		Elongation at Break, %	
	with US	without US	with US	without US
PC 20 BBE 0	11	9.7	160	90
PC 40 BBE 0	9.1	7.8	58	19
PC 60 BBE 0	7.1	6.3	25	9
PC 20 BBE 2	10.5	9.2	158	90
PC 40 BBE 2	8.8	7.1	56	15
PC 60 BBE 2	6.7	5.9	24	8
PC 20 BBE 5	10.1	8.5	155	86
PC 40 BBE 5	8.1	6.7	42	13
PC 60 BBE 5	6.3	5.4	24	8
PC 20 BBE 8	9.7	7.9	140	67
PC 40 BBE 8	6.9	5.9	36	12
PC 60 BBE 8	5.9	4.6	10	6
PC 20 BBE 12	9.2	7.3	114	60
PC 40 BBE 12	6.7	5.4	33	12
PC 60 BBE 12	5.3	4.3	9	5

From the results obtained, it is clearly seen that the ultrasonic treatment increases the melt flow rate by approximately 2 times compared with the control samples obtained without ultrasonic treatment.

It is clearly seen that the introduction of up to 5% BBE has little effect on the physico-mechanical properties of PCM. The introduction of thermoplastic starch is of great importance; its amount reduces the breaking stress and elongation at break. It has been established that the processing of PCM melts increases the physico-mechanical properties of the materials, which is especially noticeable when comparing the elongation at break. The values of this indicator are approximately 1.5–2 times greater for PCMs obtained with ultrasonic treatment of the melt as compared to control samples. To determine the optimal concentration of BBE in PCM, studies were conducted to determine the shelf life of food products in packages. Dependence of the relative increase in shelf life on the concentration of BBE in polymer compositions containing thermoplastic starch is shown in Figure 10.

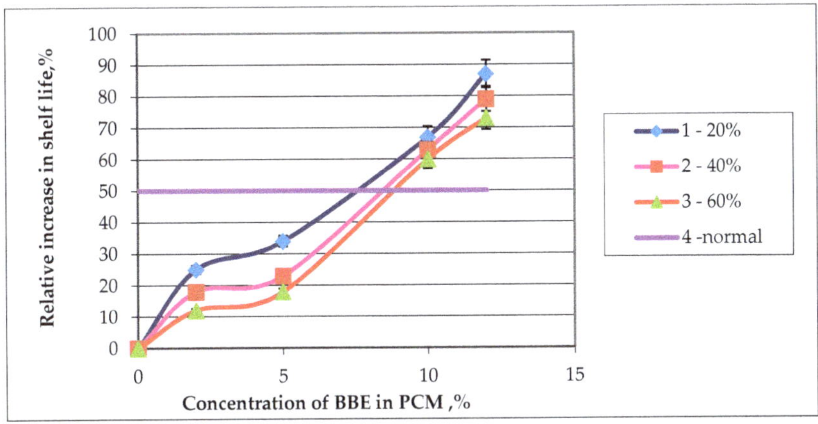

Figure 10. Dependence of the relative increase in shelf life of the food product on the contents of birch bark extract (BBE) in composition material (PCM): (1—content of thermoplastic starch 20% in PCM; 2—content of thermoplastic starch 30% in PCM; 3—content of thermoplastic starch 60% in PCM; 4—desired value to increase product storage).

Based on the data obtained, BBE concentration of 8% was selected.

Next, the research was conducted using the PCM method of composting (Table 10). As the criterion for evaluating the properties, the change in elongation at break after 6 months of composting was used.

Table 10. Change in relative elongation at break of composite polymeric materials (PC) containing birch bark extract (BBE), received with (with US) and without the use of ultrasonic treatment (without US).

Polymer Composition	Change in Elongation at Break after Composting for 6 months, %	
	with US	without US
PC 20 BBE 0	20–23	16–18
PC 40 BBE 0	39–43	28–32
PC 60 BBE 0	78–81	67–69
PC 20 BBE 8	17–19	13–15
PC 40 BBE 8	32–35	24–28
PC 60 BBE 8	60–68	46–54

It is worth noting that the samples obtained with ultrasonic treatment showed more noticeable change in relative elongation at break after composting than those without ultrasonic treatment. This indicates that the effect of ultrasound treatment accelerates the decomposition of polymer materials in the environment due to the uniform distribution of the filler in the polymer matrix and the increase in oxygen-containing groups in polymers, which was noted in the [27]. It should be noted that the introduction of BBE in PCM leads to the decrease in this indicator. This can probably be attributed to the action of BBE as the antimicrobial supplement.

Special attention should be paid to the material based on PE containing 60% starch and 8% BBE, in which, after 6 months of composting, the elongation at break changed by 60–68%. This is a good criterion for obtaining biodegradable PCMs.

4. Conclusions

The influence of ultrasonic treatment of melts of the polymer compositions based on PE and BBE, as well as PE, thermoplastic starch and BBE, was investigated. Packaging materials in the form of films based on polymer compositions were obtained. A comparison of identical PCM compositions obtained with and without ultrasonic treatment of the melts revealed that the ultrasonic treatment increases the fluidity of the melts of polymer compositions. Using electron microscopy, it was found that ultrasonic treatment of the melts of polymer compositions contributes to the production of materials with the uniform distribution of the components of the composition, as shown by the example of a PE sample containing 2% and 5% BBE. When studying oxygen and water vapor permeability of polymeric compositions based on PE and BBE, it was found that the introduction of the filler increases vapor permeability and oxygen permeability by about 8–12% compared to control samples without the addition of BBE. Ultrasonic treatment of polymer compositions does not affect these indicators. It was established that the processing of PCM melts increases the physico-mechanical properties of the materials, which is especially noticeable when comparing the elongation at break. The values of this indicator are approximately 1.5–2 times higher for PCMs obtained with ultrasonic treatment of the melt as compared to control samples. It is predetermined that the BBE content of 1.0% and above in the composition of PE provides packaging materials with antimicrobial properties. The optimal concentration of BBE in compositions based on PE and thermoplastic starch was determined. To obtain biodegradable materials with antimicrobial properties based on PE and thermoplastic starch, it is advisable to introduce BBE in the amount of greater than 8%. It was found that the extension of the shelf life of the food products stored in packaging materials based on PE and BBE in the amount of 2.0% and PE containing thermoplastic starch (60%) and BBE (8%), was greater than 50%, when compared with the control samples without additives. It was revealed that ultrasonic treatment of polymer compositions leads to the acceleration of their biodegradation. The introduction of BBE into PCM leads to the decrease in the biodegradation time, which is associated with the action of BBE as the antimicrobial supplement.

Author Contributions: Writing—review and edit, I.K.; writing—preparation of the original project, Y.F.; conceptualization, O.B. (Olga Beznaeva); verification, D.M.; research, I.T. and V.R.; methodology, O.B. (Olga Bannikova); visualization, D.Z. All authors have read and agreed to the published version of the manuscript.

Funding: The study was carried out as part of the implementation of the federal target program, the unique project identifier RFMEFI57418X0191.

Conflicts of Interest: The authors declare no conflict of interest.

References

1. Han, J.W.; Ruiz-Garcia, L.; Qian, J.P.; Yang, X.T. Food packaging: A comprehensive review and future trends. *Compr. Rev. Food Sci. Food Saf.* **2018**, *17*, 860–877. [CrossRef]
2. Galford, G.L.; Peña, O.; Sullivan, A.K.; Nash, J.; Gurwick, N.; Pirolli, G.; Richards, M.; White, J.; Wollenberg, E. Agricultural development addresses food loss and waste while reducing greenhouse gas emissions. *Sci. Total Environ.* **2020**, *699*, 134318. [CrossRef]
3. Batt, C.A.; Tortorello, M.L. *Encyclopedia of Food Microbiology*; Academic Press: London, UK, 2014; pp. 1–1014.
4. Matthews, K.R.; Kniel, K.E.; Montville, T.J. *Food Microbiology: An Introduction*; ASM Press: Washington, DC, USA, 2019.
5. Abrunhosa, L.; Morales, H.; Soares, C.; Calado, T.; Vila-Chã, A.S.; Pereira, M.; Venâncio, A. A review of mycotoxins in food and feed products in Portugal and estimation of probable daily intakes. *Crit. Rev. Food Sci. Nutr.* **2016**, *56*, 249–265. [CrossRef]

6. Lee, K.; Watanabe, M.; Sugita-Konishi, Y.; Hara-Kudo, Y.; Kumagai, S. *Penicillium camemberti* and *Penicillium roqueforti* enhance the growth and survival of Shiga toxin producing *Esherihia coli* O157 under mild acidic conditions. *J. Food Sci.* **2012**, *77*, 102–107. [CrossRef] [PubMed]

7. Zahra, S.A.; Butt, Y.N.; Nasar, S.; Akram, S.; Fatima, Q.; Ikram, J. Food Packaging in Perspective of Microbial Activity: A Review. *J. Microbiol. Biotechnol. Food Sci.* **2016**, *6*, 752–757. [CrossRef]

8. Pobiega, K.; Kraśniewska, K.; Gniewosz, M. Application of propolis in antimicrobial and antioxidative protection of food quality–A review. *Trends Food Sci. Technol.* **2019**, *83*, 53–62. [CrossRef]

9. Cazón, P.; Velazquez, G.; Ramírez, J.A.; Vázquez, M. Polysaccharide-based films and coatings for food packaging: A review. *Food Hydrocoll.* **2017**, *68*, 136–148. [CrossRef]

10. Fang, Z.; Zhao, Y.; Warner, R.D.; Johnson, S.K. Active and intelligent packaging in meat industry. *Trends Food Sci. Technol.* **2017**, *61*, 60–71. [CrossRef]

11. Jideani, V.A.; Vogt, K. Antimicrobial packaging for extending the shelf life of bread—A review. *Crit. Rev. Food Sci. Nutr.* **2016**, *56*, 1313–1324. [CrossRef]

12. Frolova, Y.V.; Kirsh, I.A.; Beznaeva, O.V.; Pomogova, D.A.; Tikhomirov, A.A. Creation of polymeric packaging materials with antimicrobial properties. *Izv. Vuzov. Prikl. Khimia I Biotekhnologiya [Proc. Univ. Appl. Chem. Biotechnol.]* **2017**, *7*, 145–152. (in Russian). [CrossRef]

13. Kirsh, I.A.; Frolova, Yu.V.; Myalenko, D.M. Packaging materials for food products with an antimicrobial component of natural origin. *Pishchevaya Promyshlennost'* **2018**, *1*, 24–25. (in Russian).

14. Gaikwad, K.K.; Singh, S.; Lee, Y.S. Antimicrobial and improved barrier properties of natural phenolic compound-coated polymeric films for active packaging applications. *J. Coat. Technol. Res.* **2019**, *16*, 147–157. [CrossRef]

15. Khaneghah, A.M.; Hashemi, S.M.B.; Limbo, S. Antimicrobial agents and packaging systems in antimicrobial active food packaging: An overview of approaches and interactions. *Food Bioprod. Process.* **2018**, *111*, 1–19. [CrossRef]

16. Huang, T.; Qian, Y.; Wei, J.; Zhou, C. Polymeric Antimicrobial Food Packaging and Its Applications. *Polymers* **2019**, *11*, 560. [CrossRef] [PubMed]

17. Otoni, C.G.; Espitia, P.J.; Avena-Bustillos, R.J.; McHugh, T.H. Trends in antimicrobial food packaging systems: Emitting sachets and absorbent pads. *Food Res. Int.* **2016**, *83*, 60–73. [CrossRef]

18. Malhotra, B.; Keshwani, A.; Kharkwal, H. Antimicrobial food packaging: Potential and pitfalls. *Front. Microbiol.* **2015**, *6*, 611. [CrossRef]

19. Zhong, Y.; Godwin, P.; Jin, Y.; Xiao, H. Biodegradable polymers and green-based antimicrobial packaging materials: A mini-review. *Adv. Ind. Eng. Polym. Res.* **2019**, in press. [CrossRef]

20. Liu, Y.; Liang, X.; Wang, S.; Qin, W.; Zhang, Q. Electrospun Antimicrobial Polylactic Acid/Tea Polyphenol Nanofibers for Food-Packaging Applications. *Polymers* **2018**, *10*, 561. [CrossRef]

21. Dzubak, P.; Hajduch, M.; Vydra, D.; Hustova, A.; Kvasnica, M.; Biedermann, D.; Markova, L.; Urban, M.; Sarek, J. Pharmacological activities of natural triterpenoids and their therapeutic implications. *Nat. Prod. Rep.* **2006**, *23*, 394–411. [CrossRef]

22. Amiri, S.; Dastghaib, S.; Ahmadi, M.; Mehrbod, P.; Khadem, F.; Behrooj, H.; Aghanoori, M.R.; Machaj, F.; Ghamsari, M.; Rosik, J.; et al. Betulin and its derivatives as novel compounds with different pharmacological effects. *Biotechnol. Adv.* **2019**, in press. [CrossRef]

23. Chen, C.Y.; Gu, J.; Weng, Y.X.; Huang, Z.G.; Qiu, D.; Shao, S.X. Optimization of the preparation process of biodegradable masterbatches and characterization of their rheological and application properties. *Polym. Test.* **2018**, *70*, 526–532. [CrossRef]

24. Sreekumar, P.A.; Elanamugilan, M.; Singha, N.K.; Al-Harthi, M.A.; De, S.K.; Al-Juhani, A. LDPE filled with LLDPE/Starch masterbatch: Rheology, morphology and thermal analysis. *Arab. J. Sci. Eng.* **2014**, *39*, 8491–8498. [CrossRef]

25. Kirsh, I.A.; Chalykh, T.I.; Pomogova, D.A. Modification of polymers and mixtures of incompatible polymers by exposure of their melts to ultrasound. *J. Charact. Dev. Nov. Mater.* **2016**, *8*, 119.

26. Kirsh, I.A.; Babin, Yu.V.; Ananiev, V.V.; Tveriynikova, I.S.; Romanova, V.A.; Bannikova, O.A.; Beznaeva, O.V. Establishing the Dependence of the Effect of Ultrasound on Pkm Melts and their Functional Technological Characteristics. *Izv. Vyss. Uchebnyh Zaved. Tekhnologiya Tekst. Promyshlennosti.* **2019**, *2*, 85–90. (in Russian).

27. Jimenez, A.; Aneli, J.N.; Kubica, S. *Chemistry and Physics of Modern Materials: Processing, Production and Applications*; Apple Academic Press: Toronto, ON, Canada, 2013; pp. 235–251.

28. Kirsh, I.A.; Ananiev, V.V.; Chalykh, T.I.; Sogrina, D.A.; Pomogova, D.A. Study of the effect of ultrasonic treatment on the rheological properties of polymers during their multiple recycling. *Plast. Massy* **2014**, *11–12*, 45–48. (in Russian).
29. ASTM D1653-2013. *Standard Test Methods for Water Vapor Transmission of Organic Coating Films*; ASTM International: West Conshohocken, PA, USA, 2013.
30. SAC GB/T 1038-2000 (China National Standards 2000). *Plastics—Film and Sheeting—Determination of Gas Transmission—Differential—Pressure Method*; Standards Press of China: Beijing, China, 2000.
31. ISO 16929:2019 (International Standard 2019). *Plastics—Determination of the Degree of Disintegration of Plastic Materials under Defined Composting Conditions in a Pilot-Scale Test*; ISO: Geneva, Switzerland, 2019.
32. ISO 4833-1:2013 (International Standard 2013). *Microbiology of the Food Chain—Horizontal Method for the Enumeration of Microorganisms—Part 1: Colony Count at 30 Degrees C by the Pour Plate Technique*; ISO: Geneva, Switzerland, 2013.

Article

On the Use of Gallic Acid as a Potential Natural Antioxidant and Ultraviolet Light Stabilizer in Cast-Extruded Bio-Based High-Density Polyethylene Films

Luis Quiles-Carrillo [1,*], Sergi Montava-Jordà [1], Teodomiro Boronat [1], Chris Sammon [2], Rafael Balart [1] and Sergio Torres-Giner [3,*]

[1] Technological Institute of Materials (ITM), Universitat Politècnica de València (UPV), Plaza Ferrándiz y Carbonell 1, 03801 Alcoy, Spain; sermonjo@mcm.upv.es (S.M.-J.); tboronat@dimm.upv.es (T.B.); rbalart@mcm.upv.es (R.B.)

[2] Materials and Engineering Research Institute, Sheffield Hallam University, Howard Street, Sheffield S1 1WB, UK; C.Sammon@shu.ac.uk

[3] Novel Materials and Nanotechnology Group, Institute of Agrochemistry and Food Technology (IATA), Spanish National Research Council (CSIC), Calle Catedrático Agustín Escardino Benlloch 7, 46980 Paterna, Spain

[*] Correspondence: luiquic1@epsa.upv.es (L.Q.-C.); storresginer@iata.csic.es (S.T.-G.); Tel.: +34-966-528-433 (L.Q.-C.); +34-963-900-022 (S.T.-G.)

Received: 27 November 2019; Accepted: 19 December 2019; Published: 23 December 2019

Abstract: This study originally explores the use of gallic acid (GA) as a natural additive in bio-based high-density polyethylene (bio-HDPE) formulations. Thus, bio-HDPE was first melt-compounded with two different loadings of GA, namely 0.3 and 0.8 parts per hundred resin (phr) of biopolymer, by twin-screw extrusion and thereafter shaped into films using a cast-roll machine. The resultant bio-HDPE films containing GA were characterized in terms of their mechanical, morphological, and thermal performance as well as ultraviolet (UV) light stability to evaluate their potential application in food packaging. The incorporation of 0.3 and 0.8 phr of GA reduced the mechanical ductility and crystallinity of bio-HDPE, but it positively contributed to delaying the onset oxidation temperature (OOT) by 36.5 °C and nearly 44 °C, respectively. Moreover, the oxidation induction time (OIT) of bio-HDPE, measured at 210 °C, was delayed for up to approximately 56 and 240 min, respectively. Furthermore, the UV light stability of the bio-HDPE films was remarkably improved, remaining stable for an exposure time of 10 h even at the lowest GA content. The addition of the natural antioxidant slightly induced a yellow color in the bio-HDPE films and it also reduced their transparency, although a high contact transparency level was maintained. This property can be desirable in some packaging materials for light protection, especially UV radiation, which causes lipid oxidation in food products. Therefore, GA can successfully improve the thermal resistance and UV light stability of green polyolefins and will potentially promote the use of natural additives for sustainable food packaging applications.

Keywords: bio-HDPE; GA; natural additives; thermal resistance; UV stability; food packaging

---

1. Introduction

The scarcity of petroleum and the great awareness about plastic waste have recently generated a great interest in the use of biopolymers for packaging applications [1]. Biopolymers include bio-based polymers, biodegradable polymers, and polymers featuring both characteristics. Bio-based polymers can successfully save fossil resources by using biomass that regenerates annually and provides the unique potential of carbon neutrality [2]. Bio-based polyethylene, also called "green" polyethylene, is a highly crystalline polyolefin produced by addition polymerization of ethylene obtained by catalytic dehydration of bioethanol [3]. Bio-based high-density polyethylene (bio-HDPE) has the same physical properties than its counterpart petrochemical resin, that is, high-density polyethylene (HDPE), showing good mechanical strength, high ductility, and improved water resistance [4,5]. In 2018, bio-based but non-biodegradable polyethylenes represented approximately 9.5% of the global bioplastics' production capacity, reaching nearly 200,000 tons/year [6].

Polyolefins are excellent materials as the base of industrial plastic formulations due to their excellent balance between performance and processability by conventional processing routes such as extrusion and injection molding [7]. However, they are highly sensitive to degradation when exposed to oxidant atmospheres or ultraviolet (UV) light [8]. Polyethylene may undergo degradation, with subsequent increase in fragility, both during processing conditions by extrusion, that is, typically around 140–160 °C [9], or injection molding, that is, above 200 °C [10], and in the presence of light, heat, and chemicals. Hence, the addition of antioxidants and/or UV light stabilizers is habitually required to preserve its original physical properties for long periods. In this regard, phenolic compounds have been extensively used to extend the life service of low-density polyethylene (LDPE) [11,12]. Nevertheless, several synthetic polymer additives have been associated with toxicity effects on human health and the environment as well as other side effects such as carcinogenesis, which has led to some restraint in their use in plastics [13,14]. For instance, synthetic antioxidants such as polyphenol, organophosphate, and thioester compounds can potentially induce some toxicity derived from their migration into food products [15].

While scientific evidence on the exact implications is not conclusive, especially due to the difficulty of assessing complex long-term exposure, there are sufficient indications that warrant further research of natural additives for packaging manufacturers. For instance, tocopherol, plant extracts, and essential oils from herbs and spices have been proposed as natural antioxidants in polyolefins [16–18]. Other published works have reported the use of dihydromyricetin (DHM), quercetin or rosmarinic acid as UV light stabilizers [19,20]. Gallic acid (GA), that is, 3,4,5-trihydroxybenzoic acid, is a naturally occurring polyphenol commonly found in a variety of fruits and vegetables such as grapes, green tea, tea leaves or tomatoes [21,22]. Bioactive phenolic compounds can be effectively obtained by classical solid–liquid extraction employing organic solvents in heat-reflux systems [23] as well as other novel techniques including the use of supercritical fluids, high pressure processes, microwave-assisted extraction (MAE), and ultrasound-assisted extraction [24,25]. Therefore, GA is a good candidate to be applied as a natural polymer additive due to its natural origin, inherently low toxicity, and high bioactive activity such as antioxidant, anti-inflammatory, anticarcinogenic, and antifungal properties [26,27].

This study originally focuses on the use of the GA natural antioxidant to protect bio-HDPE from thermal and UV degradation. To this end, two contents of GA were melt-mixed during extrusion with bio-HDPE and the resultant materials were shaped into films by cast extrusion. The films were characterized in terms of their mechanical, morphological, and thermal performance as well as UV light stability to ascertain their potential in packaging applications.

2. Experimental

2.1. Materials

Bio-HDPE, SHA7260 grade, was manufactured by Braskem (São Paulo, Brazil) and supplied in pellet form by FKuR Kunststoff GmbH (Willich, Germany). This resin has a density of 0.955 g·cm^{-3} and a melt flow index (MFI) of 20 (2.16 kg, 190 °C). It has been developed for injection molding applications and its minimum bio-based content is 94%, determined by ASTM D6866. GA, with commercial reference G7384, having 97.5%–102.5% (titration) and 170.12 g·mol^{-1}, was supplied in powder form by Sigma-Aldrich S.A. (Madrid, Spain). This is a water-soluble phenolic acid obtained from grapes and the leaves of different plants.

2.2. Manufacturing of Films

Different mixtures of bio-HDPE and GA were manually premixed in a zipper bag and melt-compounded in a co-rotating twin-screw extruder from Construcciones Mecánicas Dupra, S.L. (Alicante, Spain). This extruder has a ratio of length (L) to diameter (D) ratio, that is, L/D, of 24, whereas its screws have a diameter of 25 mm. The speed of the screws was set at 20 rpm and the temperature profile was adjusted as follows: 145 °C (hopper)–150 °C–160 °C–165 °C (die). The extruded materials were cooled in air and then pelletized using an air-knife unit. GA was added at 0.3 and 0.8 parts per hundred resin (phr) of bio-HDPE, whereas a neat bio-HDPE sample was prepared in the same conditions as the control sample.

The compounded pellets were, thereafter, cast-extruded into films using a cast-roll machine MINI CAST 25 from EUR.EX.MA (Venegono, Italy). The extrusion speed was set at 25 rpm and the temperature profile was 150 °C (feeding)–155 °C–160 °C–165 °C–165 °C–170 °C–170 °C (head). Bio-HDPE films with an average thickness of approximately 150 µm were obtained by adjusting the speed of the calendar and the drag.

2.3. Color Measurements

A Hunter Mod. CFLX-DIF-2 colorimeter (Hunterlab, Murnau, Germany) was used to determine the color coordinates of the film samples. The values of L^* (lightness), a^* (red to green), and b^* (yellow to blue) parameters were determined while the color difference between two samples (ΔE_{ab}^*) was calculated using Equation (1):

$$\Delta E_{ab}^* = \sqrt{\Delta L^{*2} + \Delta a^{*2} + \Delta b^{*2}} \tag{1}$$

where ΔL^*, Δa^*, and Δb^* represent the differences in L^* and the a^* and b^* coordinates, respectively, between the neat bio-HDPE film and the GA-containing bio-HDPE films. At least five readings were taken for each film and the average values were reported. The following assessment was used to evaluate the color change of the films based on the ΔE_{ab}^* values: below 1 indicates an unnoticeable difference in color; 1–2 a slight difference that can only be noticed by an experienced observer; 2–3.5 a noticeable difference by an unexperienced observer; 3.5–5 a clear noticeable difference; and above 5, different colors are noticeable [28].

2.4. Mechanical Tests

A universal test machine Elib 50 from S.A.E. Ibertest (Madrid, Spain) was used to perform the tensile tests in the bio-HDPE film samples following the guidelines of ISO 527-1:2012. The selected load cell was 5 kN and the cross-head speed was set at 10 mm·min^{-1}. Standard tensile samples (type 2) with a total length and width of 160 and 10 mm, respectively, were tested as indicated in ISO 527-3. Tests were performed at room conditions and at least six samples per film were analyzed.

2.5. Thermal Characterization

The main thermal transitions of the bio-HDPE film samples were obtained by differential scanning calorimetry (DSC) in a Mettler-Toledo 821 calorimeter (Mettler-Toledo, Schwerzenbach, Switzerland). Samples with a total weight of about 5–10 mg were placed into aluminum crucibles. Two types of DSC tests were carried out to evaluate the antioxidant efficiency of GA. The first test was based on a dynamic program from 30 to 350 °C in an air atmosphere at a heating rate of 5 °C·min^{-1} where the oxidative degradation was identified as the onset oxidation temperature (OOT). The second test consisted of a heating ramp from 30 to 210 °C in an air atmosphere at a heating rate of 5 °C·min^{-1}, followed by an isotherm at 210 °C for a whole period of 400 min. The latter test allowed for the oxidation induction time (OIT) to be obtained. Furthermore, the degree of crystallinity (X_C) was calculated following Equation (2):

$$X_C = \left[\frac{\Delta H_m - \Delta H_{CC}}{\Delta H_m^0 \Delta (1 - w)} \right] \cdot 100 \tag{2}$$

where ΔH_m ($J \cdot g^{-1}$) and ΔH_{CC} ($J \cdot g^{-1}$) correspond to the melt and cold crystallization enthalpies, respectively. $\Delta H_m{}^0$ ($J \cdot g^{-1}$) stands for the melt enthalpy of a theoretically fully crystalline of bio-HDPE with a value of 293.0 $J \cdot g^{-1}$ [29] and the term *1-w* represents the weight fraction of bio-HDPE.

Thermal stability was also determined by thermogravimetric analysis (TGA) in a Mettler-Toledo TGA/SDTA 851 thermobalance (Mettler-Toledo, Schwerzenbach, Switzerland). Samples with an average weight of 5–7 mg were placed in standard alumina crucibles (70 µL) and subjected to a heating program from 30 to 700 °C in air atmosphere at heating rates of 20 °C·min^{-1}. All the thermal tests were performed in triplicate.

2.6. Aging Treatment

The aging treatment of materials was performed by means of a high-pressure mercury lamp, with 1000 W and 350 nm wavelength, model UVASPOT 1000RF2 (Honle Spain S.A., Barcelona, Spain) in a closed chamber under ambient conditions. Samples were exposed for a period of up to 10 h and tests were carried out in triplicate.

2.7. Infrared Spectroscopy

Attenuated total reflection–Fourier transform infrared (ATR-FTIR) spectroscopy was used to perform chemical analysis of the films. A Vector 22 from Bruker S.A. (Madrid, Spain) coupling a PIKE MIRacle™ ATR accessory from PIKE Technologies (Madison, WI, USA) was used to record the FTIR spectra. Ten scans were averaged from 4000 to 450 cm^{-1} at a resolution of 4 cm^{-1}. Film samples that were UV treated at 30 min intervals were used to collect variable time FTIR spectra for a whole span time of 10 h.

2.8. Microscopy

The morphology of the fracture surfaces of the UV-treated films of bio-HDPE was observed by field emission scanning electron microscopy (FESEM) in a ZEISS ULTRA 55 from Oxford Instruments (Abingdon, UK). Samples were obtained by cryo-fracture and an acceleration voltage of 2 kV was applied during FESEM observation. The surfaces were previously coated with a gold-palladium alloy in an EMITECH sputter coating SC7620 model from Quorum Technologies, Ltd. (East Sussex, UK).

3. Results and Discussion

3.1. Optical Properties of the GA-Containing Bio-HDPE Films

Figure 1 shows the surface view of the bio-HDPE films varying the GA content. Simple naked eye examination of these images indicated that all of the biopolymer films showed a high contact transparency. Indeed, bio-HDPE is highly transparent due to its high crystalline nature [30]. All the film samples exhibited a smooth, defect-free, and uniform surface, in which GA yielded a yellow color and also certain opacity. The latter effect can be ascribed to the presence of the GA particles, which reduced the transparency properties by blocking the passage of ultraviolet–visible (UV–Vis) light and scattering light. A similar yellowing effect was observed by Al-Malaica et al. [31], who reported the effect of changing the concentration of tocopherol and Irganox 1010 (a commercial phenolic antioxidant) on the color stability of polypropylene (PP). At low additive concentrations, both antioxidants showed low influence on the color sample, expressed in terms of differences in yellow index, whereas higher concentrations of tocopherol led to noticeable color changes. In order to quantify the optical parameters, Table 1 gathers the values of L^*, a^*, and b^* of all the bio-HDPE films and also the $\Delta E_{ab}{}^*$ values of the bio-HDPE films containing GA. One can observe that as the GA content increased, the luminance of the film decreased, confirming that the bio-HDPE films became less transparent. It could also be observed that the a^* coordinate slightly changed from negative values (green) to nearly neutral values, while the b^* coordinate also changed remarkably from negative values (blue) to positive values (yellow) [32]. Therefore, the incorporation of the here-tested GA loadings induced an increase in both opacity and the hue of yellow color, which could restrict the use of biopolymer films for transparent applications. Furthermore, the development of a different color in the bio-HDPE film after the GA addition was noticeable ($\Delta E_{ab}{}^* \geq 5$). However, the GA-containing bio-HDPE films can also offer some advantages for certain packaging applications. For instance, this optical property can be desirable for the protection of foodstuff from light, especially UV radiation, which can cause lipid oxidation in food products [33,34]. Examples include snack products that are made with refined vegetable oils and dried soups such as chicken soup that are sensitive to UV light because they contain highly sensitive unsaturated fatty acids or dry broccoli cream soup that is sensitive to visible light because it contains the photosensitizers chlorophyll from broccoli and riboflavin from dairy ingredients. Another potential application of the here-developed films is to avoid the discoloration of sliced sausage, which is a well-known adverse effect of light that often occurs even if the product is packed under vacuum [35].

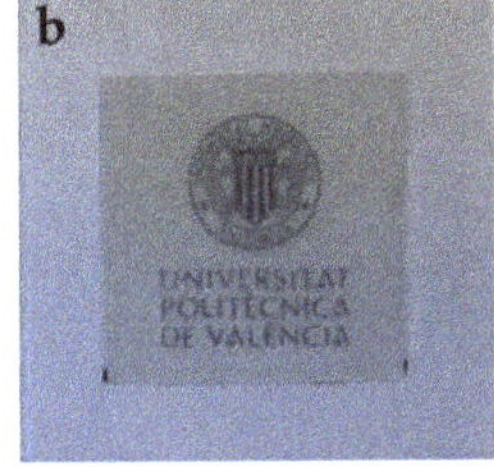
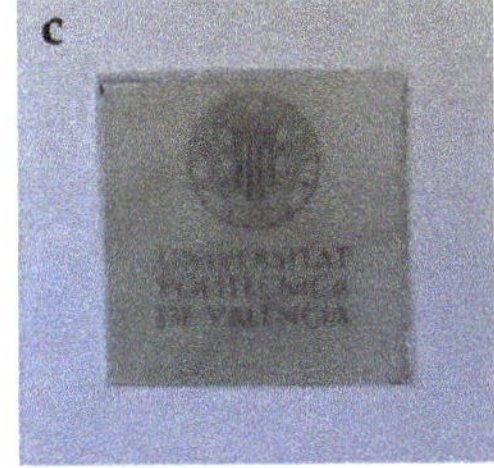

Figure 1. Visual appearance of the bio-based high-density polyethylene (bio-HDPE) films containing different amounts of gallic acid (GA): (**a**) Bio-HDPE; (**b**) Bio-HDPE + 0.3GA; (**c**) Bio-HDPE + 0.8GA.

Table 1. Color parameters (L^*, a^*, b^*, and $\Delta E_{ab}{}^*$) of the bio-based high-density polyethylene (bio-HDPE) films containing different amounts of gallic acid (GA).

Film	L^*	a^*	b^*	$\Delta E_{ab}{}^*$
Bio-HDPE	82.9 ± 1.0	-1.9 ± 0.1	-2.8 ± 0.3	-
Bio-HDPE + 0.3GA	75.3 ± 0.9	-0.7 ± 0.3	8.8 ± 0.4	13.9 ± 0.9
Bio-HDPE + 0.8GA	70.6 ± 0.5	-0.1 ± 0.1	10.9 ± 0.2	18.5 ± 0.4

3.2. Mechanical Properties of the GA-Containing Bio-HDPE Films

Tensile tests were carried out in order to analyze the mechanical properties of the GA-containing bio-HDPE films. Table 2 summarizes the values of tensile modulus ($E_{tensile}$), maximum tensile strength (σ_{max}), and elongation at break (ε_b). One can observe that $E_{tensile}$ of the neat bio-HDPE film was 292.5 MPa and this value was reduced to 222.1 and 243.6 MPa with the incorporation of 0.3 and 0.8 phr of GA, respectively. The value of σ_{max} was in the 20–21 MPa range for all of the bio-HDPE film samples, which is similar to the values reported by other authors [36]. In relation to ε_b, the neat bio-HDPE film showed a value of 45.2%, which was also reduced to 18.6% and 20.2% after the incorporation of 0.3 phr and 0.8 phr of GA, respectively. Significantly higher ε_b values, around 450%–550%, have been reported for injection-molded articles of bio-HDPE [5,37], which can be ascribed to the testing conditions, processing method, and differences in the percentage of crystallinity as well as crystal orientation during manufacturing. Therefore, the incorporation of GA resulted in a reduction in both elasticity and ductility of bio-HDPE. In this regard, crystallinity can play a significant role in the mechanical and durability performance in rigid applications. A decrease in the polymer's crystallinity can lead to a reduction of $E_{tensile}$ and σ_{max}, which are parameters ascribed to mechanical strength [38]. For instance, Jamshidian et al. [39] showed that the use of different antioxidants in polylactide (PLA) films yielded lower values of $E_{tensile}$, σ_{max}, and ε_b. Whereas the reduction in mechanical strength was related to an effect of reduced crystallinity, the ductility impairment observed was ascribed to a phenomenon of stress concentration by the presence of additives with a low interfacial adhesion with the biopolymer matrix. Similarly, work performed by Jamshidian et al. [40] demonstrated that the addition of antioxidants in PLA films yielded a reduction in their mechanical performance due to the additive not being homogenously distributed throughout the entire polymer structure, which could lead to polymer inconstancy and be another reason for decreased mechanical parameters. In general, the incorporation of antioxidants and other polymer additives can alter the film continuity and then decrease the movement of the polymer chains, leading to a ductility decrease [41]. In the particular case of HDPE, its mechanical performance reduction has been attributed to the direct reaction of the antioxidant with oxygen that lowers the efficiency of the inhibitor, pro-oxidant transformation products that may be formed during the processing operations and can participate in oxidative degradation and, more importantly, to limitations in the solubility of antioxidants in the polyolefin matrix [42]. This effect differs from that of UV stabilizers, which tend to be more soluble in low-molecular weight (M_W) organic solvents [43].

Table 2. Tensile properties of the bio-based high-density polyethylene (bio-HDPE) films containing different amounts of gallic acid (GA) in terms of tensile modulus ($E_{tensile}$), maximum tensile strength (σ_{max}), and elongation at break (ε_b).

Film	$E_{tensile}$ (MPa)	σ_{max} (MPa)	ε_b (%)
Bio-HDPE	292.5 ± 22.1	21.3 ± 1.2	45.2 ± 3.5
Bio-HDPE + 0.3GA	222.1 ± 24.2	20.1 ± 0.6	18.6 ± 2.1
Bio-HDPE + 0.8GA	243.6 ± 31.5	20.8 ± 0.9	20.2 ± 2.3

3.3. Thermal Properties of the GA-Containing Bio-HDPE Films

Both DSC and TGA tests were carried out in order to ascertain the influence of the GA addition on the thermal stability of the bio-based polyolefin. Figure 2 shows the dynamical DSC curves of the cast-extruded bio-HDPE films, whereas Table 3 summarizes the main thermal parameters obtained from the curves. One can observe that, in all cases, the polyolefin melted sharply in a single peak at approximately 132 °C. A similar melting profile has been observed previously for this polyolefin, regardless of the origin and the methodology followed to prepare the articles [37,44]. It can also be seen that the ΔH_m values of bio-HDPE slightly reduced as the GA content in the green polyolefin increased. In particular, the crystallinity degree, that is, X_C, was slightly reduced from 54.8% for the neat bio-HDPE film to 53.5% and 52% for the bio-HDPE films containing 0.3 phr and 0.8 phr of GA, respectively. This result suggests that the presence of the GA antioxidant decreased the lamellae size of the bio-HDPE crystals by inducing imperfections [45]. For instance, Lopez-de-Dicastillo et al. [46] similarly reported that the incorporation of ascorbic acid, ferulic acid, quercetin, or green tea extract induced a lower and more deficient crystallinity structure for poly(ethylene–co–vinyl alcohol) (EVOH).

Table 3. Thermal parameters of the bio-based high-density polyethylene (bio-HDPE) films containing different amounts of gallic acid (GA) in terms of melting temperature (T_m), normalized melting enthalpy (ΔH_m), degree of crystallinity (X_C), onset oxidation temperature (OOT), and oxidation induction time (OIT).

Film	T_m (°C)	ΔH_m (J·g^{-1})	X_C (%)	OTT (°C)	OIT (min)
Bio-HDPE	132.1 ± 0.3	160.6 ± 1.5	54.8± 0.8	226.3 ± 1.5	4.9 ± 0.3
Bio-HDPE + 0.3GA	132.4 ± 0.5	156.5 ± 1.4	53.4± 0.7	262.8 ± 2.1	60.8 ± 0.5
Bio-HDPE + 0.8GA	132.2 ± 0.7	152.7 ± 2.0	52.0 ± 0.9	270.2 ± 1.9	244.7 ± 1.0

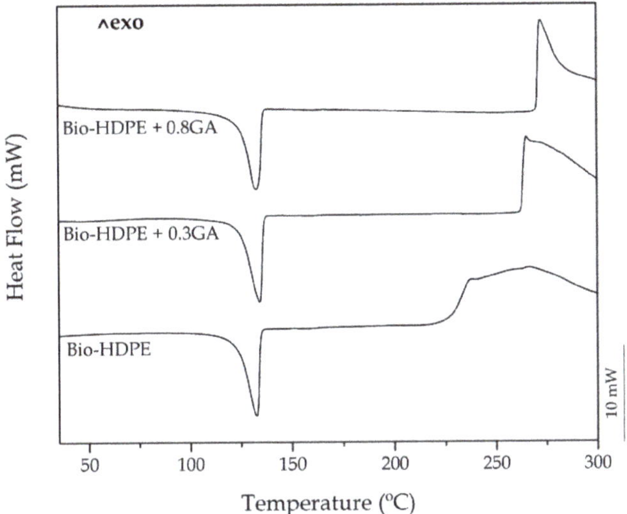

Figure 2. Differential scanning calorimetry (DSC) heating curves of the bio-based high-density polyethylene (bio-HDPE) films containing different amounts of gallic acid (GA).

More interestingly, the DSC plots also revealed the significant oxidative retardant effect of GA on bio-HDPE. It can be observed that the onset of thermal degradation (T_{onset}), also called OOT when the DSC run is carried out in an oxygen-rich environment, started at 226.3 °C in the neat bio-HDPE film. This value is relatively similar to that reported by Jorda-Vilaplana et al. [47], who showed that bio-HDPE started thermal degradation at approximately 232.5 °C. The value of T_{onset} then increased by 36.5 °C and nearly 44 °C in the bio-HDPE films containing 0.3 phr and 0.8 phr of GA, respectively. Similar results were obtained by Samper et al. [17] where 0.5 wt % silibinin and quercetin acted as oxidative retardants for PP as both natural additives successfully delayed the onset of thermal oxidation. In this sense, Dopico-Garcia et al. [48] showed that the use of natural antioxidants could successfully result in polyolefins with enhanced stabilization against thermal-oxidation degradation. The criteria for the antioxidant activity is based on the o-dihydroxy structure of their B-ring, which confers higher stability to the radical form and participates in electron delocalization for effective radical scavenging.

Figure 3 shows the isothermal DSC curves of the cast-extruded bio-HDPE films measured at 210 °C for a span time of 400 min. It can be observed that, after a heating ramp of 36 min, all the DSC pans reached 210 °C and the green polyolefin samples already melted and then showed similar curves in which, thereafter, oxidation occurred at different times. The OIT value, that is, the time between melting and the decomposition onset in isothermal conditions, was seen as an exothermic peak. One can notice that in the neat bio-HDPE film, oxidation initiated at approximately 5 min. The addition of 0.3 phr and 0.8 phr of GA successfully delayed oxidative thermal degradation of bio-HDPE by approximately 56 min and 240 min. It is worth noting that the high performance achieved herein using GA, a natural antioxidant, in comparison to other antioxidants. For instance, the use of other phenolic compounds such as the natural antioxidants naringin or silbinin at 0.5 wt % resulted in an OIT value of 17 min at 210 °C in PP [17]. In relation to synthetic antioxidants, Li et al. [49] showed that the incorporation of 0.1 wt % of dendritic antioxidant delayed the oxidation of PP and LDPE by 40 min and 50 min, respectively.

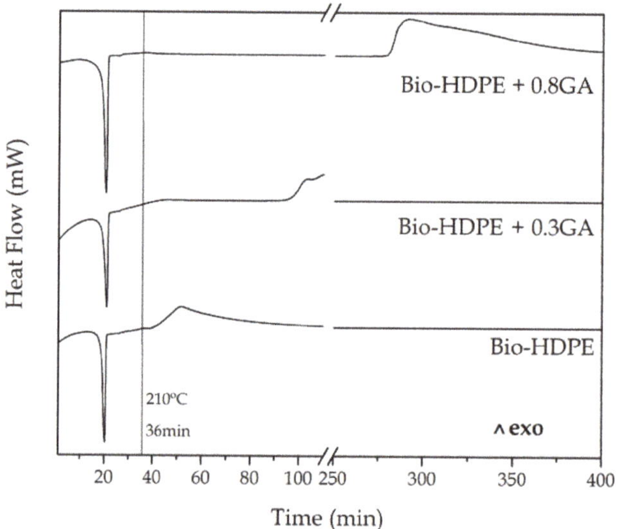

Figure 3. Differential scanning calorimetry (DSC) isothermal curves measured at 210 °C for a span time of 400 min of the bio-based high-density polyethylene (bio-HDPE) films containing different amounts of gallic acid (GA).

The improvement attained with the incorporation of the GA can also be related to the good dispersion of GA achieved within the bio-HDPE matrix. Thus, the additive chemically makes better contact with the peroxyradical on the polymer chains to inhibit the oxidation reaction. In this regard, Koontz et al. [50] showed that the addition of tocopherol improved the OIT value of linear low-density polyethylene (LLDPE) in 68 min when it was uniformly dispersed in the polyolefin matrix. Furthermore, the particular chemical structure of GA provides a great antioxidant capacity, which has been widely reported in food technology, medicine, pharmacy, etc. [51–53]. Consistent with most polyphenolic antioxidants, both the configuration and total number of hydroxyl groups can substantially influence its antioxidant activity mechanism [54,55]. In particular, GA is a free radical scavenger that is based on the high reactivity of the hydroxyl substituents ($F - OH$) that participate in the next reaction [56]:

$$F - OH + R^{\cdot} \rightarrow F - O^{\cdot} + RH$$

Hence, when a free radical ($R^{\cdot}$) is formed by thermo-oxidation, the phenolic compounds move toward this unstable point to block further degradation and produce a stabilization effect. Thus, $F - OH$ donates hydrogen to become peroxyl ($F - O^{\cdot}$), stabilizing the free radical. Among structurally homologous flavones and flavanones, peroxyl and hydroxyl scavenging increases linearly and curvy-linearly, respectively, according to the total number of hydroxyl groups [57].

Figure 4 shows the TGA curves (Figure 4a) and DTG curves (Figure 4b) of the cast-extruded bio-HDPE films while Table 4 summarizes the main thermal parameters obtained from the curves. The neat bio-based polyolefin presented an onset degradation temperature (T_{onset}) of 256.9 °C. The temperature of maximum degradation (T_{deg}), which corresponds to the temperature with the maximum degradation rate, was 427.8 °C. Although the thermal degradation of the green polyolefin was produced in a single step, a lower decomposition rate was observed up to approximately 370 °C, which can be seen as a shoulder in the DTG curve of the neat bio-HDPE. In this thermal range, the decomposition of the C–C covalent bond started and free radicals were generated. At higher temperatures, the free radicals formed led to sequential thermal degradation and breakdown of the main polyolefin chain [58]. Finally, all the film samples showed a similar residual mass of 0.2%–0.3%, indicating full thermal decomposition at 700 °C. In this sense, Montanes et al. [59] observed a similar thermal degradation profile for this green polyolefin, which was based on a one-step weight loss that ranged between 390 and 508 °C. The addition of 0.3 and 0.8 phr of GA successfully induced an improvement in the bio-based HDPE film of approximately 27 and 35 °C in the T_{onset} value, respectively, and suppressed the formation of the above-described free radicals. This thermal stability enhancement was relatively similar to that obtained above by DSC, as shown in previous Table 3, which is related to the intrinsic antioxidant activity of the natural polyphenol. In comparison to previous works using synthetic antioxidants, Zeinalov et al. [60] showed that thermal degradation of neat polystyrene (PS) started at around 270 °C, while the addition of 1% of antioxidant (Fullerene C60) delayed it up to 300 °C. In relation to other works reporting the use of GA, Luzi et al. [61] recently described that the addition of 5 wt % GA successfully increased the thermal stability of EVOH films by nearly 20 °C.

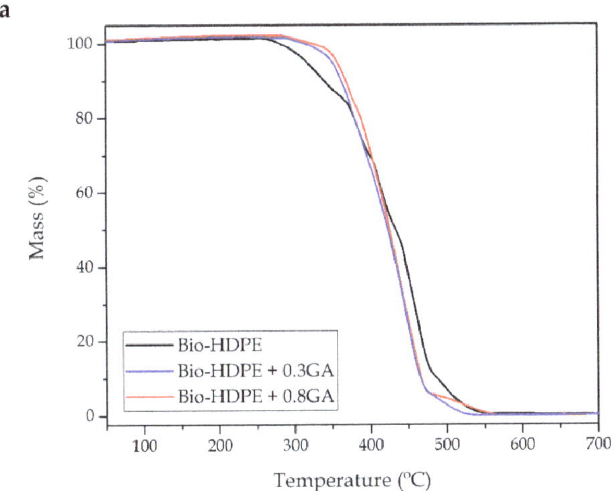

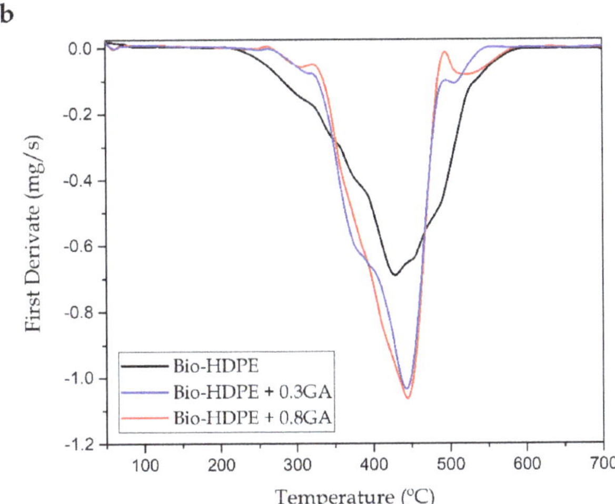

Figure 4. (**a**) Thermogravimetric analysis (TGA) curves and (**b**) first derivative (DTG) of the bio-based high-density polyethylene (bio-HDPE) films containing different amounts of gallic acid (GA).

Table 4. Thermal decomposition parameters of the bio-based high-density polyethylene (bio-HDPE) films containing different amounts of gallic acid (GA) in terms of onset degradation temperature (T_{onset}), temperature of maximum degradation (T_{deg}), and residual mass at 700 °C.

Film	T_{onset} (°C)	T_{deg} (°C)	Residual mass (%)
Bio-HDPE	256.9 ± 1.8	427.8 ± 1.3	0.22 ± 0.05
Bio-HDPE + 0.3GA	283.9 ± 2.0	442.9 ± 1.1	0.20 ± 0.04
Bio-HDPE + 0.8GA	291.6 ± 2.1	444.6 ± 1.2	0.17± 0.05

It is also worthy to note that the GA addition increased the values of T_{deg} by approximately 15 °C, but it also increased the of mass loss rate during thermal degradation. In general, the incorporation of different natural antioxidants can significantly improve the thermal stability of polymers. In particular, some authors have reported similar results with other natural antioxidants [62,63]. For instance, España et al. [64] showed that the incorporation of phenolic compounds successfully improved the T_{deg} values of green composites made of a mixture of lignin and organic coconut fibers (CFs) with an excellent stabilization provided by tannic acid.

3.4. Chemical Characterization of the GA-Containing Bio-HDPE Films

The chemical changes in the bio-HDPE films after the GA addition were analyzed by means of FTIR spectroscopy. Figure 5 shows the FTIR absorbance spectra of the GA, in powder form, and the neat bio-HDPE film and GA-containing bio-HDPE films. The main peaks of GA were observed at the 3100–3500 cm^{-1} region and, more intensely, from 1650 to 560 cm^{-1}. The peak located at 3491 cm^{-1} is ascribed to the O–H stretching vibration of the hydroxyl groups of polyphenols [65]. Xu et al. [66] showed that the strong absorption peak at around 1614 cm^{-1} and bands between 1400 cm^{-1} and 1200 cm^{-1} are characteristic of polysaccharides. The peaks in the 1200–1000 cm^{-1} region originated from ring vibrations that overlapped with the stretching vibrations of C–OH side groups and the C–O–C glycosidic band vibration [67]. In relation to the green polyethylene, the intense peaks at 2925, 2850, 1460, and 725 cm^{-1} were respectively assigned to stretching vibrations and bending and rocking deformations of the methylene (CH$_2$) groups [68]. Furthermore, the low-intense bands located between at 1377 and 1351 cm^{-1} were assignable to the wagging deformation and symmetric deformation of the CH$_2$ and methyl (CH$_3$) groups, respectively. The peaks in the 1700–1800 cm^{-1} and 1200–1300 cm^{-1} regions have been ascribed to carbonyl compounds formed in the oxidation products of polyethylene [69].

The incorporation of GA into the bio-HDPE film generated the appearance of a series of peaks, particularly noticeable for the film containing 0.8 phr, which confirmed the presence of the natural antioxidant in the polyolefin. Briefly, the hydroxyl groups of GA altered the bands related to the CH$_2$ groups of bio-HDPE in the 3100–3500 cm^{-1} region. In addition, the formation of a new weak band at 1607 cm^{-1} can be ascribed to the stretching and bending vibrations of the aromatic ring of GA [65]. One can also observe the formation of a new low-intensity peak centered around 1030 cm^{-1}. Bands formed between 1021 and 1037 cm^{-1} have been ascribed to the formation of dimers or oligomers of GA that can result from the stretching vibration of C–C and C–O bonds [65,70]. Indeed, GA is known to be auto-oxidized to its semiquinone free radicals, which can consequently generate hydroquinone [71].

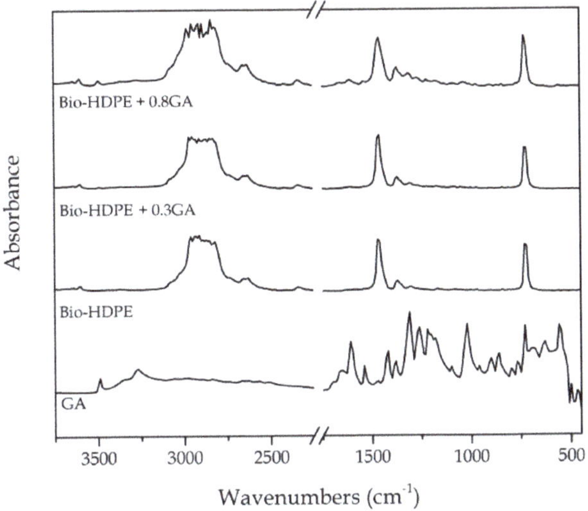

Figure 5. Fourier transform infrared (FTIR) spectra, from bottom to top, of gallic acid (GA) powder, bio-based high-density polyethylene (bio-HDPE) film, and bio-HDPE films containing 0.3 and 0.8 parts per hundred resin (phr) of GA.

3.5. UV Light Stability of the GA-Containing Bio-HDPE Films

The bio-HDPE films were subjected to UV light for a span time of up to 10 h, herein referred to as the aging time, in order to ascertain the influence of the GA addition on their UV light stability. Based on the above spectra, FTIR spectroscopy was used to analyze the chemical changes on the samples after being exposed to UV light. Figure 6 shows the 3D plots of the FTIR spectra taken across the exposure time to UV light. In the case of the neat bio-HDPE film, the UV exposure greatly increased the relative intensity of the strongest peaks observed at 2919 and 2851 cm^{-1}, which are assigned to the CH$_2$ antisymmetric and symmetric stretching, respectively [68]. Furthermore, the peaks centered at 1463 and 720 cm^{-1}, which are respectively ascribed to bending and rocking deformations in polyethylene [68], also increased. These chemical changes suggested an increase of the CH$_2$ groups in the film sample, which can be related to the partial breakdown of the polyolefin chain by UV light exposure and the formation of more terminal groups. It is also noteworthy that UV degradation was already noticeable after 30 min of UV light treatment, whereas it increased slightly in the whole aging time tested. No further changes were observed in the bands related to oxidized groups since changes were very subtle and are also known to appear after longer UV exposure periods and higher temperatures [72]. The fast degradation changes observed in bio-HDPE can be ascribed to the above-reported mechanism based on free radicals with high reactivity. Interestingly, these absorbance bands related to CH$_2$ compounds of both GA-containing bio-HDPE films remained nearly constant and a slight increase was observed after 4 h of UV light exposure. Therefore, the addition of GA successfully improved the oxidation stability of the green polyolefin, also offering long-term UV stability. As also explained above during the thermal analysis, the chemical configuration of GA and the significant number of hydroxyl groups could successfully stabilize the free radicals formed during UV light exposure.

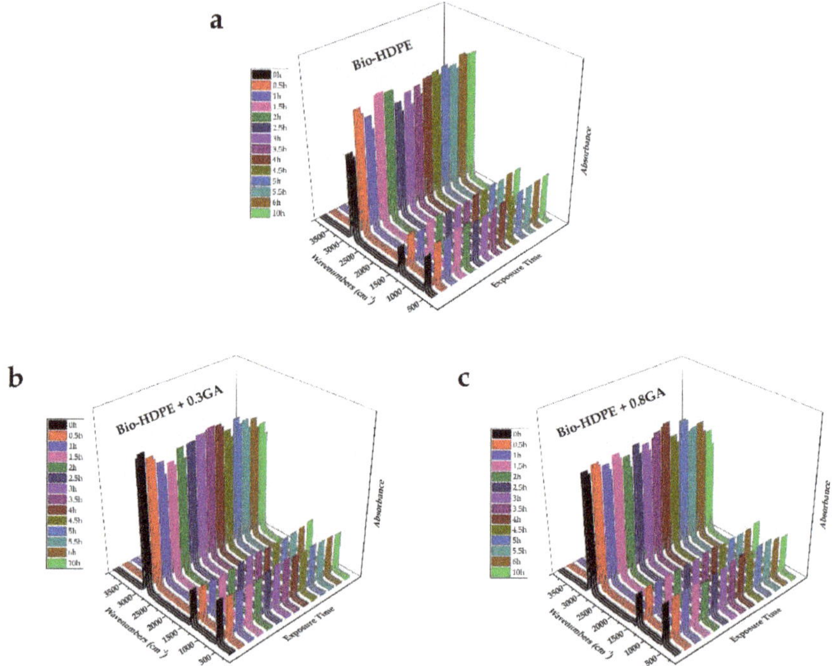

Figure 6. Fourier transform infrared (FTIR) spectra taken across the exposure time to ultraviolet (UV) light of the bio-based high-density polyethylene (bio-HDPE) films containing different amounts of gallic acid (GA): (**a**) Bio-HDPE; (**b**) Bio-HDPE + 0.3GA; (**c**) Bio-HDPE + 0.8GA.

Finally, Figure 7 shows the FESEM images of the bio-HDPE films exposed to 5 h UV light. One can observe that, prior to UV light exposure, all of the films presented a similar fracture surface without any cracks or wrinkles. After 1 h of UV light exposure, the films developed an increase in roughness on their fracture surfaces. However, the GA-containing films generated a smoother surface, which is representative for a slower or negligible UV light aging. One can also observe that the fracture surface of the neat bio-HDPE suffered a remarkable modification after 2.5 h of UV light exposure. Indeed, the life time of an article made of HDPE without stabilizers can be as low as one year since the polyolefin decomposes rapidly by UV light action [73]. This phenomenon is related to the presence of impurities that are formed during their synthesis such as carbonyl, peroxide, hydroxyl, hydroxyperoxide, or any substances with unsaturated groups, which absorb light at higher wavelengths and thus yield the generation of free radicals. The incorporation of 0.3 phr GA significantly reduced the UV degradation of bio-HDPE and the fracture surfaces remained similar for up to 2.5 h, time at which some wrinkles were formed. Furthermore, 0.8 phr GA successfully kept the film samples stable up to 5 h of UV light exposure, showing fracture surfaces free of cracks. Therefore, the morphological analysis correlates well with the FTIR spectroscopy results shown above and confirmed the UV light stability provided by GA to the bio-HDPE films that can thus improve the shelf life of the green polyolefin. The antioxidant effect of GA is considered to also protect the free residues that are generated during polymer synthesis with improved degradation stability [74]. Indeed, UV stabilizers have always been categorized as a subgroup within the antioxidant additive group. Similarly, Du et al. [75] showed that HDPE/wood flour composites containing pigments presented fewer cracks on the surface than composites without pigment after accelerated UV weathering. The authors suggested that pigments can mask some UV radiation and prevent HDPE against UV radiation damage. Similar results were previously reported by Samper et al. [17] through the use of quercetin and silibinin as UV light stabilizers for PP.

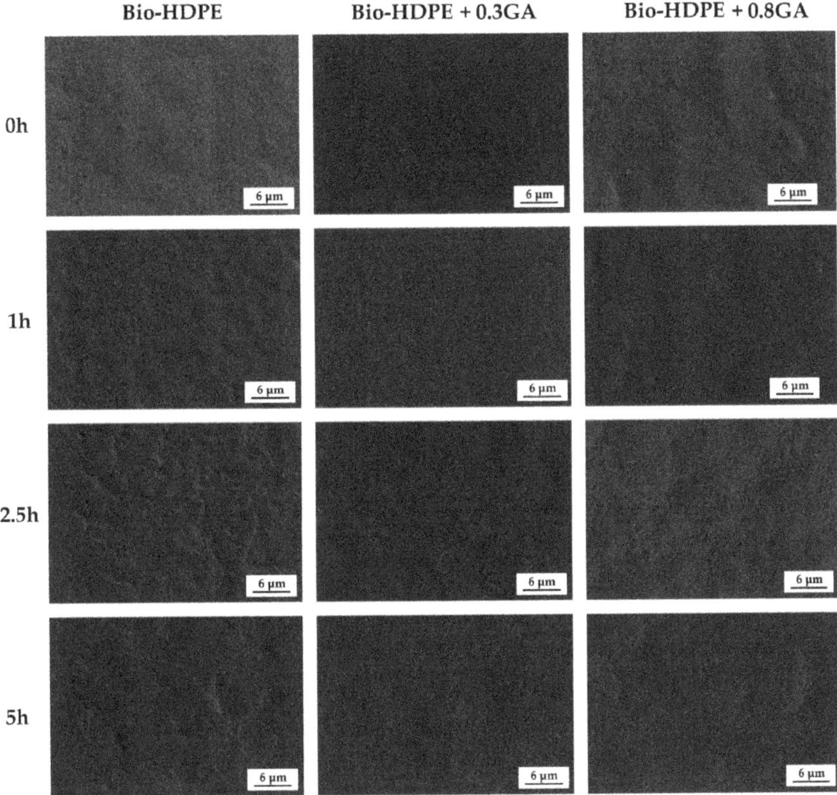

Figure 7. Field emission scanning electron microscopy (FESEM) micrographs of the bio-based high-density polyethylene (bio-HDPE) films containing different amounts of gallic acid (GA) exposed to different ultraviolet (UV) light exposure times.

4. Conclusions

This work describes the development and characterization of cast-extruded bio-HDPE films containing the natural antioxidant GA in order to ascertain their potential application in food packaging. The incorporation of GA at 0.3 and 0.8 phr contents induced a mechanical elasticity and ductility impairment and also a crystallinity reduction of bio-HDPE due to its limited solubility in the green polyolefin matrix. The bio-HDPE films also developed a low-intense yellow color but were still contact transparent. Interestingly, the OOT values was delayed by 36.5 °C and nearly 44 °C while the OIT values were reduced by approximately 56 and 240 min in the bio-HDPE films containing 0.3 and 0.8 phr GA, respectively. Furthermore, the UV light stability of bio-HDPE was significantly improved after the GA addition for an aging time monitored by FTIR spectroscopy of 10 h. The enhancement attained was ascribed to the high capacity of the phenolic compounds present in the natural antioxidant to stabilize the free radicals formed during degradation of the green polyolefin. As a result, GA can be regarded as a natural antioxidant and UV light stabilizer that can potentially replace synthetic additives in biopolymer formulations for food packaging applications following the Bioeconomy principles. Nevertheless, future studies should be addressed to increase the ductility of the resultant biopolymer films by, for instance, the addition of natural plasticizers, while the analysis of their barrier properties and the performance of specific migration tests will also be required according to the targeted application.

Author Contributions: Conceptualization, R.B., L.Q.-C. and S.T.-G.; Methodology, L.Q., S.M.-J. and C.S.; Validation, C.S., R.B. and S.T.-G.; Formal analysis, L.Q.-C., T.B. and R.B.; Investigation, C.S. and S.T.-G.; Data curation, L.Q.-C., T.B. and S.M.-J.; Writing—original draft preparation, L.Q.-C. and S.T.-G.; Writing—review and editing, R.B. and S.T.-G.; Supervision, R.B. and S.T.-G.; Project administration, R.B. and S.T.-G. All authors have read and agreed to the published version of the manuscript.

Funding: This research work was funded by the Spanish Ministry of Science, Innovation, and Universities (MICIU) project numbers RTI2018-097249-B-C21 and MAT2017-84909-C2-2-R.

Acknowledgments: L.Q.-C. wants to thank Generalitat Valenciana (GVA) for his FPI grant (ACIF/2016/182) and the Spanish Ministry of Education, Culture, and Sports (MECD) for his FPU grant (FPU15/03812). S.T.-G. is the recipient of a Juan de la Cierva contract (IJCI-2016-29675) from MICIU. Microscopy services at UPV are acknowledged for their help in collecting and analyzing the FESEM images.

Conflicts of Interest: The authors declare no conflicts of interest.

References

1. Torres-Giner, S.; Gil, L.; Pascual-Ramírez, L.; Garde-Belza, J. Packaging: Food waste reduction. In *Encyclopedia of Polymer Applications*; Mishra, M., Ed.; CRC Press: Boca Raton, FL, USA, 2018.

2. Babu, R.P.; O'connor, K.; Seeram, R. Current progress on bio-based polymers and their future trends. *Prog. Biomater.* **2013**, *2*, 8. [CrossRef]

3. Chen, G.; Li, S.; Jiao, F.; Yuan, Q. Catalytic dehydration of bioethanol to ethylene over tio2/γ-al2o3 catalysts in microchannel reactors. *Catal. Today* **2007**, *125*, 111–119. [CrossRef]

4. Quiles-Carrillo, L.; Montanes, N.; Jorda-Vilaplana, A.; Balart, R.; Torres-Giner, S. A comparative study on the effect of different reactive compatibilizers on injection-molded pieces of bio-based high-density polyethylene/polylactide blends. *J. Appl. Polym. Sci.* **2019**, *136*, 47396. [CrossRef]

5. Torres-Giner, S.; Torres, A.; Ferrándiz, M.; Fombuena, V.; Balart, R. Antimicrobial activity of metal cation-exchanged zeolites and their evaluation on injection-molded pieces of bio-based high-density polyethylene. *J. Food Saf.* **2017**, *37*. [CrossRef]

6. Vasile, C.; Râpă, M.; Ştefan, M.; Stan, M.; Macavei, S.; Darie-Niţă, R.; Barbu-Tudoran, L.; Vodnar, D.; Popa, E.; Ştefan, R. New pla/zno: Cu/Ag bionanocomposites for food packaging. *Express Polym. Lett.* **2017**, *11*, 531–544. [CrossRef]

7. Carbonell-Verdú, A.; García-García, D.; Jordá, A.; Samper, M.; Balart, R. Development of slate fiber reinforced high density polyethylene composites for injection molding. *Compos. Part B Eng.* **2015**, *69*, 460–466. [CrossRef]

8. Araújo, J.; Waldman, W.; De Paoli, M. Thermal properties of high density polyethylene composites with natural fibres: Coupling agent effect. *Polym. Degrad. Stab.* **2008**, *93*, 1770–1775. [CrossRef]

9. Wang, J.; Du, Z.; Lian, T. Extrusion–calendering process of single-polymer composites based on polyethylene. *Polym. Eng. Sci.* **2018**, *58*, 2156–2165. [CrossRef]

10. Quiles-Carrillo, L.; Montanes, N.; Fombuena, V.; Balart, R.; Torres-Giner, S. Enhancement of the processing window and performance of polyamide 1010/bio-based high-density polyethylene blends by melt mixing with natural additives. *Polym. Int.* **2019**. [CrossRef]

11. Gao, X.; Hu, G.; Qian, Z.; Ding, Y.; Zhang, S.; Wang, D.; Yang, M. Immobilization of antioxidant on nanosilica and the antioxidative behavior in low density polyethylene. *Polymer* **2007**, *48*, 7309–7315. [CrossRef]

12. Yu, W.; Reitberger, T.; Hjertberg, T.; Oderkerk, J.; Costa, F.; Englund, V.; Gedde, U.W. Chlorine dioxide resistance of different phenolic antioxidants in polyethylene. *Polym. Degrad. Stab.* **2015**, *111*, 1–6. [CrossRef]

13. Ito, N.; Fukushima, S.; Haqlwara, A.; Shibata, M.; Ogiso, T. Carcinogenicity of butylated hydroxyanisole in F344 rats. *J. Natl. Cancer Inst.* **1983**, *70*, 343–352.

14. Reddy, V.; Urooj, A.; Kumar, A. Evaluation of antioxidant activity of some plant extracts and their application in biscuits. *Food Chem.* **2005**, *90*, 317–321. [CrossRef]

15. Gómez-Estaca, J.; López-de-Dicastillo, C.; Hernández-Muñoz, P.; Catalá, R.; Gavara, R. Advances in antioxidant active food packaging. *Trends Food Sci. Technol.* **2014**, *35*, 42–51. [CrossRef]

16. Peltzer, M.; Wagner, J.; Jiménez, A. Thermal characterization of uhmwpe stabilized with natural antioxidants. *J. Therm. Anal. Calorim.* **2007**, *87*, 493–497. [CrossRef]

17. Samper, M.; Fages, E.; Fenollar, O.; Boronat, T.; Balart, R. The potential of flavonoids as natural antioxidants and UV light stabilizers for polypropylene. *J. Appl. Polym. Sci.* **2013**, *129*, 1707–1716. [CrossRef]

18. Tovar, L.; Salafranca, J.; Sánchez, C.; Nerín, C. Migration studies to assess the safety in use of a new antioxidant active packaging. *J. Agric. Food Chem.* **2005**, *53*, 5270–5275. [CrossRef]

19. Kirschweng, B.; Bencze, K.; Sárközi, M.; Hégely, B.; Samu, G.; Hári, J.; Tátraaljai, D.; Földes, E.; Kállay, M.; Pukánszky, B. Melt stabilization of polyethylene with dihydromyricetin, a natural antioxidant. *Polym. Degrad. Stab.* **2016**, *133*, 192–200. [CrossRef]

20. Doudin, K.; Al-Malaika, S.; Sheena, H.; Tverezovskiy, V.; Fowler, P. New genre of antioxidants from renewable natural resources: Synthesis and characterisation of rosemary plant-derived antioxidants and their performance in polyolefins. *Polym. Degrad. Stab.* **2016**, *130*, 126–134. [CrossRef]

21. Graham, H.N. Green tea composition, consumption, and polyphenol chemistry. *Prev. Med.* **1992**, *21*, 334–350. [CrossRef]

22. Yilmaz, Y.; Toledo, R.T. Major flavonoids in grape seeds and skins: Antioxidant capacity of catechin, epicatechin, and gallic acid. *J. Agric. Food Chem.* **2004**, *52*, 255–260. [CrossRef]

23. Martins, S.; Aguilar, C.N.; de la Garza-Rodriguez, I.; Mussatto, S.I.; Teixeira, J.A. Kinetic study of nordihydroguaiaretic acid recovery from larrea tridentata by microwave-assisted extraction. *J. Chem. Technol. Biotechnol.* **2010**, *85*, 1142–1147. [CrossRef]

24. Valdés, A.; Vidal, L.; Beltran, A.; Canals, A.; Garrigós, M.C. Microwave-assisted extraction of phenolic compounds from almond skin byproducts (prunus amygdalus): A multivariate analysis approach. *J. Agric. Food Chem.* **2015**, *63*, 5395–5402. [CrossRef]

25. Martins, S.; Mussatto, S.I.; Martínez-Avila, G.; Montañez-Saenz, J.; Aguilar, C.N.; Teixeira, J.A. Bioactive phenolic compounds: Production and extraction by solid-state fermentation. A review. *Biotechnol. Adv.* **2011**, *29*, 365–373. [CrossRef]

26. Kim, J.H.; Kang, N.J.; Lee, B.K.; Lee, K.W.; Lee, H.J. Gallic acid, a metabolite of the antioxidant propyl gallate, inhibits gap junctional intercellular communication via phosphorylation of connexin 43 and extracellular-signal-regulated kinase1/2 in rat liver epithelial cells. *Mutat. Res./Fundam. Mol. Mech. Mutagenesis* **2008**, *638*, 175–183. [CrossRef]

27. Quiles-Carrillo, L.; Montanes, N.; Lagaron, J.M.; Balart, R.; Torres-Giner, S. Bioactive multilayer polylactide films with controlled release capacity of gallic acid accomplished by incorporating electrospun nanostructured coatings and interlayers. *Appl. Sci.* **2019**, *9*, 533. [CrossRef]

28. Agüero, A.; Morcillo, M.d.C.; Quiles-Carrillo, L.; Balart, R.; Boronat, T.; Lascano, D.; Torres-Giner, S.; Fenollar, O. Study of the influence of the reprocessing cycles on the final properties of polylactide pieces obtained by injection molding. *Polymers* **2019**, *11*, 1908. [CrossRef]

29. Castro, D.; Ruvolo-Filho, A.; Frollini, E. Materials prepared from biopolyethylene and curaua fibers: Composites from biomass. *Polym. Test.* **2012**, *31*, 880–888. [CrossRef]

30. Paradkar, R.; Sakhalkar, S.; He, X.; Ellison, M. Estimating crystallinity in high density polyethylene fibers using online raman spectroscopy. *J. Appl. Polym. Sci.* **2003**, *88*, 545–549. [CrossRef]

31. Al-Malaika, S.; Goodwin, C.; Issenhuth, S.; Burdick, D. The antioxidant role of α-tocopherol in polymers II. Melt stabilising effect in polypropylene. *Polym. Degrad. Stab.* **1999**, *64*, 145–156. [CrossRef]

32. Fenollar, O.; García, D.; Sánchez, L.; López, J.; Balart, R. Optimization of the curing conditions of pvc plastisols based on the use of an epoxidized fatty acid ester plasticizer. *Eur. Polym. J.* **2009**, *45*, 2674–2684. [CrossRef]

33. Figueroa-Lopez, K.J.; Vicente, A.A.; Reis, M.A.M.; Torres-Giner, S.; Lagaron, J.M. Antimicrobial and antioxidant performance of various essential oils and natural extracts and their incorporation into biowaste derived poly(3-hydroxybutyrate-*co*-3-hydroxyvalerate) layers made from electrospun ultrathin fibers. *Nanomaterials* **2019**, *9*, 144. [CrossRef]

34. Melendez-Rodriguez, B.; Figueroa-Lopez, K.J.; Bernardos, A.; Martínez-Máñez, R.; Cabedo, L.; Torres-Giner, S.; Lagaron, J.M. Electrospun antimicrobial films of poly(3-hydroxybutyrate-*co*-3-hydroxyvalerate) containing eugenol essential oil encapsulated in mesoporous silica nanoparticles. *Nanomaterials* **2019**, *9*, 227. [CrossRef]

35. Møller, J.K.S.; Bertelsen, G.; Skibsted, L.H. Photooxidation of nitrosylmyoglobin at low oxygen pressure. Quantum yields and reaction stoechiometries. *Meat Sci.* **2002**, *60*, 421–425. [CrossRef]

36. Zhong, Y.; Janes, D.; Zheng, Y.; Hetzer, M.; De Kee, D. Mechanical and oxygen barrier properties of organoclay-polyethylene nanocomposite films. *Polym. Eng. Sci.* **2007**, *47*, 1101–1107. [CrossRef]

37. Ferrero, B.; Fombuena, V.; Fenollar, O.; Boronat, T.; Balart, R. Development of natural fiber-reinforced plastics (NFRP) based on biobased polyethylene and waste fibers from posidonia oceanica seaweed. *Polym. Compos.* **2015**, *36*, 1378–1385. [CrossRef]

38. Harris, A.M.; Lee, E.C. Improving mechanical performance of injection molded pla by controlling crystallinity. *J. Appl. Polym. Sci.* **2008**, *107*, 2246–2255. [CrossRef]

39. Jamshidian, M.; Tehrany, E.A.; Cleymand, F.; Leconte, S.; Falher, T.; Desobry, S. Effects of synthetic phenolic antioxidants on physical, structural, mechanical and barrier properties of poly lactic acid film. *Carbohydr. Polym.* **2012**, *87*, 1763–1773. [CrossRef]

40. Jamshidian, M.; Tehrany, E.A.; Imran, M.; Akhtar, M.J.; Cleymand, F.; Desobry, S. Structural, mechanical and barrier properties of active pla–antioxidant films. *J. Food Eng.* **2012**, *110*, 380–389. [CrossRef]

41. Jongjareonrak, A.; Benjakul, S.; Visessanguan, W.; Tanaka, M. Antioxidative activity and properties of fish skin gelatin films incorporated with BHT and α-tocopherol. *Food Hydrocoll.* **2008**, *22*, 449–458. [CrossRef]

42. Chirinos Padrón, A.J.; Colmenares, M.A.; Rubinztain, Z.; Albornoz, L.A. Influence of additives on some physical properties of high density polyethylene-i. Commercial antioxidants. *Eur. Polym. J.* **1987**, *23*, 723–727. [CrossRef]

43. Chirinos Padrón, A.J.; Rubinztain, Z.; Colmenares, M.A. Influence of additives on some physical properties of high density polyethylene-II. Commercial u.V. Stabilizers. *Eur. Polym. J.* **1987**, *23*, 729–732. [CrossRef]

44. López-Naranjo, E.J.; Alzate-Gaviria, L.M.; Hernández-Zárate, G.; Reyes-Trujeque, J.; Cupul-Manzano, C.V.; Cruz-Estrada, R.H. Effect of biological degradation by termites on the flexural properties of pinewood residue/recycled high-density polyethylene composites. *J. Appl. Polym. Sci.* **2013**, *128*, 2595–2603. [CrossRef]

45. Xu, T.; Lei, H.; Xie, C. The effect of nucleating agent on the crystalline morphology of polypropylene (PP). *Mater. Des.* **2003**, *24*, 227–230. [CrossRef]

46. López-de-Dicastillo, C.; Gómez-Estaca, J.; Catalá, R.; Gavara, R.; Hernández-Muñoz, P. Active antioxidant packaging films: Development and effect on lipid stability of brined sardines. *Food Chem.* **2012**, *131*, 1376–1384. [CrossRef]

47. Jordá-Vilaplana, A.; Carbonell-Verdú, A.; Samper, M.-D.; Pop, A.; Garcia-Sanoguera, D. Development and characterization of a new natural fiber reinforced thermoplastic (NFRP) with cortaderia selloana (pampa grass) short fibers. *Compos. Sci. Technol.* **2017**, *145*, 1–9. [CrossRef]

48. Dopico-García, M.; Castro-López, M.; López-Vilariño, J.; González-Rodríguez, M.; Valentao, P.; Andrade, P.; García-Garabal, S.; Abad, M. Natural extracts as potential source of antioxidants to stabilize polyolefins. *J. Appl. Polym. Sci.* **2011**, *119*, 3553–3559. [CrossRef]

49. Li, C.; Wang, J.; Ning, M.; Zhang, H. Synthesis and antioxidant activities in polyolefin of dendritic antioxidants with hindered phenolic groups and tertiary amine. *J. Appl. Polym. Sci.* **2012**, *124*, 4127–4135. [CrossRef]

50. Koontz, J.L.; Marcy, J.E.; O'Keefe, S.F.; Duncan, S.E.; Long, T.E.; Moffitt, R.D. Polymer processing and characterization of LLDPE films loaded with α-tocopherol, quercetin, and their cyclodextrin inclusion complexes. *J. Appl. Polym. Sci.* **2010**, *117*, 2299–2309. [CrossRef]

51. Brewer, M. Natural antioxidants: Sources, compounds, mechanisms of action, and potential applications. *Compr. Rev. Food Sci. Food Saf.* **2011**, *10*, 221–247. [CrossRef]

52. Fogliano, V.; Verde, V.; Randazzo, G.; Ritieni, A. Method for measuring antioxidant activity and its application to monitoring the antioxidant capacity of wines. *J. Agric. Food Chem.* **1999**, *47*, 1035–1040. [CrossRef]

53. Choubey, S.; Varughese, L.R.; Kumar, V.; Beniwal, V. Medicinal importance of gallic acid and its ester derivatives: A patent review. *Pharm. Pat. Anal.* **2015**, *4*, 305–315. [CrossRef]

54. Burda, S.; Oleszek, W. Antioxidant and antiradical activities of flavonoids. *J. Agric. Food Chem.* **2001**, *49*, 2774–2779. [CrossRef]

55. Pannala, A.S.; Chan, T.S.; O'Brien, P.J.; Rice-Evans, C.A. Flavonoid b-ring chemistry and antioxidant activity: Fast reaction kinetics. *Biochem. Biophys. Res. Commun.* **2001**, *282*, 1161–1168. [CrossRef]

56. Heim, K.E.; Tagliaferro, A.R.; Bobilya, D.J. Flavonoid antioxidants: Chemistry, metabolism and structure-activity relationships. *J. Nutr. Biochem.* **2002**, *13*, 572–584. [CrossRef]

57. Cao, G.; Sofic, E.; Prior, R.L. Antioxidant and prooxidant behavior of flavonoids: Structure-activity relationships. *Free Radic. Biol. Med.* **1997**, *22*, 749–760. [CrossRef]

58. Jana, R.N.; Mukunda, P.G.; Nando, G.B. Thermogravimetric analysis of compatibilized blends of low density polyethylene and poly(dimethyl siloxane) rubber. *Polym. Degrad. Stab.* **2003**, *80*, 75–82. [CrossRef]

59. Montanes, N.; Garcia-Sanoguera, D.; Segui, V.; Fenollar, O.; Boronat, T. Processing and characterization of environmentally friendly composites from biobased polyethylene and natural fillers from thyme herbs. *J. Polym. Environ.* **2018**, *26*, 1218–1230. [CrossRef]

60. Zeinalov, E.B.; Koßmehl, G. Fullerene C60 as an antioxidant for polymers. *Polym. Degrad. Stab.* **2001**, *71*, 197–202. [CrossRef]

61. Luzi, F.; Puglia, D.; Dominici, F.; Fortunati, E.; Giovanale, G.; Balestra, G.; Torre, L. Effect of gallic acid and umbelliferone on thermal, mechanical, antioxidant and antimicrobial properties of poly(vinyl alcohol-*co*-ethylene) films. *Polym. Degrad. Stab.* **2018**, *152*, 162–176. [CrossRef]

62. Cerruti, P.; Malinconico, M.; Rychly, J.; Matisova-Rychla, L.; Carfagna, C. Effect of natural antioxidants on the stability of polypropylene films. *Polym. Degrad. Stab.* **2009**, *94*, 2095–2100. [CrossRef]

63. Ambrogi, V.; Cerruti, P.; Carfagna, C.; Malinconico, M.; Marturano, V.; Perrotti, M.; Persico, P. Natural antioxidants for polypropylene stabilization. *Polym. Degrad. Stab.* **2011**, *96*, 2152–2158. [CrossRef]

64. España, J.; Fages, E.; Moriana, R.; Boronat, T.; Balart, R. Antioxidant and antibacterial effects of natural phenolic compounds on green composite materials. *Polym. Compos.* **2012**, *33*, 1288–1294. [CrossRef]

65. Neo, Y.P.; Ray, S.; Jin, J.; Gizdavic-Nikolaidis, M.; Nieuwoudt, M.K.; Liu, D.; Quek, S.Y. Encapsulation of food grade antioxidant in natural biopolymer by electrospinning technique: A physicochemical study based on zein–gallic acid system. *Food Chem.* **2013**, *136*, 1013–1021. [CrossRef]

66. Xu, W.; Zhang, F.; Luo, Y.; Ma, L.; Kou, X.; Huang, K. Antioxidant activity of a water-soluble polysaccharide purified from pteridium aquilinum. *Carbohydr. Res.* **2009**, *344*, 217–222. [CrossRef]

67. Kacurakova, M.; Capek, P.; Sasinkova, V.; Wellner, N.; Ebringerova, A. FT-IR study of plant cell wall model compounds: Pectic polysaccharides and hemicelluloses. *Carbohydr. Polym.* **2000**, *43*, 195–203. [CrossRef]

68. Gulmine, J.V.; Janissek, P.R.; Heise, H.M.; Akcelrud, L. Polyethylene characterization by FTIR. *Polym. Test.* **2002**, *21*, 557–563. [CrossRef]

69. Sugimoto, M.; Shimada, A.; Kudoh, H.; Tamura, K.; Seguchi, T. Product analysis for polyethylene degradation by radiation and thermal ageing. *Radiat. Phys. Chem.* **2013**, *82*, 69–73. [CrossRef]

70. Markarian, S.A.; Zatikyan, A.L.; Bonora, S.; Fagnano, C. Raman and ft ir atr study of diethylsulfoxide/water mixtures. *J. Mol. Struct.* **2003**, *655*, 285–292. [CrossRef]

71. Gil-Longo, J.; González-Vázquez, C. Vascular pro-oxidant effects secondary to the autoxidation of gallic acid in rat aorta. *J. Nutr. Biochem.* **2010**, *21*, 304–309. [CrossRef]

72. Qin, H.; Zhao, C.; Zhang, S.; Chen, G.; Yang, M. Photo-oxidative degradation of polyethylene/montmorillonite nanocomposite. *Polym. Degrad. Stab.* **2003**, *81*, 497–500. [CrossRef]

73. Grigoriadou, I.; Paraskevopoulos, K.; Chrissafis, K.; Pavlidou, E.; Stamkopoulos, T.-G.; Bikiaris, D. Effect of different nanoparticles on HDPE UV stability. *Polym. Degrad. Stab.* **2011**, *96*, 151–163. [CrossRef]

74. Scott, G. The antioxidant role of uv stabilisers. *Pure Appl. Chem.* **1980**, *52*, 365–387. [CrossRef]

75. Du, H.; Wang, W.; Wang, Q.; Zhang, Z.; Sui, S.; Zhang, Y. Effects of pigments on the UV degradation of wood-flour/HDPE composites. *J. Appl. Polym. Sci.* **2010**, *118*, 1068–1076. [CrossRef]

MDPI
St. Alban-Anlage 66
4052 Basel
Switzerland
Tel. +41 61 683 77 34
Fax +41 61 302 89 18
www.mdpi.com

Polymers Editorial Office
E-mail: polymers@mdpi.com
www.mdpi.com/journal/polymers